INTRODUCTION TO MAGNETIC RANDOM-ACCESS MEMORY

IEEE Press
445 Hoes Lane
Piscataway, NJ 08854

INTRODUCTION TO MAGNETIC RANDOM-ACCESS MEMORY

Edited by

BERNARD DIENY
RONALD B. GOLDFARB
KYUNG-JIN LEE

IEEE MAGNETICS

IEEE PRESS

WILEY

Published by John Wiley & Sons, Inc., Hoboken, New Jersey. All rights reserved
Published simultaneously in Canada

Library of Congress Cataloging-in-Publication Data is available.

ISBN: 978-1-119-00974-0

Printed in the United States of America

10 9 8 7 6 5 4 3 2 1

CONTENTS

ABOUT THE EDITORS

Bernard Dieny has conducted research in magnetism for 30 years. He played a key role in the pioneering work on spin-valves at IBM Almaden Research Center in 1990–1991. In 2001, he co-founded SPINTEC in Grenoble, France, a public research laboratory devoted to spin-electronic phenomena and components. Dieny is co-inventor of 70 patents and has co-authored more than 340 scientific publications. He received an outstanding achievement award from IBM in 1992 for the development of spin-valves, the European Descartes Prize for Research in 2006, and two Advanced Research Grants from the European Research Council in 2009 and 2015. He is co-founder of two companies, one dedicated to magnetic random-access memory, Crocus Technology, the other to the design of hybrid CMOS/magnetic circuits, EVADERIS. In 2011 he was elected Fellow of the Institute of Electrical and Electronics Engineers.

Ronald B. Goldfarb was leader of the Magnetics Group at the National Institute of Standards and Technology in Boulder, Colorado, USA, from 2000 to 2015. He has published over 60 papers, book chapters, and encyclopedia articles in the areas of magnetic measurements, superconductor characterization, and instrumentation. In 2004 he was elected Fellow of the Institute of Electrical and Electronics Engineers (IEEE). From 1995 to 2004 he was editor-in-chief of *IEEE Transactions on Magnetics*. He is the founder and chief editor of *IEEE Magnetics Letters*, established in 2010. He received the IEEE Magnetics Society Distinguished Service Award in 2016.

Kyung-Jin Lee is a professor in the Department of Materials Science and Engineering, and an adjunct professor of the KU-KIST Graduate School of Converging Science and Technology, at Korea University. Before joining the university, he worked for Samsung Advanced Institute of Technology in the areas of magnetic recording and magnetic random-access memory. His current research is focused on understanding the underlying physics of current-induced magnetic excitations and exploring new spintronic devices based on spin-transfer torque. He is co-inventor of 20 patents and has more than 100 scientific publications in the areas of magnetic random-access memory, spin-transfer torque, and spin–orbit torque. He received an outstanding patent award from the Korea Patent Office in 2005 and an award for Excellent Research on Basic Science from the Korean government in 2010. In 2013 he was recognized by the National Academy of Engineering of Korea as a leading scientist in spintronics, "one of the top 100 technologies of the future."

PREFACE

A PERSPECTIVE ON NONVOLATILE MAGNETIC MEMORY TECHNOLOGY

As context for this book, computer memory hierarchy ranges from cache (fastest and most expensive) to main memory, to mass storage (slowest and least expensive). Cache memory, immediately accessible by the central processing unit, is usually static random-access memory (SRAM). Main memory is usually dynamic random-access memory (DRAM); like SRAM, it requires power to maintain its memory state, but additionally must be electrically refreshed, typically every 64 ms (1). Mass storage is exemplified by nonvolatile flash memory (of the NAND and NOR varieties) and magnetic hard-disk drives.

Although magnetic random-access memory (MRAM), in particular spin-transfer torque MRAM (STT-MRAM), has the potential to serve as a universal computer memory—cache, main memory, and mass storage—it will likely be most cost-effective as main memory. MRAM is already a commercial product, albeit expensive and of low density relative to DRAM, and several variations of STT-MRAM are in development. Particular interest is in the superior performance of STT-MRAM with perpendicular magnetic anisotropy. Still under research are three-terminal spin–orbit torque MRAM (2,3), which has high endurance and separate paths for read and write current, and voltage-controlled magnetoelectric MRAM (4–7), which has low energy requirements.

More generally, MRAM, the subject of this book, is characterized by nonvolatility, low energy dissipation, high endurance (repeated writing), scalability to advanced (sub-20 nm) technology nodes, compatibility with complementary metal–oxide semiconductor (CMOS) processing, resistance to radiation damage, and short read and write times. The most intriguing of these, nonvolatility and low energy dissipation, are the main drivers of the technology.

Introduction to Magnetic Random-Access Memory, First Edition. Edited by Bernard Dieny, Ronald B. Goldfarb, and Kyung-Jin Lee.

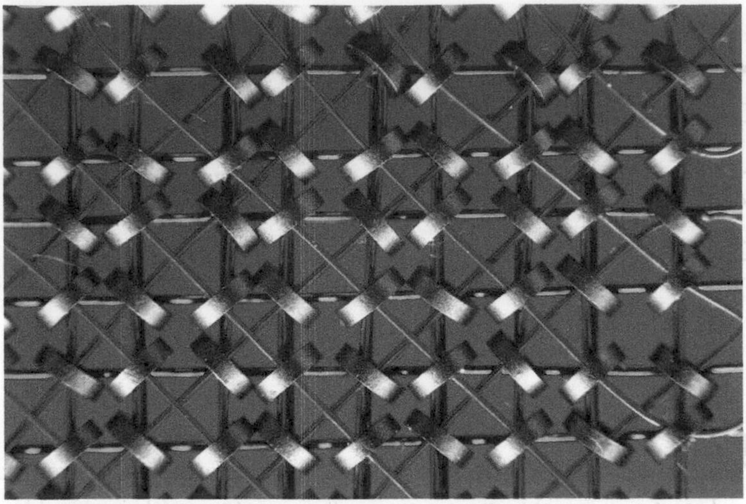

Details of a 1024 bit core plane memory module (11). When magnetic-core memory was introduced in the mid-1950s, toroid cores were about 2 mm in outer diameter.

Nonvolatile MRAM is not really new. It was first developed in the early 1950s by Jay W. Forrester at Massachusetts Institute of Technology (8,9). As envisioned by Forrester, a three-dimensional magnetic-core memory module consisted of circumferentially magnetized toroids strung with x-, y-, and z-plane select wires and a fourth inductively driven output-signal wire. Switching times of the original material studied by Forrester, grain-oriented $Ni_{50}Fe_{50}$ "Deltamax," were on the order of 10 ms. He noted that nonmetallic magnetic ferrites would switch in less than 1 μs based on materials research by William N. Papian (10), and that is how the technology developed. The estimated cost per bit was $1 ($9 today, adjusted for monetary inflation).

Another form of nonvolatile magnetic memory was developed by Andrew H. Bobeck and colleagues at Bell Laboratories in the late 1960s and early 1970s for mass storage: magnetic bubble memory (12,13). It was based on sequential, not random, access. Magnetic bubble domains in synthetic garnets with perpendicular magnetic anisotropy were stabilized by bias fields from permanent magnets. The presence or absence of a bubble—a logic "1" or "0"—was detected with magnetoresistive sensors. Bubbles could be generated, propagated, transferred, replicated, stored, and annihilated. Two orthogonal drive coils provided an in-plane rotating magnetic field to control the magnetization of Ni-Fe bubble-propagation elements (14).

The advent of DRAM in the 1970s, which sacrificed nonvolatility for reduced size, higher speed, and reduced cost, made core memory obsolete for main memory. By the early 1980s, storage density advances and cost reductions in hard-disk drives made bubble memory obsolete for mass storage (although bubble memory continued to be used in military and aerospace applications that required ruggedness).

Besides MRAM, other forms of nonvolatile memory are subjects of intense research. They are based on binary state variables that include "spin, phase, multipole orientation, mechanical position, polarity, orbital symmetry, magnetic flux quanta, molecular configuration, and other quantum states" (15). Mechanisms

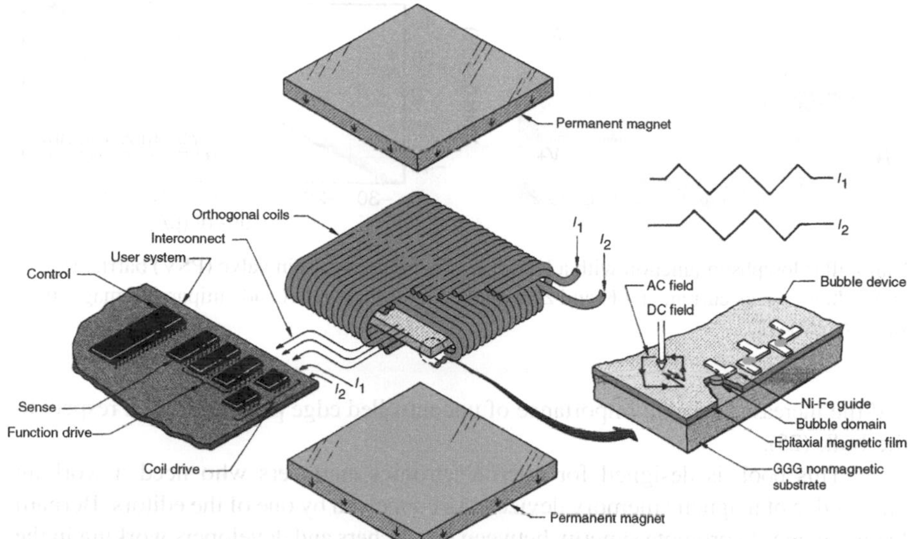

Schematic diagram of the first commercial magnetic bubble memory module, TIB0103, manufactured in 1977 (14). Shown are the bias magnets, drive coils, control and interface circuits, bubble chip, and Ni-Fe "T-bar" bubble-propagation pattern. (An asymmetric chevron pattern was used instead in the TIB0203 in 1978.) The magnetic bubble diameter was 5 μm. The chip had 92,000 bits, a storage density of 155,000 bits/cm^2, an access time of 2–4 ms, and a data transfer rate of 50,000 bits/s.

with great potential include resistive RAM ("memristors") (16) and phase-change RAM (17). One of these may eventually become the dominant technology for cache memory, main memory, or mass storage, but for now, MRAM seems the most promising for main memory. Nevertheless, alternatives to MRAM should not be discounted (18,19).

Coincident with the memory revolution is a rethinking of computer logic and architecture, particularly to address problems of energy consumption in supercomputers and massive data centers. For example, in the United States, the Intelligence Advanced Research Projects Activity (IARPA) has sponsored the development of a prototype cryogenic computer under its "Cryogenic Computing Complexity" program (20). Dramatic reductions are projected in both energy consumption and size (21). Such a computer would combine superconducting, single flux quantum (SFQ) logic with hybrid superconducting/magnetic RAM. Hybrid superconducting/magnetic Josephson junctions switched by spin-transfer torque, resulting in measureable changes in critical current, have been demonstrated (22).

Recent accelerated growth in data centers and their demand for energy are bringing the need for new computer logic and memory to a head. World Wide Web search engines have resorted to storing most of their data in energy-inefficient DRAM (23) because retrieval from mass storage is too slow. There is a need to prototype, test, and benchmark (24) the energy dissipation, high-speed performance, reliability, dimensional scalability, temperature margins, and fabrication reproducibility of MRAM materials, devices, and circuits. Inevitably, new physical phenomena arise as nanostructures shrink in size, and failures will be determined by unknown variables

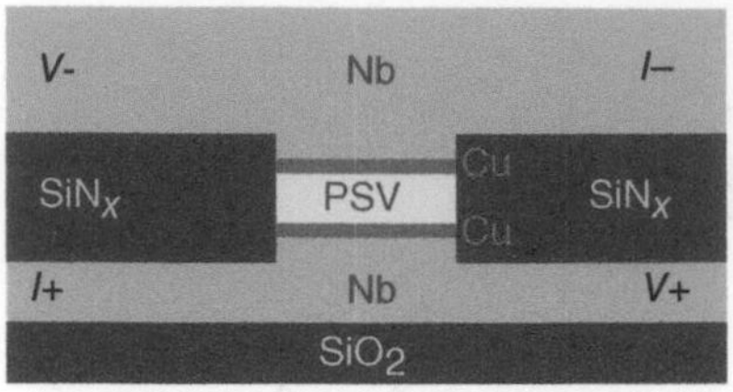

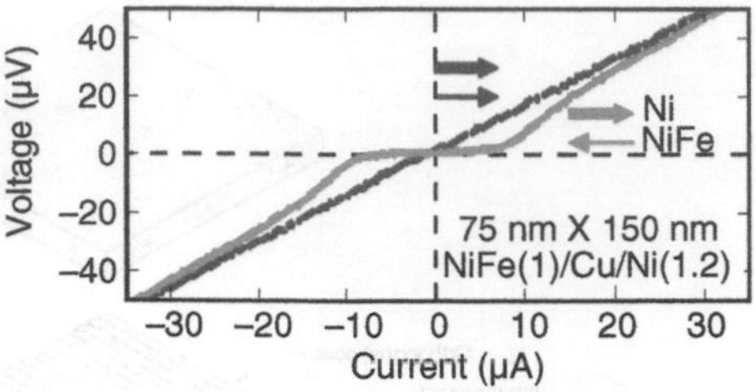

Nanopillar Josephson junction with a $Ni_{0.8}Fe_{0.2}$/Cu/Ni pseudo-spin valve (PSV) barrier (left) and voltage versus current at 4 K and zero applied field for parallel and antiparallel magnetic states (right) (22).

and the increased relative importance of uncontrolled edge properties with respect to the bulk (25).

This book is designed for microelectronics engineers who need a working knowledge of magnetic memory devices. As conceived by one of the editors, Bernard Dieny, it aims to promote synergy between researchers and developers working in the field of electron devices and those in magnetics and information storage.

The chapters in this volume cover basic concepts in spin electronics (spintronics); magnetic properties of materials; micromagnetic modeling; dynamics of magnetic precession and damping; different implementations of MRAM; the integration of MRAM with CMOS; and future hybrid logic-in-memory architectures.

The authors are leaders in their respective fields. Nicolas Locatelli and Vincent Cros are known for their major contributions in the field of spin torque nano-oscillators and nanodevices assembled in novel computer architectures (26,27). Shinji Yuasa is noted for his pioneering experimental work (28,29) on giant magnetoresistance in magnetic tunnel junctions with MgO barriers (30,31). Liliana Buda-Prejbeanu is expert in micromagnetic modeling and computational magnetics (32). Bill Bailey is an authority on magnetization dynamics and spin–orbit coupling (33). Bernard Dieny is famous for his key role in the discovery of giant magnetoresistance in spin valve structures (34,35). Lucian Prejbeanu is known for his work on thermally assisted MRAM (36). Michael Gaidis is a microwave engineer who has specialized in "back-end-of-line" integration of MRAM with CMOS (37). The chapter on non-volatile logic-in-memory (38,39) by Takahiro Hanyu, Hideo Ohno, and the team at Tohoku University represents one of the most complete compilations on this topical subject.

This book was sponsored by the IEEE Magnetics Society and published under the Wiley-IEEE Press imprint. The editor at Wiley-IEEE Press was Mary Hatcher.

RONALD B. GOLDFARB

National Institute of Standards and Technology
Boulder, Colorado, USA

REFERENCES

1. I. Bhati, M.-T. Chang, Z. Chishti, S.-L. Lu, and B. Jacob, "DRAM refresh mechanisms, penalties, and trade-offs," *IEEE Trans. Comput.* 65, pp. 108–121 (2016); doi: 10.1109/TC.2015.2417540.
2. I. M. Miron, K. Garello, G. Gaudin, P.-J. Zermatten, M. V. Costache, S. Auffret, S. Bandiera, B. Rodmacq, A. Schuhl, and P. Gambardella, "Perpendicular switching of a single ferromagnetic layer induced by in-plane current injection," *Nature* 476, pp. 189–194 (2011); doi: 10.1038/nature10309.
3. L.-Q. Liu, C.-F. Pai, Y. Li, H. W. Tseng, D. C. Ralph, and R. A. Buhrman, "Spin-torque switching with the giant spin Hall effect of tantalum," *Science* 336, pp. 555–558 (2012); doi: 10.1126/science.1218197.
4. S. Kanai, M. Yamanouchi, S. Ikeda, Y. Nakatani, F. Matsukura, and H. Ohno, "Electric field-induced magnetization reversal in a perpendicular-anisotropy CoFeB-MgO magnetic tunnel junction," *Appl. Phys. Lett.* 101, 122403 (2012); doi: 10.1063/1.4753816.
5. W.-G. Wang, M. Li, S. Hageman, and C. L. Chien, "Electric-field-assisted switching in magnetic tunnel junctions," *Nat. Mater.* 11, pp. 64–68 (2012); doi: 10.1038/nmat3171.
6. Y. Shiota, T. Nozaki, F. Bonell, S. Murakami, T. Shinjo, and Y. Suzuki, "Induction of coherent magnetization switching in a few atomic layers of FeCo using voltage pulses," *Nat. Mater.* 11, pp. 39–43 (2012); doi: 10.1038/nmat3172.
7. P. K. Amiri, J. G. Alzate, X. Q. Cai, F. Ebrahimi, Q. Hu, K. Wong, C. Grèzes, H. Lee, G. Yu, X. Li, M. Akyol, Q. Shao, J. A. Katine, J. Langer, B. Ocker, and K. L. Wang, "Electric-field-controlled magnetoelectric RAM: progress, challenges, and scaling," *IEEE Trans. Magn.* 51, 3401507 (2015); doi: 10.1109/TMAG.2015.2443124.
8. J. W. Forrester, "Digital information in three dimensions using magnetic cores," *J. Appl. Phys.* 22, pp. 44–48 (1951); doi: 10.1063/1.1699817.
9. J. W. Forrester, "Multicoordinate digital information storage device," U.S. Patent 2,736,880 (filed May 11, 1951, published February 28, 1956); https://www.google.com/patents/US2736880.
10. W. N. Papian, "A coincident-current magnetic memory cell for the storage of digital information," *Proc. IRE* 40, pp. 475–478 (1952); doi: 10.1109/JRPROC.1952.274045.
11. K. Lunzet, CC BY-SA 3.0, http://wikimedia.org/w/index.php?curid=7025574 (2009).
12. A. H. Bobeck, "Properties and device applications of magnetic domains in orthoferrites," *Bell Syst. Tech. J.* 46, pp. 1901–1925 (1967); doi: 10.1002/j.1538-7305.1967.tb03177.x.
13. A. H. Bobeck and H. E. D. Scovil, "Magnetic bubbles," *Sci. Am.* 224 (6), pp. 78–90 (1971); http://www.nature.com/scientificamerican/journal/v224/n6/pdf/scientificamerican0671-78.pdf.
14. D. Toombs, "An update: CCD and bubble memories," *IEEE Spectr.* 15 (4), pp. 22–30 (1978); doi: 10.1109/MSPEC.1978.6367665.
15. G. I. Bourianoff, P. A. Gargini, and D. E. Nikonov, "Research directions in beyond CMOS computing," *Solid-State Electron.* 51, pp. 1426–1431 (2007); doi: 10.1016/j.sse.2007.09.018.
16. E. Gale, "TiO_2-based memristors and ReRAM: materials, mechanisms and models (a review)," *Semicond. Sci. Technol.* 29, 104004 (2014); doi: 10.1088/0268-1242/29/10/104004.
17. M. Wuttig and N. Yamada, "Phase-change materials for rewriteable data storage," *Nat. Mater.* 6, pp. 824–832 (2007); doi: 10.1038/nmat2009.
18. Y. Fujisaki, "Current status of nonvolatile semiconductor memory technology," *Jpn. J. Appl. Phys.* 49, 100001 (2010); doi: 10.1143/JJAP.49.100001.
19. J. S. Meena, S. M. Sze, U. Chand, and T.-Y. Tseng, "Overview of emerging nonvolatile memory technologies," *Nanoscale Res. Lett.* 9, 526 (2014); doi: 10.1186/1556-276X-9-526.
20. Cryogenic Computing Complexity (C3) (2014), http://www.iarpa.gov/index.php/research-programs/c3 (accessed Sept. 15, 2016).
21. D. S. Holmes, A. L. Ripple, and M. A. Manheimer, "Energy-efficient superconducting computing: power budgets and requirements," *IEEE Trans. Appl. Supercond.* 23, 1701610 (2013); doi: 10.1109/TASC.2013.2244634.
22. B. Baek, W. H. Rippard, M. R. Pufall, S. P. Benz, S. E. Russek, H. Rogalla, and P. D. Dresselhaus, "Spin-transfer torque switching in nanopillar superconducting-magnetic hybrid Josephson junctions," *Phys. Rev. Appl.* 3, 011001 (2015); doi: 10.1103/PhysRevApplied.3.011001.

23. J. Ousterhout, "The volatile future of storage," *IEEE Spectr.* 52 (11), pp. 34 ff. (2015); doi: 10.1109/MSPEC.2015.7335899.
24. D. E. Nikonov and I. A. Young, "Benchmarking of beyond-CMOS exploratory devices for logic integrated circuits," *IEEE J. Exploratory Solid-State Comput. Devices Circuits* 1, pp. 3–11 (2015); doi: 10.1109/JXCDC.2015.2418033.
25. H. T. Nembach, J. M. Shaw, T. J. Silva, W. L. Johnson, S. A. Kim, R. D. McMichael, and P. Kabos, "Effects of shape distortions and imperfections on mode frequencies and collective linewidths in nanomagnets," *Phys. Rev. B* 83, 094427 (2011); doi: 10.1103/PhysRevB.83.094427.
26. N. Locatelli, V. Cros, and J. Grollier, "Spin-torque building blocks," *Nat. Mater.* 13, pp. 11–20 (2014); doi: 10.1038/NMAT3823.
27. N. Locatelli, V. V. Naletov, J. Grollier, G. de Loubens, V. Cros, C. Deranlot, C. Ulysse, G. Faini, O. Klein, and A. Fert, "Dynamics of two coupled vortices in a spin valve nanopillar excited by spin transfer torque," *Appl. Phys. Lett.* 98, 062501 (2011); doi: 10.1063/1.3553771.
28. S. Yuasa, T. Nagahama, A. Fukushima, Y. Suzuki, and K. Ando, "Giant room-temperature magnetoresistance in single-crystal Fe/MgO/Fe magnetic tunnel junctions," *Nat. Mater.* 3, pp. 868–871 (2004); doi: 10.1038/nmat1257.
29. S. Yuasa and D. D. Djayaprawira, "Giant tunnel magnetoresistance in magnetic tunnel junctions with a crystalline MgO(0 0 1) barrier," *J. Phys. D Appl. Phys.* 40, pp. R337–R354 (2007); doi: 10.1088/0022-3727/40/21/R01.
30. W. H. Butler, X.-G. Zhang, T. C. Schulthess, and J. M. MacLaren, "Spin-dependent tunneling conductance of Fe|MgO|Fe sandwiches," *Phys. Rev. B* 63, 054416 (2001); doi: 10.1103/PhysRevB.63.054416.
31. S. S. P. Parkin, C. Kaiser, A. Panchula, P. M. Rice, B. Hughes, M. Samant, and S.-H. Yang, "Giant tunneling magnetoresistance at room temperature with MgO(1 0 0) tunnel barriers," *Nat. Mater.* 3, pp. 862–867 (2004); doi: 10.1038/nmat1256.
32. L. D. Buda, I. L. Prejbeanu, U. Ebels, and K. Ounadjela, "Micromagnetic simulations of magnetisation in circular cobalt dots," *Comput. Mater. Sci.* 24, pp. 181–185 (2002); doi: 10.1016/S0927-0256(02)00184-2.
33. W. E. Bailey, L. Cheng, D. J. Keavney, C. C. Kao, E. Vescovo, and D. S. Arena, "Precessional dynamics of elemental moments in a ferromagnetic alloy," *Phys. Rev. B* 70, 172403 (2004); doi: 10.1103/PhysRevB.70.172403.
34. B. Dieny, V. S. Speriosu, S. S. P. Parkin, B. A. Gurney, D. R. Wilhoit, and D. Mauri, "Giant magnetoresistance in soft ferromagnetic multilayers," *Phys. Rev. B* 43, pp. 1297–1300 (1991); doi: 10.1103/PhysRevB.43.1297.
35. B. Dieny, "Giant magnetoresistance in spin-valve multilayers," *J. Magn. Magn. Mater.* 136, pp. 335–359 (1994); doi: 10.1016/0304-8853(94)00356-4.
36. I. L. Prejbeanu, M. Kerekes, R. C. Sousa, H. Sibuet, O. Redon, B. Dieny, and J. P. Nozieres, "Thermally assisted MRAM," *J. Phys Condens. Matter* 19, 165218 (2007); doi: 10.1088/0953-8984/19/16/165218.
37. M. C. Gaidis, E. J. O'Sullivan, J. J. Nowak, Y. Lu, S. Kanakasabapathy, P. L. Trouilloud, D. C. Worledge, S. Assefa, K. R. Milkove, G. P. Wright, and W. J. Gallagher, "Two-level BEOL processing for rapid iteration in MRAM development," *IBM J. Res. Dev.* 50, pp. 41–54 (2006); doi: 10.1147/rd.501.0041.
38. S. Matsunaga, J. Hayakawa, S. Ikeda, K. Miura, H. Hasegawa, T. Endoh, H. Ohno, and T. Hanyu, "Fabrication of a nonvolatile full adder based on logic-in-memory architecture using magnetic tunnel junctions," *Appl. Phys. Exp.* 1, 091301 (2008); doi: 10.1143/APEX.1.091301.
39. T. Hanyu, D. Suzuki, N. Onizawa, S. Matsunaga, M. Natsui, and A. Mochizuki,"Spintronics-based nonvolatile logic-in-memory architecture towards an ultra-low-power VLSI computing paradigm," *Proceedings of the Design, Automation & Test in Europe Conference* (2015), pp. 1006–1011; http://dl.acm.org/citation.cfm?id=2757048.

CHAPTER 1

BASIC SPINTRONIC TRANSPORT PHENOMENA

Nicolas Locatelli[1,2] *and Vincent Cros*[2]

[1]Unité Mixte de Physique, CNRS, Thales, Univ. Paris-Sud, Université Paris-Saclay, 91767 Palaiseau, France

[2]Centre de Nanosciences et de Nanotechnologies, CNRS, Univ. Paris-Sud, Université Paris-Saclay, 91405 Orsay France

Spintronics is a merger of magnetism and electronics. Conventional electronics uses only the charge of the electrons. For instance, transistors are based on the modulation of the density of electrons in a semiconductor channel by an electric field. Semiconductor-based memory (e.g., DRAM, Flash) stores information in the form of an amount of charge stored in a capacitor. In contrast, spintronics uses the spin of the electrons in addition to their charge to obtain new properties and use these properties in innovative devices. The spin of the electrons is an elementary magnetic moment carried by each electron. It has a quantum mechanical origin. Magnetic materials can be used as polarizers or analyzers for electron spins. This is why most spintronic devices combine magnetic and nonmagnetic materials, which can be metals, semiconductors, or insulators.

Magnetism has been used for a long time for data storage applications. Indeed, information can be stored in some magnetic materials in the form of a magnetization orientation. This was developed for storage on magnetic tapes as well as in magnetic hard disk drives (HDDs). The increase in the demand for storage capacity has stimulated an increase by eight orders of magnitude in the areal density of information stored in HDDs over the past 50 years; the bit area has decreased by the same factor. In 2014, bit sizes are typically on the order of 40 nm × 15 nm. This decrease in bit size has required continual improvements in the storage medium, in the write head used to switch the magnetization in the medium, and in the read head used to read out the magnetic state. This field of magnetic recording has benefited strongly from research and development in the field of spintronics. In particular, the discoveries of giant magnetoresistance in 1988 (1) and tunnel magnetoresistance at room temperature in

Introduction to Magnetic Random-Access Memory, First Edition. Edited by Bernard Dieny, Ronald B. Goldfarb, and Kyung-Jin Lee.

1995 (2,3) have been major breakthroughs from a scientific point of view, but they also helped recording technology keep moving forward. In 2010, a total of around 12,000 PB (10^{15} bytes) of storage capacity contained in 674.6 million HDDs were shipped worldwide.

Another type of spintronic devices that was proposed in the late 1990s is magnetic random-access memory (MRAM) (4). Indeed, solid state memory is of primary importance both for storage (the introduction of solid state drives in personal computers, tablets, and handheld devices) and for fast working memory between logic units and hard disk drives. In these applications, random-access memory based on devices involving magnetic materials, called magnetic tunnel junctions, are among the most promising technologies for future nonvolatile data storage, and may replace, in the near future, semiconductor-based memory (i.e., DRAM and SRAM), which represents a huge market.

This chapter is an introduction to the physical concepts required to understand how information is stored in a magnetic data cell, how this information can be detected, and how it is possible to modify the information by switching the magnetization from one state to another. First we introduce the basics of electronic transport in magnetic materials, a concept that is required for the comprehension of the physical mechanisms at the origin of magnetoresistive properties: giant magnetoresistance (GMR) and tunneling magnetoresistance (TMR), which is the magnetoresistive effect at play in spin-transfer torque (STT) MRAM devices. Then we describe how a spin-polarized current can exert a STT on the magnetizations in nanostructured spintronic devices by the interaction with local magnetic moments. Finally, we show how this novel effect can be used to modify the state of a magnetic element, leading to current-induced magnetization switching as the writing process in STT-MRAM.

1.1 GIANT MAGNETORESISTANCE

After introducing the basic concepts of electronic transport in ferromagnetic metals, a simple model of the GMR effect is presented, the so-called "two-current model" that was proposed to describe the dependence of the electrical resistance of magnetic multilayered stacks on their magnetic configuration. This model is helpful for the understanding of the basic principles of spin-dependent transport. Finally, the main applications of GMR are discussed.

1.1.1 Basics of Electronic Transport in Magnetic Materials

Magnetism, as produced by magnetite (Fe_3O_4), has been known from at least ancient Greek times. It was described as a force, either attractive or repulsive, that can act at distance. The origin of this force is due to a magnetic field that is created by some materials (called magnets), or is induced by the motion of electrons, that is, electrical currents. In magnetic materials, such as iron (Fe) or cobalt (Co), sources of magnetization are mainly the electrons' intrinsic magnetic moment associated with spin angular momentum, or simply "spin," and also to the electrons' orbital angular momentum. In Nature, other sources of magnetism are due to nuclear magnetic moments of the nuclei, typically thousands of times smaller than the electrons' magnetic moments. Consequently, these nuclear moments are negligible in the context of the magnetization of materials. However, they play an important role in nuclear magnetic resonance (NMR) and magnetic resonance imaging (MRI).

The spin magnetic moment $\vec{\mu}_S$ and spin angular momentum $\vec{S}$ are linked through the relationship $\vec{\mu}_S = -g\mu_B\vec{S}$, where g is a dimensionless number called the g-factor (or Landé factor) and $\mu_B = e\hbar/2m_e$ is the Bohr magneton. In this expression, $e = 1.60 \times 10^{-19}$C is the electron charge, $m_e = 9.31 \times 10^{-31}$ kg is the electron mass, and $\hbar = 1.05 \times 10^{-34}$ m^2 kg/s is the reduced Planck constant. Due to the quantum mechanical nature of the spin, measurement of the projection of the electron spin $\vec{S}$ on any direction can take only two values: +1/2 and −1/2.

In the context of spintronics, the main question is to understand how this fundamental characteristic property of the electrons, that is, the spin, influences the mobility of the electrons in materials. In fact, although it was suggested by N. Mott in 1936, the influence of spin on the transport properties in a ferromagnetic material was clearly demonstrated experimentally and described theoretically only in the late 1960s (for a review, see Ref. 5). The property of spin-dependent transport is at the heart of not only the GMR effect but also all related effects that have allowed the development of spintronic devices.

The ferromagnetic transition metals, such as Fe, Co, Ni, and their alloys, which are the key compounds in today's spintronic devices, have a specific electronic band structure compared to normal (nonmagnetic) metals. In transition metals, the two highest filled energy bands, which are the conduction bands, are occupied by $3d$ and $4s$ electrons. This nomenclature refers to atomic orbitals of the electrons. They are labeled s-orbital, p-orbital, d-orbital, and f-orbital referring to orbitals with angular momentum quantum number $l = 0$, 1, 2, and 3, respectively. These nomenclatures indicate the orbital shape and are used to describe the electron configuration. In a crystal, electrons with similar orbitals associate to fill a band of energy. More precise description on this topic can be found in any quantum mechanics textbook. In the case of ferromagnetic transition metals, each of these bands splits in two subbands corresponding to each spin configuration (see Fig. 1.1a). And in these magnetic materials, the interaction between spins, called the *exchange interaction*, energetically favors a parallel orientation of the electrons' spins.

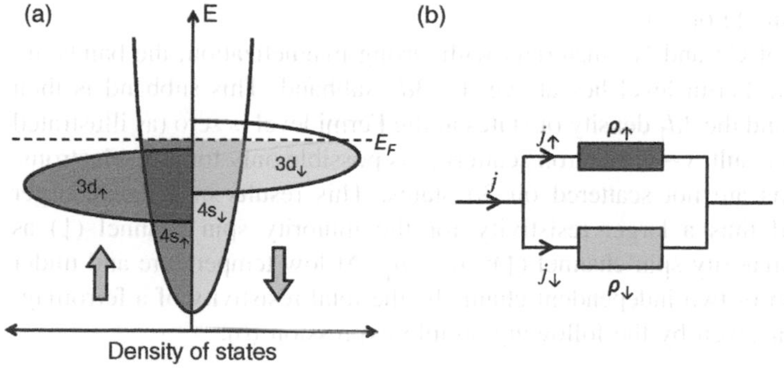

Figure 1.1 (a) Schematic representation of the band structure of a transition metal with strong ferromagnetic properties such as Co or Ni. (b) Equivalent circuit for the two-spin subbands in the "two-current" model.

In the following, we will refer to electrons with a magnetic moment aligned parallel to the local magnetization as "spin-up" (↑) electrons, and to electrons with a magnetic moment aligned antiparallel to the local magnetization as "spin-down" (↓) electrons.

As for ferromagnetic metals, similar to a normal metal, the 4s band contains almost an equal number of spin-up and spin-down electrons. But the specificity of ferromagnetic metals lies in the structure of their 3*d* bands, for which the resulting lowest energy situation corresponds to a shift of the two subbands $3d_{\uparrow}$ and $3d_{\downarrow}$ (Fig. 1.1a). This offset generates an asymmetry for the number of electrons of each orientation, also responsible for the spontaneous magnetization. Consequently, they are also known as majority spin (↑) and minority spin (↓) electrons. It finally creates, for each spin orientation, a difference between spin-up and spin-down densities of states at the Fermi energy E_F. We remind here that the density of states (DOS) *D(E)* of a system describes the number of states *dn(E)* per interval of energy *dE* around each energy level *E* that are available to be occupied by electrons: $dn(E)=D(E)dE$. The Fermi level E_F corresponds to the highest energy level occupied by electrons in a system at a temperature $T=0$ K. Electrons involved in the transport process lie at (or close to) the Fermi level.

In the low-temperature limit, one considers that the electron's spin is conserved during most scattering events. Under this assumption, transport properties associated with spin-up and spin-down electrons can be represented by two independent parallel conduction channels (Fig. 1.1b), and the mixing of these two conduction channels is then considered as negligible. In ferromagnetic metals, these two channels have different resistivities $\rho_{\uparrow}$ and $\rho_{\downarrow}$, which depend on whether the electron magnetic moment is parallel (↑) or antiparallel (↓) to the direction of the local magnetization.

In a first, simple approximation, we can consider that 4*s* electrons, which are fully delocalized in the metal because they belong to outer electronic shells, constitute the conduction electrons that carry most of the current. In contrast, 3*d* electrons are more localized and responsible for the magnetic properties of the metal. The overlapping of *s* and *d* bands at the Fermi level allows current-carrying 4*s* electrons to be scattered on the localized 3*d* states, on the condition that they have the same energy and the same spin. The difference between the density of states of (↑) and (↓) 3*d* electrons at the Fermi level, therefore, results in different scattering probabilities for 4*s* electrons with spin (↑) or (↓).

In the case of Co and Ni, materials with strong magnetization, the bands are filled such that the Fermi level lies above the $3d_{\uparrow}$ subband. This subband is then completely filled and the $3d_{\uparrow}$ density of states at the Fermi level is zero (as illustrated in Fig. 1.1a). As a result, $s\rightarrow d$ electron scattering is possible only for s (↓) electrons, while (↑) electrons are not scattered on 3*d* states. This results in a much larger diffusion rate and thus a larger resistivity for the minority spin channel (↓) as compared to the majority spin channel (↑): $\rho_{\uparrow}<\rho_{\downarrow}$. At low temperature and under this approximation of two independent channels, the total resistivity of a ferromagnetic metal is then given by the following simple expression (6):

$$\rho=\frac{\rho_{\uparrow}\rho_{\downarrow}}{\rho_{\uparrow}+\rho_{\downarrow}} \tag{1.1}$$

At high temperatures, some additional scattering of conduction electrons, for instance, by spin waves (propagating perturbations in the magnetic materials), can cause *spin-flip* events, that is, a mixing of the two conduction channels, but those can be ignored in first approximation up to room temperature.

Two definitions for the spin asymmetry coefficient of a given ferromagnetic material are used in the literature, $\alpha = \rho_\downarrow/\rho_\uparrow$ or $\beta = (\rho_\uparrow - \rho_\downarrow)/(\rho_\uparrow + \rho_\downarrow)$. In strong ferromagnets such as Co or Ni, $\rho_\uparrow < \rho_\downarrow$ and thus $\alpha > 1$.

A major consequence of the resistivity difference between conduction channels of minority and majority spin is that most of the current flows through the low resistivity spin (↑) channel. Consequently, an asymmetry in the current densities associated with (↑) and (↓) electrons appears. Hence, the current flowing in the ferromagnetic material is *spin polarized*. Calling $j_\uparrow$ and $j_\downarrow$ the current densities of spin (↑) and (↓) electrons, respectively, and p the current spin polarization, p is defined by $p = (j_\uparrow - j_\downarrow)/(j_\uparrow + j_\downarrow)$. Note that $p = \beta$ at low temperature.

1.1.2 A Simple Model to Describe GMR: The "Two-Current Model"

Historically, the *two-current model*, proposed by Mott and then by Fert and Campbell, was developed to explain the spin-dependent resistivity in materials doped with magnetic impurities. It allows the anticipation, in a rather simple way, of the GMR effect in magnetic multilayers. For this, we consider an archetypal multilayered stack consisting of thin layers of alternating ferromagnetic metals (F) and nonmagnetic (NM) metals. The magnetization of the ferromagnetic layers is supposed to be uniform within each layer. We also assume that the relative orientation of the magnetization in the successive F layers can somehow be changed from parallel (*P*) to antiparallel (*AP*) magnetic configuration, as illustrated in Fig. 1.3. The way this is achieved will be explained in more detail in the next section.

Two geometries to evaluate the resistance of this multilayered structure can be considered: either with current flowing parallel to the plane of the layers (known as "current-in-plane GMR," CIP-GMR) or with current flowing in the direction perpendicular to the plane of the layers (known as "current-perpendicular-to-plane GMR," CPP-GMR). The same model can be used to evaluate the magnetoresistive properties for both geometries, provided that the layers' thickness remains small compared to a characteristic length associated with each geometry.

For the CIP case, the characteristic length is actually the mean free path λ. For the CPP case, it is the spin-flip length or spin diffusion length l_{sf} (7).

As illustrated in Fig. 1.2, during their Brownian motion throughout the structure with an average drift along the electrical field direction, the electrons traverse successive ferromagnetic layers. We denote $r/2$ the resistance associated with traversing a F layer for the majority spin channel (same direction as the magnetization) and $R/2$ the corresponding resistance for the minority spin channel (opposite direction to the magnetization), with $r < R$. $r/2$ and $R/2$ are associated with the average resistance sensed by the electrons as they spend half of their total path, respectively, in the majority or minority spin channels. For the sake of simplicity, let us also assume that the resistance of the nonmagnetic separating layer is much smaller

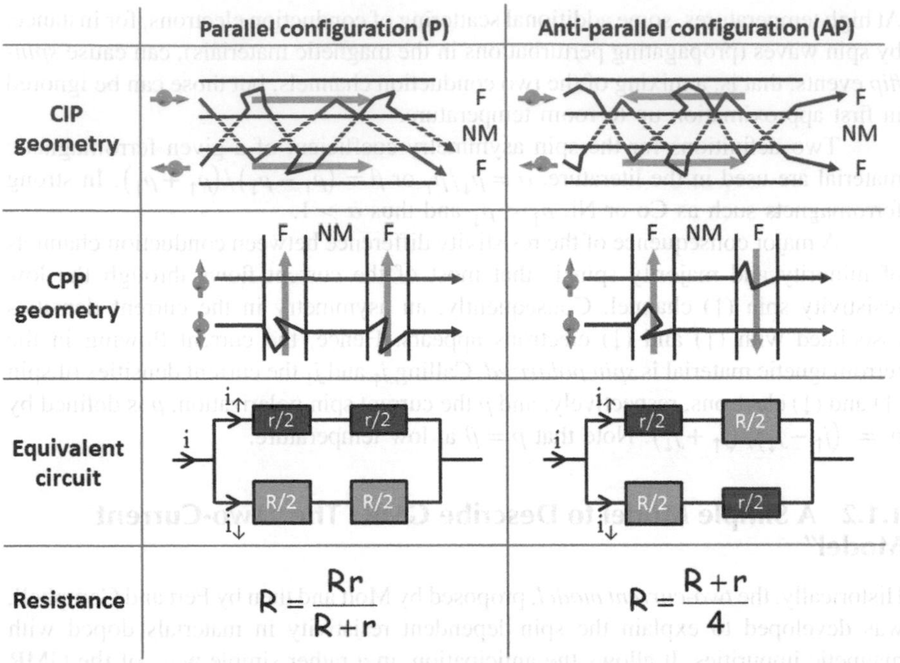

Figure 1.2 Illustration of the two-current model. The conduction paths of spin-up and spin-down electrons in a ferromagnetic metal/normal metal/ferromagnetic metal (F/N/F) multilayer, in the two cases of CIP and CPP transport. Conduction electrons with spin magnetic moment aligned antiparallel (blue paths) to the local magnetization experience more scattering events than those with parallel spin (red paths). The equivalent resistance circuit is represented for the two magnetic configurations: parallel and antiparallel.

than r and R. Then, in the P configuration, spin (↑) and (↓) electrons behave, respectively, as majority and minority electrons in all magnetic layers. As a result, the respective resistances of the two spin channels are $r_\uparrow = r$ and $r_\downarrow = R$. Since these two channels conduct the current in parallel, the equivalent resistance of the F/NM/F stack can be written as $r_P = rR/(r+R)$. In the case of materials with large spin asymmetry ($\alpha \gg 1$ and $r \ll R$), the multilayer can be considered short-circuited by the spin (↑) channel; its equivalent resistance is $r_P \approx r$.

For the AP configuration, the electrons alternatively behave as majority or minority electrons as they propagate from one ferromagnetic layer to another. As a result, they are alternatively weakly and strongly scattered. Thus, the short-circuit effect previously mentioned in P configuration is here suppressed. In the AP configuration, the two channels have the same resistance $(R+r)/2$. The F/NM/F equivalent resistance is then $r_{AP} = (R+r)/4$, which is, in general, much larger than $r_P = r$.

Following this model, one can finally derive a simple expression for the amplitude of the GMR ratio:

$$\mathrm{GMR} = \frac{r_{AP} - r_P}{r_P} = \frac{(R-r)^2}{4Rr}. \tag{1.2}$$

Let us mention that another definition of the GMR ratio is sometimes used in the literature, notably in theoretical articles. It consists of normalizing the resistance variation between *P* and *AP* configurations by the resistance in the *AP* configuration: $GMR = (r_{AP} - r_P)/r_{AP}$. In this definition, the GMR amplitude has a maximum value of 100%, whereas the commonly used definition often leads to magneto-resistive ratios over 100%. The GMR ratio is of prime importance for the characterization of the resistance variation, which is measured to determine the magnetic state of the stack.

1.1.3 Discovery of GMR and Early GMR Developments

The characteristic length scale of spin-dependent diffusion in thin films is on the order of a few nanometers in magnetic materials and tens of nanometers in nonmagnetic materials. These numbers explain why it took almost 20 years between the first basic studies on spin-dependent transport carried out on magnetic alloys in the late 1960s and the GMR discovery. GMR could be observed only in multilayered stacks consisting of nanometer thick layers. The growth of such multilayers became possible in the 1980s, thanks to the development of a new growth technique adapted from semiconductor industry: molecular beam epitaxy (MBE). GMR was actually discovered in magnetic metallic multilayers consisting of alternating layers of iron and chromium (Fe/Cr). This discovery by Albert Fert in Orsay, France, and Peter Grünberg in Jülich, Germany, in 1988, consisted of a very large variation in the CIP electrical resistance of these stacks under the application of an external magnetic field. Due to an antiferromagnetic coupling that exists between the successive Fe layers across the Cr spacers, the magnetization in the successive Fe layers spontaneously orient themselves in an *AP* configuration in zero magnetic field, as represented in Fig. 1.3. Upon application of a large enough magnetic field to overcome this antiferromagnetic coupling, the magnetization of all Fe layers can be saturated in the direction of the field, resulting in a *P* magnetic configuration. The GMR consists in a very large drop of resistance of 80% at 4 K (50% at 300 K) between the *AP* and *P* configurations. In 1988, it has been named "giant magnetoresistance" because the GMR amplitude was much larger than all known magnetoresistance effects at room temperature at that time. This discovery of GMR is considered the starting point of spinelectronics or spintronics. Almost immediately, GMR attracted enormous interest both from the point of view of fundamental physics and also for its possible applications, especially in the fields of data storage and magnetic field sensors. Fert and Grünberg were awarded the Nobel Prize in Physics in 2007 for this discovery.

Research on magnetic multilayers and GMR rapidly became a very active topic. It is not our aim here to provide an exhaustive review of all experimental and theoretical results that followed the initial discovery. A more complete review can be found in Ref. 6. Here we will rather introduce some of the key advances in GMR that occurred in the first years of spintronics.

Parkin et al. first demonstrated in 1990 the existence of GMR in multilayers prepared by sputtering (8), a simpler and faster physical vapor deposition (PVD) technique: a technique that is compatible with industrial processes. In magnetic multilayers consisting of alternating ferromagnetic and nonmagnetic layers, they also demonstrated the existence of oscillations in the FM interlayer coupling across the

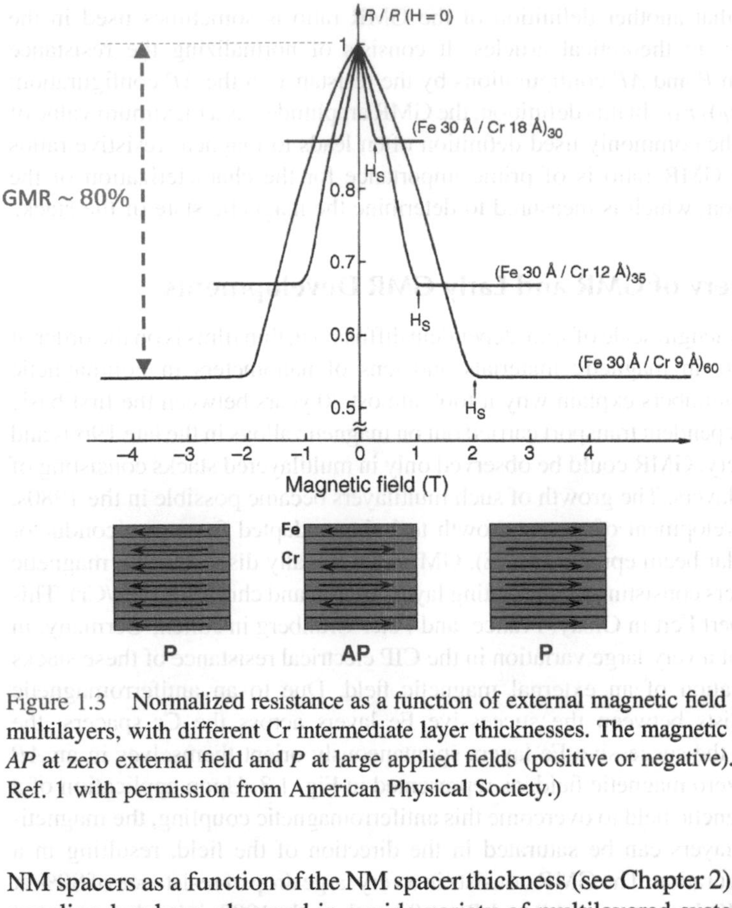

Figure 1.3 Normalized resistance as a function of external magnetic field $\mu_0 H$ for Fe/Cr multilayers, with different Cr intermediate layer thicknesses. The magnetic configuration is *AP* at zero external field and *P* at large applied fields (positive or negative). (Adapted from Ref. 1 with permission from American Physical Society.)

NM spacers as a function of the NM spacer thickness (see Chapter 2). This oscillatory coupling has been observed in a wide variety of multilayered systems, in particular (Fe/Cr) and (Co/Cu) multilayers. Another crucial step toward applications of GMR was also made in 1991 by Dieny et al., who developed trilayered ferromagnet/nonmagnetic metal/ferromagnet (F/NM/F) structures called *spin valves*, which exhibit GMR at low magnetic fields (a few millitesla instead of a few teslas for (Fe/Cr) multilayers (9). In these trilayers, the magnetization of one of the ferromagnetic layers is pinned in a fixed direction using a phenomenon called exchange bias, whereas the magnetization of the second ferromagnetic layer can be easily switched in the direction of the applied field. It is then possible to change the magnetic configuration of these spin valves from *P* to *AP* by applying a weak magnetic field (a few millitesla) parallel or antiparallel to the direction of magnetization of the pinned layer. These systems, which exhibit very high 'resistance versus field' sensitivity, were introduced as magnetoresistive heads in hard disk drives by IBM in 1998.

1.1.4 Main Applications of GMR

GMR has been primarily used as spin valve magnetoresistive heads in HDDs between 1998 and 2004. It was subsequently replaced by TMR heads, which exhibit even larger magnetoresistance amplitude. GMR sensors are being used in other types of applications in the automotive industry, robotics, as three-dimensional position sensors, and sensors for biotechnological applications.

Some attempts have been made to use CIP-GMR for memory applications (MRAM), but only low-memory densities of about 1 Mbit per chip could be achieved. This memory is used mainly for space applications because of its radiation hardness (10).

1.2 TUNNELING MAGNETORESISTANCE

The development of artificial magnetic systems based on magnetic tunnel junctions (MTJs) was a second major breakthrough in spin electronics. Magnetic tunnel junctions look like spin valves from a magnetic point of view (two ferromagnetic layers separated by a nonmagnetic spacer) but a major difference is that the nonmagnetic spacer consists of a very thin insulating layer. In these junctions, the current flows perpendicular to the plane of the layers so that the electrons have to tunnel from one ferromagnetic layer to the other one across the thin insulating barrier.

After the pioneering work of Jullière (11) on Fe–GeO–Co junctions at 4.2 K in 1973, it was not until the mid-1990s that the improvement of both growth techniques and lithography processes allowed the fabrication of reliable MTJs. The devices studied used amorphous aluminum oxide (Al_2O_3) as the insulating barrier (2,3). They led to the first measurements of large magnetoresistive effects (TMR) with ratios on the order of 10–70% at room temperature. A great advantage of TMR with respect to GMR is the much larger impedance of MTJs compared to GMR metallic structures in the CPP geometry. Indeed, in MTJs, the resistance of the structure is effectively determined by the thickness of the tunnel barrier. MTJs can be patterned in the form of deeply submicrometer pillars with resistance ranging from kilohms to megohms, depending on the barrier thickness. This makes MTJs easier to integrate with CMOS components such as transistors, which have typical resistance in conducting mode of a few kilohms. In contrast, GMR submicrometer pillars in CPP geometry have resistances on the order of a few tens of ohms, which is fine for sensor applications, but difficult to integrate with CMOS components, in particular for MRAM applications.

An important research effort was undertaken on the materials side to improve the TMR amplitude. In 2004, it was discovered that much larger TMR ratios of about 250% at room temperature could be obtained in MTJs containing a monocrystalline magnesium oxide (MgO) barrier instead of an amorphous alumina barrier (12,13), reaching up to 600% at room temperature. These improvements have allowed a large increase of the sensitivity of magnetic HDD heads, which are needed for increased bit density. Thanks to their higher impedance than CPP-GMR devices, they also enabled new types of MRAM that could be scaled down in size and ramped up in density. Magnetic tunnel junctions are today the core elements of all MRAM technology (see Chapter 5).

As already mentioned, the transport mechanism across MTJs is no longer Ohmic transport as in GMR but relies on the well-known quantum mechanical tunneling effect. We start our second section by summarizing the basics of quantum mechanical tunneling. This will allow us to introduce Jullière's model for TMR, giving an intuitive explanation of the magnetoresistive effect in magnetic tunnel junctions. This introduction will be completed by a description of a more accurate model (Slonczewski's model), and the spin filtering effect. Lastly, the important matter of voltage dependence of TMR will be discussed.

1.2.1 Basics of Quantum Mechanical Tunneling

In classical physics, charge transport through an insulating layer (even if it is ultra-thin) is forbidden. Hence, tunnel conduction through a potential barrier is a pure quantum mechanical phenomenon, called the *tunnel effect*. This effect, predicted in the early years of quantum physics, now has important applications in semiconductor devices such as the tunnel diodes. It is extensively described in elementary textbooks on quantum mechanics: we will only summarize here the main points relevant to MTJ physics.

In Fig. 1.4, the potential landscape for a (injector) metal–insulator–metal (collector) junction is schematically depicted. The main characteristics of the insulating barrier are its energy height Φ, and its thickness d. We consider an electron with an energy E, propagating in the metallic injector in the direction of the stack (perpendicular to the layers), with a wave vector amplitude $k_{\perp} = \sqrt{2m_e E/\hbar^2}$, where m_e is the electron effective mass and $\hbar$ is the reduced Planck constant. Based on the *free-electron approximation*, the resolution of Schrödinger's equation demonstrates that the electron has a nonzero probability of propagating through the insulating barrier and inside the collector electrode. We recall that in the free electron approximation, it is considered that the electrons are not subjected to any confining potential inside the metal or the barrier. The only potential variation is due to the insulating barrier. This model is well suited to 4*s* electrons, which are delocalized in the crystal.

Within the tunnel barrier, the electron wave function decays exponentially so that the probability for the electron to tunnel through the insulating barrier is given by

$$T(E) \propto e^{-2\kappa d}, \tag{1.3}$$

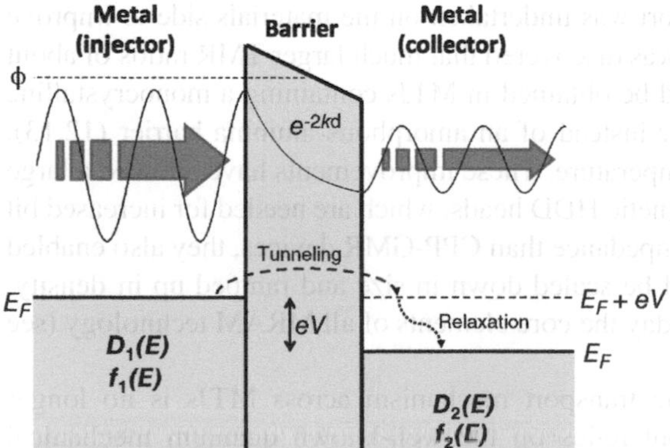

Figure 1.4 Schematics of the wave function of an electron tunneling between two metallic layers across a potential barrier.

with the decay coefficient $\kappa(E) = \sqrt{2m_e(\Phi - E)/\hbar^2}$. A first conclusion from these expressions is that in order to limit the MTJ resistance with typically used materials (MgO, Al_2O_3), the thickness of the insulating layer has to be on the order of a nanometer, corresponding to a few atomic layers. As an illustration, considering that an electron has to cross a $(\Phi - E_F) = 1\ eV$ barrier, it would experience a decay length of $1/\kappa \approx 0.2$ nm.

In MTJs at zero bias voltage, the Fermi levels of the two electrodes align so that the same number of electrons are steadily tunneling from one side of the barrier to the other and vice versa. In order to obtain a net nonzero current flow, a bias voltage has to be applied between the two metallic electrodes. When a bias voltage V is applied, the collector electrode's Fermi level is lowered by eV relative to the injector's, so that electrons tunnel from injector to collector (see Fig. 1.4). The resulting current then depends on the barrier properties, but also on the states accessible on both sides of the barrier. Indeed, according to the Fermi's golden rule, the probability for an electron having an energy E to tunnel through the barrier from metal 1 to metal 2 is proportional to the number of unoccupied electron states in 2 (collector) at this energy. In addition, the number of electrons that are candidates for tunneling is proportional to the number of occupied state in 1 (injector) at E. Therefore, the tunneling current from 1 to 2 due to electrons with energy E can be written as

$$I_{1\to2}(E) \propto D_1(E) f_1(E)\, T(E)\, D_2(E+eV)(1 - f_2(E+eV)), \qquad (1.4)$$

where $D_1(E)$ and $D_2(E+eV)$ are the density of states (DOS), respectively, in electrode 1 at the energy E and in electrode 2 at $(E+eV)$, and the functions $f_1(E)$ and $f_2(E)$ are the Fermi–Dirac distributions, which give the occupation probabilities of states in electrodes 1 and 2. Consequently, the product $D_1(E) f_1(E)$ represents the probability, in electrode 1, of having an electron with the energy E and $D_2(E+eV)(1 - f_2(E+eV))$ the probability, in electrode 2, of having an unoccupied state at the energy $(E+eV)$. Finally, the term $T(E)$ is the previously described transmission coefficient.

Using Eq. (1.4) to sum $I_{1\to2}(E) - I_{2\to1}(E)$ overall energies, the total tunneling current may be calculated. In the limit of zero temperature and small voltage V, it can be shown that only electrons at the Fermi level E_F contribute to the current. The resulting conductance is then simply proportional to the product of the densities of states at the Fermi energy in the two electrodes:

$$G_{T=0} \propto D_1(E_F) \cdot D_2(E_F). \qquad (1.5)$$

1.2.2 First Approach to Tunnel Magnetoresistance: Jullière's Model

The simplified approach described in the previous section for deriving the amplitude of the tunneling current was used by Jullière in 1975 to analyze his pioneer results of tunnel magnetoresistance in a MTJ composed of two ferromagnetic thin films, iron (Fe) and cobalt (Co), separated by a thin GeO semiconducting layer (11). The ferromagnetic character of the electrodes was taken into account by introducing different densities of states $D^{\uparrow}$ for the majority and $D^{\downarrow}$ for the minority electrons, using the following definition for the spin polarization of a ferromagnet:

$$P_0 = \frac{D^{\uparrow}(E_F) - D^{\downarrow}(E_F)}{D^{\uparrow}(E_F) + D^{\downarrow}(E_F)}. \tag{1.6}$$

As for the GMR, the TMR can then be understood within the two-current model by summing in parallel the conductances of the spin-up and spin-down channels assuming that the tunneling process is spin-conserving. As illustrated in Fig. 1.5, in the *P* magnetic configuration, majority spin (↑) electrons from the injector tunnel

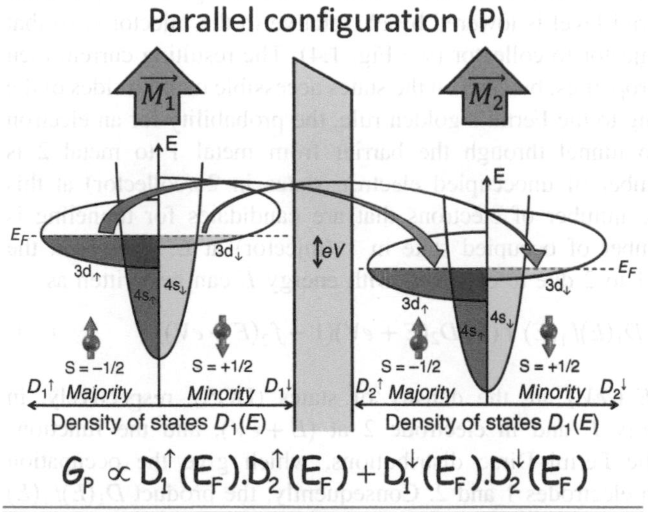

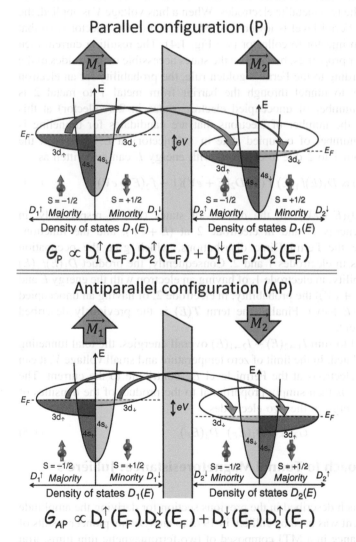

Figure 1.5 Schematics of the spin-dependent tunneling process through an insulating barrier when the magnetizations in ferromagnetic electrodes are in parallel *P* or antiparallel *AP* magnetic configurations. The process is assumed to be purely ballistic, so that no mixing of spin states occurs during the tunneling process.

toward majority spin (↑) empty states in the collector. At the same time, minority spin (↓) electrons from the injector tunnel toward minority spin (↓) empty states in the collector. The conductance in P configuration is then written as

$$G_P \propto D_1^{\uparrow}(E_F).D_2^{\uparrow}(E_F) + D_1^{\downarrow}(E_F).D_2^{\downarrow}(E_F). \tag{1.7}$$

In case of AP magnetic configuration, majority spin (↑) electrons from the injector tunnel toward minority spin (↓) empty states of the collector. At the same time, minority spin (↓) electrons from the injector tunnel toward majority spin (↑) empty states of the collector. The conductance in AP configuration is then written as

$$G_{AP} \propto D_1^{\uparrow}(E_F).D_2^{\downarrow}(E_F) + D_1^{\downarrow}(E_F).D_2^{\uparrow}(E_F). \tag{1.8}$$

Using these expressions, the TMR ratio, which characterizes the resistance difference in P and AP configurations, may be expressed as

$$\mathrm{TMR} = \frac{G_P - G_{AP}}{G_{AP}} = \frac{R_{AP} - R_P}{R_P} = \frac{2P_1P_2}{1 - P_1P_2}, \tag{1.9}$$

where $P_{1,2}$ are the spin polarization of electrode 1 and electrode 2 as defined in Eq. (1.6). Note that similar to the GMR case, a "pessimistic" definition of the TMR ratio can also be defined as $\mathrm{TMR} = ((G_P - G_{AP})/G_P) < 1$.

This model is oversimplified in the sense that it neglects all band structure effects in the magnetic electrodes and in the barrier. Nevertheless, this simple model successfully predicted the amplitude of TMR (typically around 50–70% with spin polarization on the order of 50–65% in Co–Fe-based alloys) in amorphous alumina-based magnetic tunnel junctions. However, the model failed to explain the very large TMR of magnetic tunnel junctions based on epitaxial barriers, notably of magnesium oxide (MgO) barriers. A key challenge to get an accurate prediction of TMR is to properly estimate the actual amplitude and sign of spin polarization for a given ferromagnetic material. Clearly, the larger the spin polarization, the higher the TMR amplitude. To maximize this spin polarization, a significant effort in materials science has been devoted to the search for half-metals, which are expected to be conducting for one spin direction and insulating for the other, resulting in 100% spin polarization at Fermi energy ($D^{\downarrow}(E_F) = 0$). Some magnetic oxides and Heusler alloys seem to exhibit this property in bulk form, but not when integrated in MTJs at room temperature.

Several experimental techniques have been used to measure the ferromagnet's spin polarization, such as the Meservey and Tedrow technique (14), which is based on tunneling transport in FM–insulator–superconductor junctions. A list of typical values obtained for a variety of ferromagnetic materials is presented in Table 1.1.

Other experiments in 1999, involving the study of spin-dependent tunneling through transition metal oxide barriers ($SrTiO_3$), revealed that different values and even opposite signs of spin polarization of tunneling electrons could be obtained for the same ferromagnetic electrode by varying the nature of the tunneling barrier (15).

TABLE 1.1 Values of Ferromagnet (FM) Spin Polarization P_0 Extracted from FM/Insulator/Al Tunneling Experiments, Where Al Is a Superconductor at Low Temperatures.

Ferromagnet	P_0 (%)
Ni	33
Co	42
Fe	45
$Ni_{90}Fe_{10}$	36
$Ni_{80}Fe_{20}$	48
$Co_{50}Fe_{50}$	51
$Ni_{40}Fe_{60}$	55

This technique of tunneling from a ferromagnet to a superconductor is used to determine the spin polarization in the ferromagnet (14).

These results clearly indicate that the spin polarization of the tunneling electrons is not only determined by the nature of the ferromagnetic electrodes but also by the whole trilayer ferromagnet/barrier/ferromagnet. In the following section, we introduce a more advanced model for treating the ferromagnet and the tunneling barrier configuration as a next step toward considering the whole trilayer band structure.

1.2.3 The Slonczewski Model

In 1989, Slonczewski proposed a more rigorous model based on the detailed calculations of the electron wave functions across the barrier, taking into account the exact matching conditions at the ferromagnetic/barrier and barrier/ferromagnetic interfaces (16). To derive the actual value of the transmission coefficient $T(E)$, one must consider not only the density of states of electrons at the Fermi level but also their wave vector k_F (or more simply their velocity v_F). This velocity is dependent on the type of electron one considers (s, p, and d bands). We will see below that this approach is required to understand the large values of TMR that are obtained in MgO-based MTJ devices.

1.2.3.1 The Model In Slonczewski model, several assumptions are made, similar to those of Jullière's model. First, the tunneling barrier is simply described as a square potential (energy height Φ and thickness d) above the Fermi level of the two ferromagnetic electrodes, considering the *free-electron approximation*. Second, the two ferromagnetic electrodes are taken to be identical. Finally, the model assumes that only electrons propagating perpendicular to the layers ($\vec{k} \approx \vec{k_\perp}$, $k_\parallel \approx 0$) can efficiently tunnel through the barrier. Indeed, if we consider electrons propagating in another direction, with $k_\parallel \neq 0$, then the decay coefficient inside the insulating layer becomes $\kappa(E) = \sqrt{k_\parallel^2 + (2m_e/\hbar^2)(\Phi - E)}$. Then, the higher the parallel wave vector, the lower the probability for the electron to tunnel through the barrier. By solving Schrödinger's equation in each part of the junction and applying continuity of the wave functions at the interfaces, it was analytically demonstrated that the equivalent spin polarization of the ferromagnet/barrier couple, P, can be written as

$$P = \frac{\left(k_{F,\uparrow} - k_{F,\downarrow}\right)\left(\kappa^2 - k_{F,\uparrow}k_{F,\downarrow}\right)}{\left(k_{F,\uparrow} + k_{F,\downarrow}\right)\left(\kappa^2 + k_{F,\uparrow}k_{F,\downarrow}\right)} = P_0 \frac{\left(\kappa^2 - k_{F,\uparrow}k_{F,\downarrow}\right)}{\left(\kappa^2 + k_{F,\uparrow}k_{F,\downarrow}\right)}, \tag{1.10}$$

where $k_{F,\uparrow}$ and $k_{F,\downarrow}$ are the wave vectors for majority and minority electrons at the Fermi level, respectively, κ corresponds to the decay coefficient of the electrons' tunneling probability through the insulating barrier (see Eq. (1.3)), and P_0 corresponds to the previously defined polarization for a ferromagnet in Jullière's model (see Eq. (1.6)). Moreover, instead of defining a net polarization for the whole ferromagnetic layer, this expression highlights that the polarization actually depends on the type of electron one considers (to which band it belongs). Indeed, the values of κ, $k_{F,\uparrow}$, and $k_{F,\downarrow}$ are associated with the band one considers, and P can be derived for each of these bands.

The expression for the spin polarization P in this model reveals that the attenuation of the wave function while crossing the barrier also results in a decrease of the effective spin polarization of the total current. In the limit of large barrier height Φ, and hence large decay coefficient κ, the resulting polarization is equivalent to the one defined in the Jullière's model. This model emphasizes the importance of the electrons' properties in the calculation of the polarization. One can perform a simple calculation by considering only the contribution of the lighter (s band) electrons, but a more complete calculation would sum the contributions of each type of electrons.

1.2.3.2 Experimental Observations On the experimental side, it has been noted that TMR amplitude varies considerably depending on the nature of the electrodes and the nature of the insulating barrier and interfaces. For example, Yuasa and Djayaprawira showed that in epitaxial magnetic tunnel junctions of Fe–Al_2O_3–CoFe, the TMR amplitude changes when the crystallographic orientation is changed among (211), (110), and (100) planes (17). The major role of the interfaces between the ferromagnet and the insulating layer has been highlighted by experiments demonstrating that Co/Al_2O_3 and Co/MgO provide positive polarization of tunneling electrons, whereas Co/$SrTiO_3$ provides negative polarization of tunneling electrons, at least at low bias voltage (15). These results demonstrate that it is possible to modify the TMR amplitude by choosing different ferromagnet/barrier configurations.

1.2.3.3 About the TMR Angular Dependence Denoting θ the angle between the magnetization directions in the two ferromagnetic electrodes, Slonczewski also succeeded in deriving an expression of the tunnel conductance as a function of the angle θ in the limit of large potential barrier U:

$$G(\theta) = G_0\left(1 + P^2 \cos\theta\right), \tag{1.11}$$

where the conductance in the AP configuration G_0 is given by

$$G_0 = \frac{\kappa}{\hbar d}\left[\frac{e\kappa\left(\kappa^2 + k_\uparrow k_\downarrow\right)\left(k_\uparrow + k_\downarrow\right)}{\pi\left(\kappa^2 + k_\uparrow^2\right)\left(\kappa^2 + k_\downarrow^2\right)}\right]^2 e^{-2\kappa d}, \tag{1.12}$$

which represents the actual conductance when the quantum system is considered as a whole (i.e., the two ferromagnetic layers separated by the tunnel barrier). With the effective tunnel conductance varying as $P^2\cos\theta$, the resistance dependence is then given by $R(\theta) = R_0/\left(1 + P^2\cos\theta\right)$, which can be approximated by $R \approx R_0\left(1 - P^2\cos\theta\right)$ when the electron polarization is weak, that is, for low TMR amplitude.

1.2.4 More Complex Models: The Spin Filtering Effect

Another major breakthrough in the field of spintronics was made with the discovery of much larger TMR amplitude in MTJ based on a crystalline MgO tunnel barrier rather than an amorphous alumina barrier. These MgO-based MTJs are today's standard elements in spintronic devices, whether for MRAM, read heads, or field sensors applications. The starting point of this work was the theoretical calculations carried out by Butler and Mathon in 2001, which predicted that TMR ratios as large as 1600% could be expected for epitaxially grown Co or Fe electrodes on MgO (18,19). For instance, an epitaxial MgO(001) barrier can be easily grown on a Fe bcc layer (001) since the lattice mismatch is small (about 4%). We remind that bcc stands for body-centered cubic, relative to the arrangement of atoms in the crystal. (001) later refers to a growth direction along the cube edge direction. These calculations, performed with *ab initio* methods, derived the tunneling probability of each kind of electrons, depending on their orbital symmetry.

1.2.4.1 *Incoherent Tunneling Through an Amorphous (Al_2O_3) Barrier* The electrons' wave functions in crystalline materials are described by Bloch states. Bloch states are wave function solutions of the Schrödinger equation describing the quantum mechanical state of electrons in a periodic potential, for example, in a perfect, infinite atomic crystal. In crystalline ferromagnetic metals such as Fe, Ni, Co (3*d* ferromagnetic transition metals), and their alloys, these Bloch states have some specific symmetries: Δ_1 symmetry (corresponding to *spd* hybridized states), Δ_2 symmetry (*d* states), and Δ_5 symmetry (*pd* hybridized states). Bloch states with Δ_1 symmetry usually have a large positive associated polarization at E_F, but it is not the case for other symmetries (they can even have negative associated polarizations). Jullière's model (previously introduced) assumes equal tunneling probabilities independent of the electrons' Bloch states in the electrodes. This is true only if the tunneling process is *incoherent*, meaning that the coherency of the Bloch states is not conserved during tunneling.

In the case of an amorphous barrier, there is no crystallographic symmetry in the barrier. Consequently, electron wave functions from the injector, no matter their symmetry, couple identically to any *tunneling/evanescent states* in the barrier, through which they will propagate and decay. The exponential decay of the wave function throughout the barrier is then independent of the initial symmetry of the tunneling electron wave function so that the transmission probability through the barrier is itself independent on the initial electron wave function symmetry (see Fig. 1.6). This tunneling process can be regarded as an *incoherent tunneling*. It can be well described by Jullière's model introduced earlier. It leads to comparable tunneling

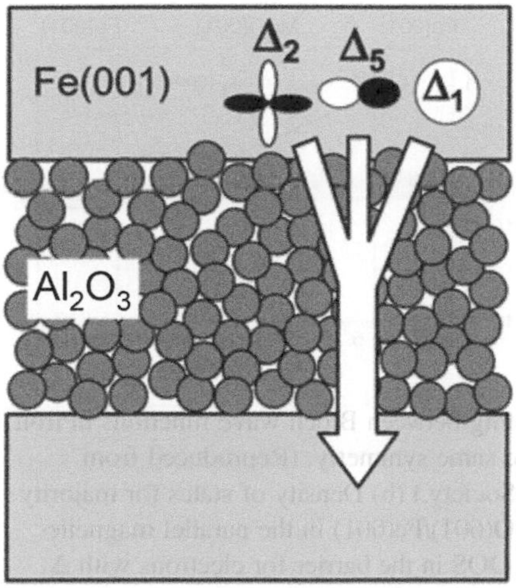

Figure 1.6 Schematic illustration of electrons tunneling through an amorphous (Al_2O_3) barrier. (From Ref. 17.)

probabilities for all Bloch states from the electrode. The *net* polarization then simply corresponds to the one calculated by summing the contributions from (↑) and (↓) DOS from all Bloch states. In this case, the net polarization is reduced due to the contributions of some Bloch states with negative spin polarization (i.e., opposite to the local magnetization).

A major difference between MTJ based on crystalline magnetic electrodes with amorphous barriers or crystalline barriers is that the symmetry of the tunneling electron wave functions may be conserved in the latter case and this symmetry can strongly influence the tunneling rates, that is, the probability of tunneling through the barrier. If one can engineer a system in which the Bloch states with large positive spin polarization have a higher tunneling probability than the Bloch states with negative spin polarization, then a very high net polarization could be expected, resulting in a very large MR ratio. This is possible in some epitaxially grown crystalline MTJs, in particular Fe/MgO/Fe, as explained in the following section.

1.2.4.2 Coherent Tunneling Through a Crystalline MgO Barrier A crystalline MgO(001) can be epitaxially grown on a bcc Fe(001) layer to prepare a crystalline Fe(001)/MgO(001)/Fe(001) MTJ. Considering the $k_\parallel = 0$ direction, there are three kinds of tunneling/evanescent states in crystalline MgO, with different associated symmetries: Δ_1, $\Delta_{2'}$, and Δ_5. In such systems, *coherent* tunneling is obtained: The electrons' wave functions in the ferromagnetic material couple with evanescent wave functions having the same symmetry in the barrier, so that electrons conserve their orbital symmetry as they tunnel. *Ab initio* calculations then predict that the tunneling probability of an electron strongly depends on its orbital symmetry, leading to effective possible *symmetry filtering of the tunneling current*. The mechanism of orbital selection for the tunnel conductance in Fe(001)/MgO/Fe(001) systems is presented in Fig. 1.7a.

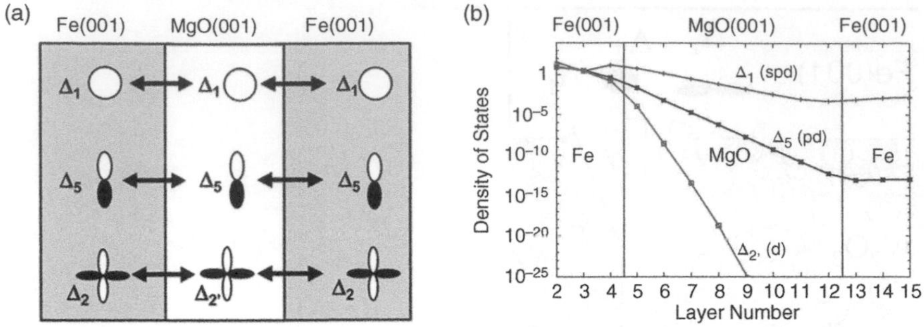

Figure 1.7 (a) Schematic illustration of the coupling between Bloch wave functions in iron and evanescent wave functions in MgO having the same symmetry. (Reproduced from Ref. 17 with permission from American Physical Society.) (b) Density of states for majority electrons tunneling ($k_{\parallel} = 0$) in a MTJ Fe(001)/MgO(001)/Fe(001) in the parallel magnetic configuration. Note the much slower decay of the DOS in the barrier for electrons with Δ_1 symmetry than for those of other symmetries. (From Ref. 17.)

In Fe, the band structure is calculated with consideration of the energy splitting between (↑) and (↓) for each Bloch state type, deducing the densities of states at the Fermi energy, and hence the polarization. Both majority and minority electrons fill many states at the Fermi energy, which corresponds to a small net polarization. But these calculations have demonstrated that only majority electrons fill Δ_1 symmetry states, implying a full polarization $P_{\Delta_1} = 1$, whereas both majority and minority Δ_2 and Δ_5 symmetry states can be found at the Fermi energy, corresponding to low polarizations. Note that for other materials such as bcc Co, only minority Δ_2 and Δ_5 symmetry states exist at the Fermi level, implying that $P_{\Delta_2} = P_{\Delta_5} = -1$.

Furthermore, calculations demonstrate that the exponential tunneling decay is much stronger for Δ_2 and Δ_5 states compared with Δ_1 states. Figure. 1.7b presents the probability for a majority electron incoming from the left electrode to propagate through the MTJ in the *P* configuration. For $d_{MgO} = 8$ monolayers (ML), which is a reasonable barrier thickness, the transmitted density of Δ_1 states is larger than that of Δ_5 states by 10 orders of magnitude.

In addition, since there are no minority $\Delta_{1\downarrow}$ states to tunnel from or to, only the $\Delta_{1\uparrow} \rightarrow \Delta_{1\uparrow}$ channel has a significant contribution to the conduction. Similarly, for the *AP* configuration, $\Delta_{1\uparrow} \rightarrow \Delta_{1\downarrow}$ and $\Delta_{1\downarrow} \rightarrow \Delta_{1\uparrow}$ channels have a potentially zero tunneling probability. There is therefore here a new mechanism of *spin filtering* of the wave functions according to their symmetry, which yields a dramatic effective increase in the net spin polarization of the tunneling current. As a result, significant conduction occurs only in the *P* configuration. It is this *spin filtering effect* that explains the very large TMR magnitudes predicted and measured in epitaxial or highly textured MgO-based MTJs.

Not only bcc Fe(001) shows this high spin polarization of the Δ_1 Bloch states, but also many other bcc ferromagnetic metals, such as cobalt (Co) and alloys based on Fe and Co. Similarly, very large TMR ratios are also predicted for other crystalline barriers such as $SrTiO_3$ (20). However, so far, MgO-based MTJs still give the best results in terms of TMR. After these theoretical predictions, a strong worldwide research effort was made to obtain epitaxial growth of structures for Fe/MgO/Fe or CoFeB/MgO/CoFeB. These efforts resulted in an extremely large TMR of about 600% obtained in 2008 in CoFeB/MgO/CoFeB MTJs at room temperature (21). The difference between the values of experimental and theoretical TMR is mainly due to imperfections in interface quality and defects in crystal growth of materials due to dislocations, vacancies, and impurities (in particular absorbed water molecules). More details on the growth of MgO-based MTJs can be found in Chapter 2 as well as in Ref. 17.

1.2.5 Bias Dependence of Tunnel Magnetotransport

In most MTJs, the magnitude of TMR decreases markedly when the applied voltage increases. This factor is critical for applications and particularly for reading out the memory magnetic state in MRAM. This voltage-induced decrease of TMR in magnetic tunnel junctions is characterized by $V_{1/2}$, which is the voltage for which the TMR ratio is reduced to half of its amplitude in the limit of zero bias voltage. Several theories have been developed to describe this bias dependence of the magnetotransport, which is often very complex because it involves several physical mechanisms of spin depolarization, as explained below.

The first of these mechanisms is based on the emission of magnons by hot electrons. When a finite bias voltage is applied across the junction, the electrons tunnel ballistically through the barrier, that is, keeping their energy so that they arrive in the collector electrode as hot electrons (Fig. 1.4). When they penetrate the collector electrode, these hot electrons very quickly lose their excess energy by inelastic relaxation to the Fermi energy. In a normal metal, the relaxation mechanisms are via electron–electron and electron–phonon interactions. In ferromagnetic materials, electrons can also reach the Fermi energy by emission of spin waves (magnons), a process that does not preserve the spin. The higher the voltage, the greater the density of emitted magnons. Both the spin polarization of the current and the TMR, directly linked to the conservation of spin information, are then reduced.

Another mechanism of spin depolarization of the current is the presence of defects inside the insulating barrier. These defects create trap states in the barrier through which electrons can co-tunnel. This means that electrons tunnel from the injector to the trap, and then from the trap to the collector, losing their spin information in the process.

Finally, the decrease of TMR with bias voltage can be described by the voltage dependence of the electronic properties involved in the tunneling process, for example, the electron effective mass, the transmission coefficients in the barrier, and the parameter of coherent reflections at interfaces. The polarization of the tunneling electrons also depends itself on the bias voltage since at a bias voltage V, electrons within a band of width eV below the Fermi energy can tunnel through the barrier. If one considers a band structure such as the one depicted in Fig. 1.1a, it is clear that the net polarization of the tunneling electrons is expected to decrease if electrons from below the Fermi energy become allowed to tunnel through the barrier.

In MgO-based MTJs, $V_{1/2}$ is typically in the range of 0.5–0.8 V, depending on the nature of the magnetic electrodes and growth conditions. The readout in MRAMs is usually performed at voltages on the order of only 0.15–0.2 V for which the drop of TMR is weak.

1.3 THE SPIN-TRANSFER PHENOMENON

In the first generation of MTJ-based MRAM, the switching of the storage layer magnetization during the write process is achieved using pulses of magnetic field (4). This field is induced by a current flowing in conducting lines located above and below the MTJ. This writing process suffers both from the large energy consumption needed to generate large enough magnetic fields and from write selectivity problems due to the spatial extension of generated fields as well as dot-to-dot distribution in switching fields (variability). A new generation of magnetic memory, called spin-transfer torque magnetic random-access memory (STT-MRAM), is based on the pioneering theoretical work of Slonczewski (22) and Berger (23), which predicted that current flowing through magnetic multilayers can directly reverse the magnetization of one of the layers.

We describe below the origin and consequences of this physical phenomenon. The details of the reversal dynamics as well as the use of this new writing mode in MRAM will be described in other chapters.

1.3.1 The Concept and Origin of the Spin-Transfer Effect

The GMR and TMR magnetoresistive effects described in the previous sections correspond to a variation of the current flow in a spin-valve device or a magnetic tunnel junction (i.e., a variation in conductance) induced by a change of the magnetic configuration of the device. This change of magnetic configuration was generally mediated through the action of an external field. The spin-transfer effect may be seen in a simple description as the reciprocal effect to GMR or TMR: via the spin-transfer torque, the current, which gets spin polarized by traversing a first magnetic layer, can exert a torque on the magnetization of a second ferromagnetic layer and thereby change the magnetic configuration of the device.

The simplest system for describing this new effect is similar to the standard magnetoresistive systems described in previous sections. It consists of two thin magnetic layers separated by a nonmagnetic layer (either a normal metal NM or an insulator I). One of the two ferromagnetic layers, F1 is presumably thick and fixed, whereas the second ferromagnetic layer F2 and the nonmagnetic one are chosen to be thin in comparison to the previously introduced characteristic length of spin-polarized transport.

1.3.1.1 The "In-Plane" Torque When electrons flow through the structure perpendicularly to the interfaces, the current spin polarization evolves to remain parallel to the direction of the local magnetization. Indeed, when the electrons penetrate into a ferromagnetic material, the spins of the conduction electrons become very rapidly aligned parallel to the local magnetization direction because of a strong exchange interaction between conducting electrons ($4s$) and the more localized electrons ($3d$) responsible for the local magnetization (see Section 1.1).

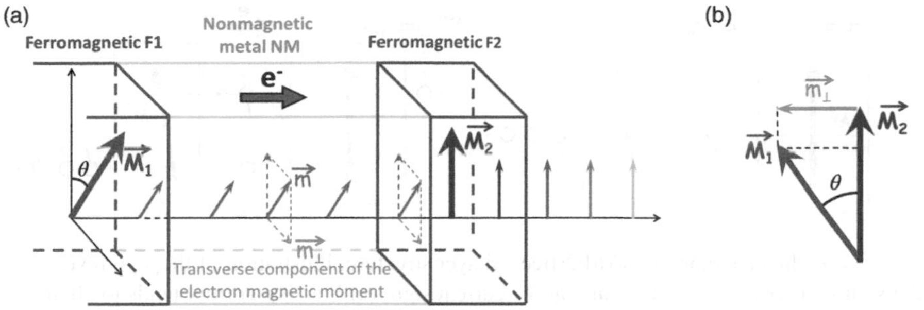

Figure 1.8 (a) Illustration of the concept of spin-transfer torque in a simple F1/NM/F2 trilayer structure; (b) Torque $\vec{m}_\perp$ exerted on the magnetization $\vec{M}_2$ of the thin magnetic layer F2, which tends to get aligned along the magnetization $\vec{M}_1$ when electrons flow from F1 to F2.

In the case where the respective magnetizations $\vec{M}_1$ and $\vec{M}_2$ are noncollinear, the spin polarization of electrons have to rotate in their paths. This occurs mainly through relaxation of the spins transverse component at the F/NM interfaces. We illustrate this phenomenon by considering the case of a spin-valve structure where the nonmagnetic spacer is metallic (NM) as illustrated in Fig. 1.8. When the electrons flow from layer F1 to F2, the current becomes spin polarized after traversing F1, through the mechanism described in Section 1.2 and thus acquires a net spin polarization along $\vec{M}_1$. This nonzero spin polarization propagates across the NM spacer as long as the spin-flip is negligible in this layer. The transmitted electrons then impinge on the NM/F2 interface with a spin polarization that is not aligned with the direction of the local magnetization $\vec{M}_2$.

If we focus on the relaxation process at the second NM/F2 interface, the incoming electrons hence have a component of their magnetic moment $\vec{m}$ transverse to the $\vec{M}_2$ direction (see Fig. 1.8a). When they penetrate into F2, their spin becomes quickly aligned toward the local magnetization direction in a very short distance after the interface (typically less than a nanometer). Hence, the electrons lose their transverse component $\vec{m}_\perp$ when they penetrate into F2 within the first nanometer from the interface. In this interaction however, the total spin angular momentum is conserved, and thus the transverse component $\vec{m}_\perp$ has been actually absorbed and transferred to the magnetization of F2.

This *transfer of spin* is therefore intrinsically an interfacial effect. For a very thick magnetic layer, with very stable magnetization, it then has a negligible influence. However, for a thin layer, this *transfer of spin* tends to modify the local magnetization direction, somehow adding the transferred spins to it. As a result, this *transfer of spin* tends to align the magnetization $\vec{M}_2$ along the direction of the spin current polarization, and hence along the magnetization $\vec{M}_1$ (see Fig. 1.9a) when electrons flow from the "reference" or "fixed" layer F1 to the "free" layer F2. It translates as a torque exerted on the magnetization $\vec{M}_2$, named *spin-transfer torque*.

If electrons now propagate in the opposite direction, that is, from layer F2 to layer F1, electrons impinging on the NM/F1 interface with spin antiparallel to the F2

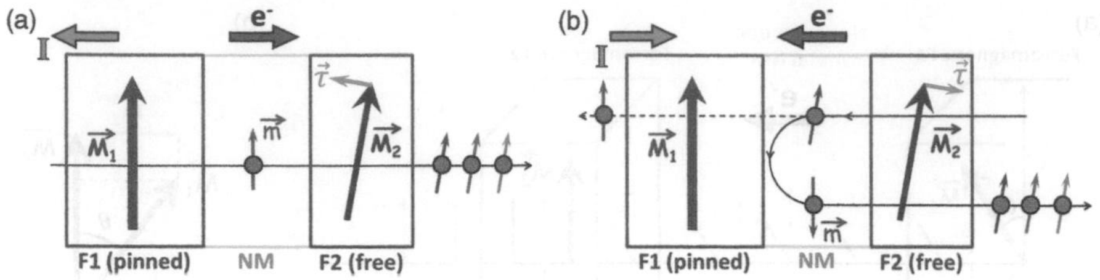

Figure 1.9 In a F1(pinned)/NM/F2(free) trilayer structure, illustration of the sign reversal of the spin-transfer torque with changing the current sign. (a) Positive current tends to align magnetization $\vec{M}_2$ parallel to $\vec{M}_1$. (b) Negative current tends to push $\vec{M}_2$ away from $\vec{M}_1$.

magnetization will have difficulty penetrating into the F1 layer or will be back-scattered because their spin is antiparallel to the local magnetization. These electrons are therefore reflected by the NM/F1 interface and impinge on the other side of the NM spacer on the NM/F2 interface with a spin antiparallel to the F1 magnetization. This flow of reflected electrons again exerts a torque on the F2 magnetization which now tends to align the F2 magnetization antiparallel to the F1 magnetization (see Fig. 1.9b). The STT therefore changes sign as a function of the current direction and can favor either *P* or *AP* magnetic configurations depending on its direction.

As the current density increases, the number of electrons crossing the magnetic layer per unit time increases, thereby proportionally increasing the spin-transfer torque. Therefore, in metallic spin valves, STT can be considered proportional to the current density.

If the system is a magnetic tunnel junction, as in STT-MRAM, the electrons are transmitted ballistically through the spacer layer (the tunnel barrier) instead of diffusely as in spin valves, but the STT effect is basically similar.

Analytically, the spin-transfer torque is equal to the sum of the absorbed transverse spin magnetic moments per unit time:

$$V\frac{d\vec{M}_2}{dt} = \text{absorbed transverse magnetic moments per unit time,} \tag{1.13}$$

considering that each electron flowing across F2 will bring a contribution $\vec{m}_\perp$ to the local magnetization:

$$\vec{m}_\perp = -\frac{g\mu_B}{2}\left(\vec{m}_2 \times \left(\vec{m}_2 \times \vec{m}_1\right)\right), \tag{1.14}$$

In this formula, the cross product $-\left(\vec{m}_2 \times \left(\vec{m}_2 \times \vec{m}_1\right)\right)$ simply corresponds to the direction of the electron's transverse magnetic moment $\vec{m}_\perp$. The free layer volume is $V = t \cdot A$ (t the thickness, A the cross section), g is the electrons Landé factor, and μ_B is the Bohr magneton. The normalized magnetization is given by $\vec{m}_i = \vec{M}_i/M_{S_i}$, where M_S is the saturation magnetization. The number of incoming electrons per second is then simply given by

$$\frac{dN}{dt} = \frac{j_{dc} \cdot A}{e}, \tag{1.15}$$

where j_{dc} is the injected current density and e the electron's charge. By summing the contributions, and considering the current spin polarization P_{spin} at the NM/F2 interface, one obtains the expression for the spin-transfer torque as

$$\left(\frac{d\vec{m}_2}{dt}\right)_{ST} = -P_{spin}\frac{j_{dc}}{2te}g\mu_B\left(\vec{m}_2 \times \left(\vec{m}_2 \times \vec{m}_1\right)\right). \tag{1.16}$$

This torque is often labeled "Slonczewski torque" (ST) or "in-plane torque" since its direction lies in the plane defined by $\vec{M}_1$ and $\vec{M}_2$. It is proportional to the current spin polarization at the interface, and to the current density. Hence, one can directly change the magnitude or even the sign of the effect by simply tuning these two parameters. Remarkably, by changing the sign of the current, it is possible to reverse the effect. One sign of the current will make the local magnetization $\vec{M}_2$ rotate to align along the magnetization $\vec{M}_1$ (thus favoring parallel magnetizations), while the other sign will make the local magnetization rotate away from the magnetization $\vec{M}_1$ (thus favoring antiparallel magnetizations). A review on in-plane torque in metallic spin valves can be found in Ref. 24.

1.3.1.2 The "Out-of-Plane" Torque The in-plane torque is always the dominant spin-transfer effect in junctions, and is the most significant torque in all metallic structures. However, an additional component of spin-torque can also become significant in tunnel junctions. It is referred to as field-like torque (FLT) or out-of-plane torque, and is associated to another source of spin-transfer, leading to a torque directed perpendicular to the $(\vec{M}_1, \vec{M}_2)$ plane. Its action is equivalent to applying an external magnetic field oriented along the $\vec{M}_1$ direction. The physical phenomena behind the effect are beyond the scope of this book; we will simply mention that

$$\left(\frac{d\vec{m}_2}{dt}\right)_{\text{Out-of-plane}} \propto -\left(\vec{m}_2 \times \vec{m}_1\right). \tag{1.17}$$

This out-of-plane torque is often considered through its ratio to the in-plane torque. It is usually negligible in metallic spin valves but its amplitude can reach about 30% of the in-plane STT in tunnel junctions. It is often neglected in first approximation but nevertheless can be responsible for undesirable behavior in the writing operation of STT-MRAM such as back-switching phenomena (25).

1.3.2 Spin-Transfer-Induced Magnetization Dynamics

After having introduced the spin-dependent transport mechanisms at the origin of spin-transfer torques, our aim is now to address the influence of these torques on the magnetization dynamics of F2. The dynamics of a magnetization in response to external torques in a solid is classically described by the Landau-Lifshitz-Gilbert (LLG) equation. Because of the STT, two terms associated to the in-plane and out-of-plane components of spin-transfer torques are added to this equation. The LLG equation and the detailed description of magnetization dynamics are the topic of another chapter. Here, we describe only some phenomenological aspects.

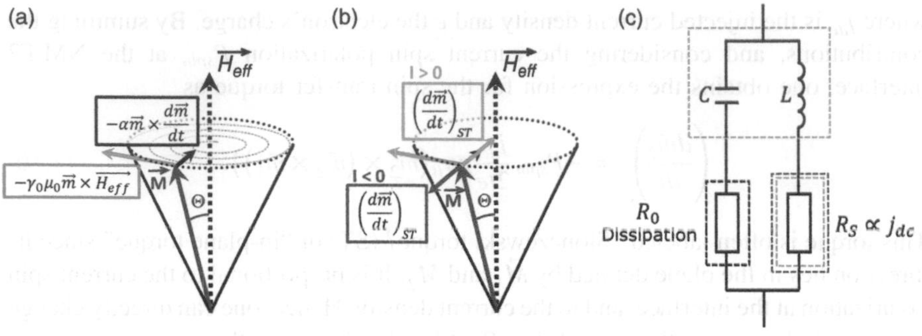

Figure 1.10 (a) Relaxation of magnetization around the effective field without spin-transfer torque. (b) Additional torque for positive or negative current. (c) Corresponding equivalent (RLC) circuit.

We return to the simple trilayer of Fig. 1.8 and consider the ferromagnetic layer F2, whose local magnetization is described by the normalized vector $\vec{m}_2$. This magnetization is aligned at equilibrium along the so-called effective field $\vec{H}_{eff}$. $\vec{H}_{eff}$ is the field resulting from the sum of all fields acting on $\vec{M}_2$, namely, the external applied magnetic field, magnetic anisotropy field, interlayer coupling field, and so on. When $\vec{M}_2$ is slightly pushed away from its stable equilibrium position, the magnetization starts processing around the effective field $\vec{H}_{eff}$, following the differential equation:

$$\left(\frac{d\vec{m}_2}{dt}\right)_{precession} = -\gamma_0\vec{m}_2 \times \mu_0\vec{H}_{eff}, \tag{1.18}$$

where γ_0 is the gyromagnetic ratio and μ_0 is the vacuum permeability. Like any other physical system with a stable equilibrium position, the magnetization will relax to this position, with a characteristic damping coefficient α, the Gilbert damping (which can range between $\sim 5\times10^{-3}$ and 0.1 for standard ferromagnets, depending on the amplitude of their spin–orbit interactions). The damping force is once again associated with a torque, written

$$\left(\frac{d\vec{m}_2}{dt}\right)_{damping} = -\alpha\vec{m}_2 \times \frac{d\vec{m}_2}{dt}, \tag{1.19}$$

which tends to bring back $\vec{m}_2$ toward $\vec{H}_{eff}$ (Fig. 1.10).

1.3.2.1 A Simple Analogy A simple analogy to understand the influence of STT on magnetization dynamics is the classical RLC resonant circuit. If the circuit is initially excited, for instance by introducing a charge on the capacity, the current in the circuit exhibits damped oscillations and gradually relaxes toward the zero equilibrium value. Because of dissipation, as long as no additional energy is supplied to the system, it has no other choice but to relax to the closest equilibrium position. However, during this relaxation process, one can act on the system to modify the relaxation rate, by reducing or increasing the damping of the system. In the RLC analogy, this would correspond to adding a supplementary positive or negative resistance in the circuit. When dealing with magnetization, a means to do this is to use spin-transfer torque. Indeed, depending on the sign of the current, the resulting STT torque can be oriented either in the same direction or in the opposite direction to the damping torque.

This simplified description allows one to understand easily the nature of the main contribution of spin-transfer force (in-plane torque), which can be described through a non-conservative force acting in the same direction as the natural damping, that is, perpendicular to the magnetization trajectory. Depending on the relative orientation of $\vec{M}_1$ and $\vec{M}_2$, as well as the sign of the injected current, the spin-transfer torque can increase or decrease the effective damping, that is, behave as an additional damping or as an antidamping.

To continue the analogy with classical systems, like the RLC circuit, one can bring it into a regime where the effective damping around equilibrium crosses zero and becomes negative ($R_S < -R_0$). In this case, the corresponding equilibrium position is no longer stable, and any deviation from equilibrium is amplified so that the oscillations diverge. For the case of unstable local magnetization, thermal excitations are sufficient to induce a small deviation from equilibrium, further amplified by spin-transfer torque. Depending on the system configuration, the magnetic system can be designed so that the oscillation divergence drives the magnetization toward another stable minimum of energy or so that steady state excitations of the magnetization can be sustained. The first behavior is the one used to write in STT-MRAM, whereas the second behavior can yield to new types of frequency-tunable RF oscillators.

1.3.2.2 Toward MRAM Based on Spin-Transfer Torque A STT-MRAM cell is effectively composed of a MTJ consisting of a pinned ferromagnetic reference layer and a ferromagnetic storage layer separated by a tunnel barrier. This system is very similar to the one described in Fig. 1.8 where the storage layer is F2 and the reference layer is F1. The MTJ is designed in such a way that the magnetization of the storage layer has two natural equilibrium positions parallel or antiparallel to the pinned layer magnetization. Using the spin-transfer torque to destabilize either one of these positions, simply by choosing the sign of the current, it is possible to induce a reversal of the magnetization, and consequently switch the value of the memory cell. When electrons flow from the pinned layer to the storage layer, parallel alignment is favored. When electrons flow from the storage layer to the reference layer, antiparallel alignment is favored. To achieve the storage layer magnetization switching, the current density must exceed a certain threshold, which corresponds to the point where the Gilbert damping becomes balanced by the STT antidamping. This current density threshold J_c can be calculated by finding the conditions for which the net effective damping is zero. The expression of the current density for switching depends in particular on whether the layers are magnetized in-plane or out-of-plane. This will be explained in more details in

Chapter 5. The LLG equation with Slonczewski in-plane torque yields the critical current density for switching:

$$J_c = \frac{2\alpha e \mu_0 M_S t}{\hbar P_{spin}} H_{eff}, \tag{1.20}$$

and the corresponding critical current:

$$I_c = \frac{2\alpha e \mu_0 M_S V}{\hbar P_{spin}} H_{eff}, \tag{1.21}$$

where M_s is the saturation magnetization, α is the Gilbert damping, t is the layer thickness, V is the volume, P_{spin} is the amplitude of the spin polarization, and H_{eff} is the effective field as previously defined. The other quantities are physical constants: e the electron charge, μ_0 the vacuum permeability, and $\hbar$ the reduced Planck constant. The predicted critical current densities are relatively large, in the range of a few 10^{11} A/m^2 in metallic spin valves and in the range of 1 to a few 10^{10} A/m^2 in MTJs. The higher STT efficiency in MTJs is associated with two phenomena: a higher spin polarization in particular in MgO-based MTJs, and the fact that in a MTJ, electrons impinging on the storage layer have mainly a perpendicular component of momentum. The electrons that tunnel the most easily through the barrier are those propagating perpendicular to the layers (with out-of-plane momentum $\vec{k} \approx \vec{k}_\perp$, $k_\parallel \approx 0$), whereas in metallic pillars, the electron momentum is broadly dispersed in all angular direction.

Very importantly for STT-MRAM considerations, STT switching is determined by a current density threshold. This means that the current required to write in STT-MRAM scales proportionally to the area of the device. For very small dimensions, for which thermal stability becomes a problem, a reduction in size must be compensated by an increase in the anisotropy field (and in H_{eff}) to maintain a desired thermal stability factor, that is, a given memory retention (see Chapter 5). As a consequence, the thermal stability limits the decrease of the critical current with the device dimensions. However, this minimum current value is in the range of 13 μA with known materials, which allows downsize scalability of STT-MRAM to sub-20 nm nodes.

The expression of critical current density provides paths to reduce the power consumption for spin-transfer-induced switching. In particular the Gilbert damping factor α plays a quite important role and must be minimized, and the spin polarization must be maximized. The other parameters (t, M_s, H_{eff}) also influence the thermal stability of the magnetization, so a trade-off must often be found between minimizing the write current density and maintaining sufficient thermal stability to achieve the specified memory retention. The goal of minimizing the critical current for writing has stimulated a strong research effort in the last few years, notably among two families of materials: magnetic oxides (26) and Heusler alloys (27).

1.3.3 Main Events Concerning Spin-Transfer Advances

A complete review of all the theoretical and experimental advances made in the last decade on spin-transfer is beyond the scope of this chapter (see Ref. 28 and references therein). However, it is worth mentioning a few key dates in spin-transfer torque research and development. The first experimental results to validate the theoretical prediction of spin-transfer torque made by Slonczewski and Berger were obtained by Tsoi et al, using a point-contact geometry for injection of a large current density into a magnetic layer (29). The experimental demonstration of the use of spin-transfer torque to reverse the magnetization of metallic spin valves without any applied field by Katine et al. in 2000 (30) showed that STT could be used as a new write scheme in MRAM instead of field-induced magnetization switching, offering a much better downsize scalability. The first demonstration of STT switching in MTJs was made in 2004 (31) once the quality of the MTJ growth became good enough to withstand the large current density required for STT switching. The first functional demonstrator of an STT-MRAM chip was developed by Sony Corp, which presented the first 4 kbit STT-RAM demonstrator in 2004 (32). A lot of further progress has been made since then on both the fundamental understanding of the STT effects as well as on the technological side (see Chapter 6). The first STT-MRAM products were announced by Everspin at the end of 2012 (see Chapter 5) and all the major microelectronic companies now have large research and development efforts, in particular aiming at DRAM replacement by STT-MRAM beyond the 20 nm technological node.

REFERENCES

1. M. Baibich, J. M. Broto, A. Fert, F. Nguyen Van Dau, F. Petroff, P. Etienne, G. Creuzet, A. Friederch, and J. Chazelas, "Giant magnetoresistance of (001)Fe/(001)Cr magnetic superlattices," *Phys. Rev. Lett.* **61**, pp. 2472–2475 (1988); doi: 10.1103/PhysRevLett.61.2472.
2. J. S. Moodera, L. R. Kinder, T. M. Wong, and R. Meservey, "Large magnetoresistance at room temperature in ferromagnetic thin film tunnel junctions," *Phys. Rev. Lett.* **74**, pp. 3273–3276 (1995); doi: 10.1103/PhysRevLett.74.3273.
3. T. Miyazaki and N. Tezuka, "Giant magnetic tunneling effect in Fe/Al_2O_3/Fe junction," *J. Magn. Magn. Mater.* **139**, pp. L231–L234 (1995); doi: 10.1016/0304-8853(95)90001-2.
4. S. Tehrani, J. M. Slaughter, E. Chen, M. Durlam, J. Shi, and M. DeHerrera, "Progress and outlook for MRAM technology," *IEEE Trans. Magn.* **35**, pp. 2814–2819 (1999); doi: 10.1109/20.800991.
5. I. A. Campbell and A. Fert, "Transport properties of ferromagnets," in *Ferromagnetic Materials*, Vol. 3, ed. E. P. Wohlfarth, North-Holland Publishing, Amsterdam, 1987, pp. 747–804.
6. A. Barthélémy, A. Fert, and F. Petroff, "Giant magnetoresistance of magnetic multilayers," in *Handbook of Magnetic Materials*, Vol. 12, ed. K. H. J. Buschow, North-Holland, Amsterdam, 1999, pp. 1–96.
7. T. Valet and A. Fert, "Theory of the perpendicular magnetoresistance in magnetic multilayers," *Phys. Rev. B* **48**, pp. 7099–7113 (1993); doi: 10.1103/PhysRevB.48.7099.
8. S. S. P. Parkin, N. More, and K. P. Roche, "Oscillations in exchange coupling and magnetoresistance in metallic superlattice structures: Co/Ru, Co/Cr, and Fe/Cr," *Phys. Rev. Lett.* **64**, pp. 2304–2307 (1990); doi: 10.1103/PhysRevLett.64.2304.
9. B. Dieny, V. S. Speriosu, S. S. P. Parkin, B. A. Gurney, D. R. Wilhoit, and D. Mauri, "Giant magnetoresistance in soft ferromagnetic multilayers," *Phys. Rev. B* **43**, pp. 1297–1300 (1991); doi: 10.1103/PhysRevB.43.1297.
10. B. Dieny, "Spin valves," in *Magnetoelectronics*, ed. M. Johnson, Elsevier, Amsterdam, 2004, pp. 67–150.
11. M. Jullière, "Tunneling between ferromagnetic films," *Phys. Lett.* **54A**, pp. 225–226 (1975); doi: 10.1016/0375-9601(75)90174-7.
12. S. Yuasa, T. Nagahama, A. Fukushima, Y. Suzuki, and K. Ando, "Giant room-temperature magnetoresistance in single-crystal Fe/MgO/Fe magnetic tunnel junctions," *Nat. Mater.* **3**, pp. 868–871 (2004); doi: 10.1038/nmat1257.

13. S. S. P. Parkin, C. Kaiser, A. Panchula, P. M. Rice, B. Hughes, M. Samant, and S-H. Yang, "Giant tunnelling magnetoresistance at room temperature with MgO(100) tunnel barriers," *Nat. Mater.* **3**, pp. 862–867 (2004); doi: 10.1038/nmat1256.
14. R. Meservey and P. M. Tedrow, "Spin-polarized electron tunneling," *Phys. Rep.* **238**, pp. 173–243 (1994); doi: 10.1016/0370-1573(94)90105-8.
15. J. M. De Teresa, A. Barthélémy, A. Fert, J. P. Contour, F. Montaigne, and P. Seneor, "Role of metal-oxide interface in determining the spin polarization of magnetic tunnel junctions," *Science* **286**, pp. 507–509 (1999); doi: 10.1126/science.286.5439.507.
16. J. Slonczewski, "Conductance and exchange coupling of two ferromagnets separated by a tunneling barrier," *Phys. Rev. B* **39**, pp. 6995–7002 (1989); doi: 10.1103/PhysRevB.39.6995.
17. S. Yuasa and D. D. Djayaprawira, "Giant tunnel magnetoresistance in magnetic tunnel junctions with a crystalline MgO(001) barrier," *J. Phys. D Appl. Phys.* **40**, pp. R337–R354 (2007); doi: 10.1088/0022-3727/40/21/R01.
18. W. H. Butler, X.-G. Zhang, T. C. Schulthess, and J. M. MacLaren, "Spin-dependent tunneling conductance of Fe/MgO/Fe sandwiches," *Phys. Rev. B* **63** 054416 (2001); doi: 10.1103/PhysRevB.63.054416.
19. J. Mathon and A. Umerski, "Theory of tunneling magnetoresistance of an epitaxial Fe/MgO/Fe (001) junction," *Phys. Rev. B* **63**, 220403 (2001); doi: 10.1103/PhysRevB.63.220403.
20. J. P. Velev, K. D. Belashchenko, D. A. Stewart, M. van Schilfgaarde, S. S. Jaswal, and E.Y. Tsymbal, "Negative spin polarization and large tunneling magnetoresistance in epitaxial Co/SrTiO3/Co magnetic tunnel junctions," *Phys. Rev. Lett.* **95**, 216601 (2005); doi: 10.1103/PhysRevLett.95.216601.
21. S. Ikeda, J. Hayakawa, Y. Ashizawa, Y. M. Lee, K. Miura, H. Hasegawa, M. Tsunoda, F. Matsukura, and H. Ohno, "Tunnel magnetoresistance of 604% at 300 K by suppression of Ta diffusion in CoFeB/MgO/CoFeB pseudo-spin-valves annealed at high temperature," *Appl. Phys. Lett.* **93**, 082508 (2008); doi: 10.1063/1.2976435.
22. J. C. Slonczewski, "Current-driven excitation of magnetic multilayers," *J. Magn. Magn. Mater.* **159**, pp. L1–L7 (1996); doi: 10.1016/0304-8853(96)00062-5.
23. L. Berger, "Emission of spin waves by a magnetic multilayer traversed by a current," *Phys. Rev. B* **54**, pp. 9353–9358 (1996); doi: 10.1103/PhysRevB.54.9353.
24. M. D. Stiles and J. Miltat, "Spin-transfer torque and dynamics," in *Spin Dynamics in Confined Magnetic Structures III*, Topics in Applied Physics, Vol. 101, eds B. Hillebrands and A. Thiaville, Springer, pp. 225–308 (2006); doi: 10.1007/10938171_7.
25. S. C. Oh, S. Y. Park, A. Manchon, M. Chshiev, J. H. Han, H. W. Lee, J. E. Lee, K. T. Nam, Y. Jo, Y. C. Kong, B. Dieny, and K. J. Lee, "Bias-voltage dependence of perpendicular spin-transfer torque in asymmetric MgO-based magnetic tunnel junctions," *Nat. Phys.* **5**, pp. 898–902 (2009); doi: 10.1038/nphys1427.
26. M. Bibes and A. Barthélémy, "Oxide spintronics," *IEEE Trans. Electron Dev.* **54**, pp. 1003–1023 (2007); doi: 10.1109/TED.2007.894366.
27. T. Graf, C. Felser, and S. S. P. Parkin, "Simple rules for the understanding of Heusler compounds," *Prog. Solid State Chem.* **39**, pp. 1–50 (2011); doi: 10.1016/j.progsolidstchem.2011.02.001.
28. D. C. Ralph and M. D. Stiles, "Spin-transfer torques," *J. Magn. Magn. Mater.* **320**, pp. 1190–1216 (2008); doi: 10.1016/j.jmmm.2007.12.019.
29. M. Tsoi, A. G. M. Jansen, J. Bass, W. C. Chiang, M. Seck, V. Tsoi, and P. Wyder, "Excitation of a magnetic multilayer by an electric current," *Phys. Rev. Lett.* **80**, pp. 4281–4284 (1998); doi: 10.1103/PhysRevLett.80.4281.
30. J. A. Katine, F. J. Albert, R. A. Buhrman, E. B. Myers, and D.C. Ralph, "Current-driven magnetization reversal and spin-wave excitations in Co/Cu/Co pillars," *Phys. Rev. Lett.* **84**, pp. 3149 (2000); doi: 10.1103/PhysRevLett.84.3149.
31. Y. Huai, F. Albert, P. Nguyen, M. Pakala, and T. Valet, "Observation of spin-transfer switching in deep submicron-sized and low-resistance magnetic tunnel junctions," *Appl. Phys. Lett.* **84**, pp. 3118–3120 (2004); doi: 10.1063/1.1707228.
32. M. Hosomi, H. Yamagishi, T. Yamamoto, K. Bessho, Y. Higo, K. Yamane, H. Yamada, M. Shoji, H. Hachino, C. Fukumoto, H. Nagao, and H. Kano,"A novel nonvolatile memory with spin-torque transfer magnetization switching: Spin-RAM," IEEE International Electron Devices Meeting, IEDM Technical Digest, December 5–7, 2005, Washington, DC, pp. 459–462; doi: 10.1109/IEDM.2005.1609379.

CHAPTER 2

MAGNETIC PROPERTIES OF MATERIALS FOR MRAM

Shinji Yuasa
National Institute of Advanced Industrial Science and Technology (AIST), Tsukuba, Japan

2.1 MAGNETIC TUNNEL JUNCTIONS FOR MRAM

A magnetic tunnel junction (MTJ) consists of an ultrathin insulating layer (tunnel barrier) sandwiched between two ferromagnetic (FM) metal layers (electrodes). A memory cell for magnetic random-access memory (MRAM) consists of an MTJ and a complementary metal–oxide semiconductor (CMOS) pass transistor, as shown in Fig. 2.1. The practical layer stacking structure of MTJ for MRAM and magnetic sensor applications is shown in Fig. 2.2. In the practical MTJ, the magnetization in one of the ferromagnetic electrodes can be reversed by an external magnetic field or a relatively large current passed through the MTJ. This layer is called "free layer" or "storage layer." The free layer should have a "uniaxial magnetic anisotropy" such as a shape magnetic anisotropy or magnetocrystalline anisotropy, which tends to direct the magnetization along a certain axis called the easy axis. Thanks to this uniaxial magnetic anisotropy, the free layer can store one bit of information in the form of the magnetization direction. The orientation of the magnetization of the other ferromagnetic electrode is fixed. This layer, called "reference layer" or "pinned layer," acts as a reference for readout of the stored information. In most common embodiments, the reference layer is composed of a FM/NM/FM trilayer, where NM is a nonmagnetic spacer. The two FM layers have antiparallel magnetization alignment, which is termed a synthetic antiferromagnetic (SAF) structure (see Fig. 2.2). The purpose of this structure is to increase the pinning of the reference layer magnetization and reduce the stray field that this SAF is creating on the storage layer. Furthermore, the magnetization direction of the top FM layer in the SAF structure is pinned by exchange bias from antiferromagnetic (AF) coupling to the bottom layer. These structures and their effects on magnetic properties are explained in detail in Section 2.2 of this chapter.

The magnetotransport response of the MTJ with practical stacking structure is schematically illustrated in Fig. 2.3. Whereas the magnetization of reference layer is

Introduction to Magnetic Random-Access Memory, First Edition. Edited by Bernard Dieny, Ronald B. Goldfarb, and Kyung-Jin Lee.

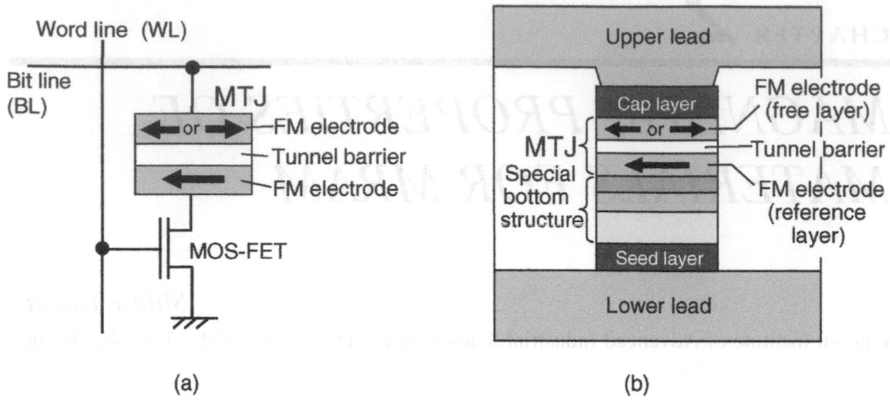

Figure 2.1 (a) Typical circuit diagram of magnetic random-access memory (MRAM). (b) Typical cross-sectional structure of magnetic tunnel junction (MTJ) with spin-valve structure.

fixed in one direction, the magnetization of the free layer can be reversed by an external magnetic field or a sufficiently large current passed through the MTJ (if spin transfer torque is used, see Chapter 1). When the free layer has a uniaxial magnetic anisotropy, the magnetization reversal process is characterized by a hysteresis loop, as shown in Fig. 2.3, which provides a nonvolatile memory function. Thus, bistable states (parallel and antiparallel magnetic alignments) appear around zero field or zero current. By applying a bipolar field or bipolar current, logic "0" or "1" can be written in the MTJ in the form of the relative orientation of the storage layer and reference layer magnetizations. The resistance of MTJ cell depends on this relative magnetization alignment (parallel or antiparallel) as a result of the tunnel magnetoresistance (TMR) effect (see Chapter 1). Thus, we can read out the bit information by passing a unipolar sense current through MTJ and measuring the MTJ resistance value. Clearly, MTJ is resistive memory, as are phase-change RAM and resistive RAM (oxide RAM, redox RAM, or conductive bridge RAM).

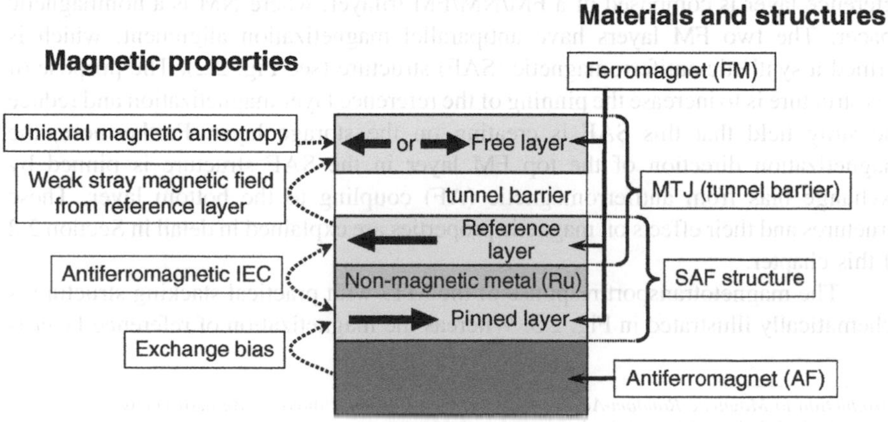

Figure 2.2 Typical stacking structure of MTJ for in-plane magnetized MRAM.

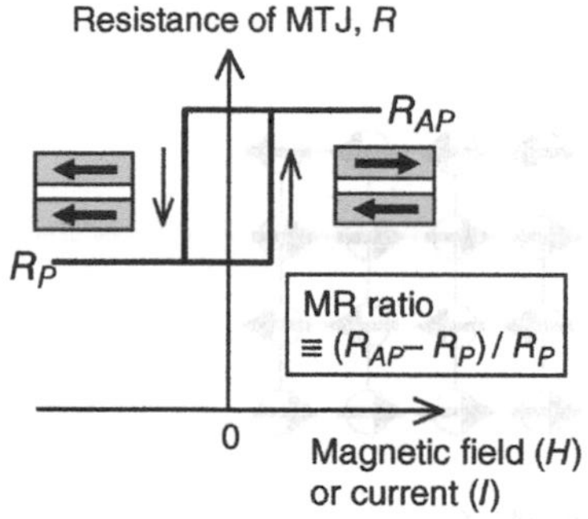

Figure 2.3 Typical magnetoresistance curve of spin-valve-type MTJ and definition of magnetoresistance (MR) ratio.

In Section 2.2, we explain basic concepts regarding magnetic materials and properties for practical MTJs, including ferromagnets, antiferromagnets, magnetic anisotropy, exchange bias, interlayer exchange coupling (IEC), and SAF structure. In Section 2.3, we explain magnetotransport properties of giant magnetoresistive device, MTJ with either amorphous aluminum oxide (AlO) or crystalline magnesium oxide (MgO) tunnel barrier.

2.2 MAGNETIC MATERIALS AND MAGNETIC PROPERTIES

2.2.1 Ferromagnet and Antiferromagnet

A ferromagnet is a material with a spontaneous magnetization (a magnetization in the absence of external magnetic field) M_s. Typical examples of ferromagnetic materials are Fe, Co, Ni (3d transition metals), Gd and Dy (rare-earth metals), and alloys mainly consisting of these ferromagnetic elements. In a ferromagnetic material, each atom of the magnetic element possesses a *localized* magnetic moment composed mainly of d or f electrons. Strictly speaking, d electrons in transition metals and alloys have itinerant nature and, therefore, are not localized exactly in an atom. However, we often use the term "localized magnetic moment" for simplicity. In ferromagnetic materials, the localized moments of the atoms tend to align parallel to each other due to a quantum mechanical ferromagnetic exchange interaction and as a result form a *net* magnetization M (magnetic moment per unit volume), as shown in Fig. 2.4a.

In an antiferromagnet, on the other hand, localized magnetic moments of neighboring atoms tend to align antiparallel because of an antiferromagnetic exchange interaction, as shown in Fig. 2.4b. An antiferromagnetic material has zero *net* magnetization.

The spontaneous magnetization in a ferromagnetic material decreases with temperature as a result of thermal activation that yields a gradual misalignment of the atomic moments. It vanishes as a result of a second-order transition from a ferromagnetic to a paramagnetic state, as shown in Fig. 2.5. The transition temperature is called the Curie temperature T_C. For a material to have stable ferromagnetic properties at room temperature (RT), its Curie temperature should be much higher than RT. For practical applications such as MRAM and magnetic sensors, stable

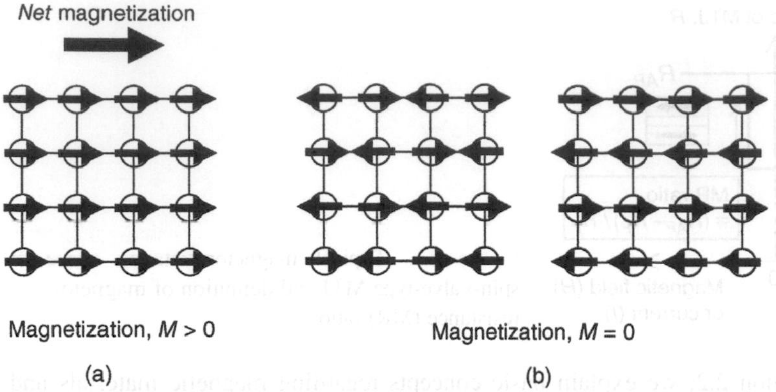

Figure 2.4 Magnetic structure of (a) ferromagnet and (b) antiferromagnet. Small arrows denote localized magnetic moments of magnetic atoms.

magnetic properties should be ensured usually for the temperature range from 0 to 80°C; this requires T_C significantly above about 300°C. Examples of Curie temperatures are 1404 K for cobalt, 1043 K for iron, and 631 K for nickel. This means that the exchange interactions are stronger in cobalt than they are in iron or nickel.

A typical magnetization process for a ferromagnet is illustrated in Fig. 2.6. Above a saturation magnetic field H_s, all the localized magnetic moments are aligned along the field direction and form a single magnetic domain (states (i) and (iii) in Fig. 2.6). The saturation magnetization is nearly equal to the spontaneous magnetization (M_s). Thus, M_s can represent both spontaneous magnetization and saturation magnetization. Below H_s, the magnetization $|M|$ can take any value between 0 and M_s because a multidomain state (state (ii) in Fig. 2.6) is formed to reduce the magnetostatic energy due to the demagnetizing field (see Section 2.2). The magnetization process in a ferromagnet is generally accompanied with a hysteresis loop. The field value for $M = 0$ is called the coercive force or the coercive field H_c. The magnetization value at zero fields is called the "remanent magnetization" M_r. When the hysteresis

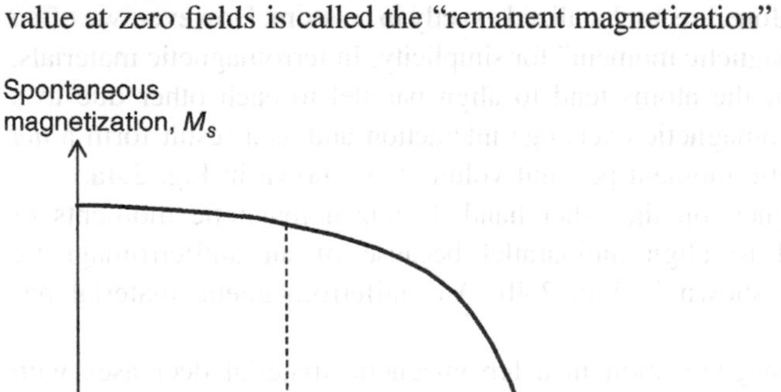

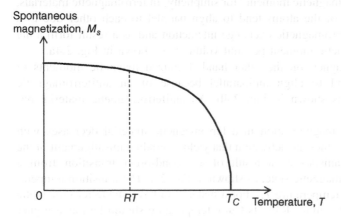

Figure 2.5 Temperature dependence of spontaneous magnetization in ferromagnet.

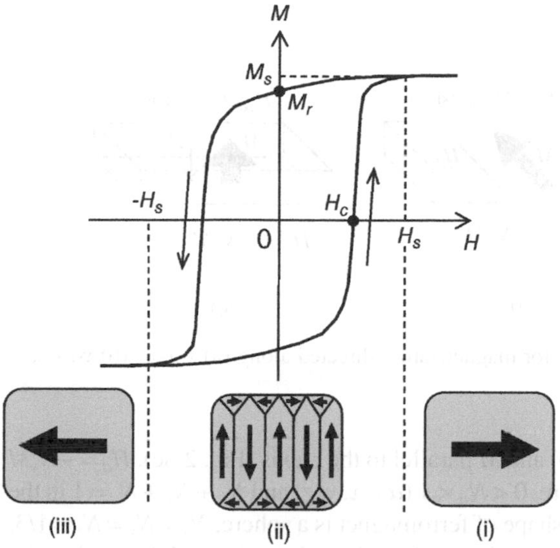

Figure 2.6 Typical magnetization process of a ferromagnet. M_s, H_s, H_c, and M_r respectively denote spontaneous (or saturation) magnetization, saturation field, coercive force, and remanent magnetization.

loop has a square shape, M_r is almost equal to M_s, and the single domain state obtained at saturation is retained even at zero field. Such a square hysteresis loop is ideal for the free layer of an MTJ in MRAM applications.

2.2.2 Demagnetizing Field and Shape Anisotropy

When the magnetization in a bulk or thin film ferromagnet is aligned parallel to an external magnetic field (H_{ext}), fictitious but mathematically described positive and negative magnetic poles are created on two sides of the ferromagnet, as shown in Fig. 2.7. These magnetic poles generate a demagnetizing field H_d, whose direction is opposite to M and H_{ext}. The effective magnetic field (H_{eff}) acting inside the ferromagnet is therefore expressed as $H_{eff}=H_{ext}-H_d$.

The following discussion is rigorously correct only for spheres and ellipsoids. Here, we approximate a rectangular thin film as an ellipsoid with a very small z-axis. A complete discussion of demagnetizing fields may be found in the work of Chen et al. (1). When the magnetization is aligned along x-axis (Fig. 2.8a), the demagnetizing field $H_d=-N_xM$. Here, N_x is the demagnetizing factor for the x-axis. For

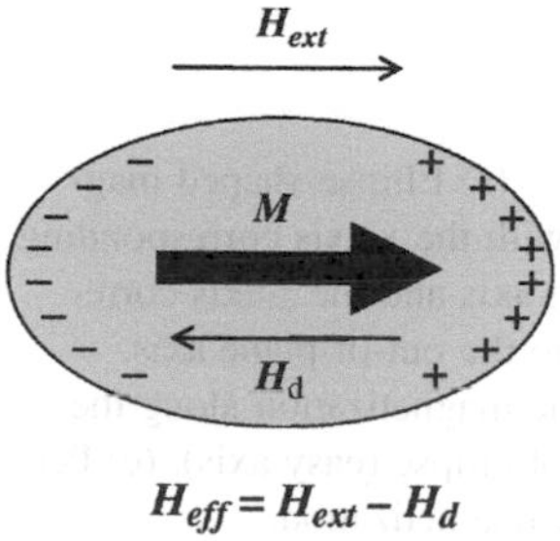

Figure 2.7 Demagnetizing field in a ferromagnet. Symbols + and – denote positive and negative magnetic charges.

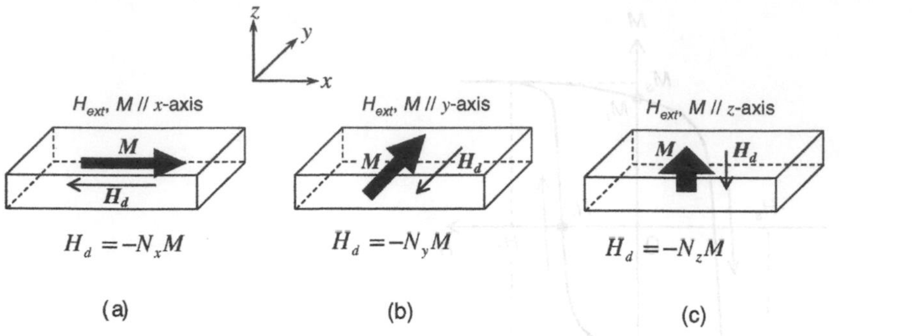

Figure 2.8 Demagnetizing field H_d for magnetization directed along (a) x-axis, (b) y-axis, and (c) z-axis.

M parallel to the y-axis (Fig. 2.8b) and M parallel to the z-axis (Fig. 2.8c), $H_d = -N_y M$ and $H_d = -N_z M$, respectively. Here, $0 < N_\alpha < 1$ ($\alpha = x, y, z$) and $N_x + N_y + N_z = 1$ in the ellipsoidal approximation. If the shape of ferromagnet is a sphere, $N_x = N_y = N_z = 1/3$. If the ferromagnet is a thin film in x–y plane, whose lateral size is much larger than the film thickness, $N_x \approx N_y \approx 0$ and $N_z \approx 1$.

Figure 2.9 shows a free layer shape in MTJ cell for MRAM. The free layer consists of a thin film with a typical thickness of a few nanometers. The free layer is usually fabricated into an ellipse or circular shape with a typical lateral size from micrometer down to a few tens of nanometers. In Fig. 2.9a, the long axis of ellipse is along x-axis. The typical dimensions $d_x > d_y \gg d_z$ yields $0 < N_x < N_y \ll N_z < 1$. The magnetostatic energy due to the demagnetizing field, E_d, when the magnetization is pointing in the direction α ($\alpha = x, y, z$) is expressed as

$$E_{d\alpha} = -\mu_0 \cdot M_s \cdot H_{d\alpha}/2 = \mu_0 \cdot N_\alpha \cdot M_s^2/2. \quad (2.1)$$

Because the magnetostatic energy is proportional to the demagnetizing factor N_α, the energy is the lowest when M is aligned to the film plane and parallel to the long axis of the ellipse (Fig. 2.9b), and the energy is the highest when M is perpendicular to the film plane (Fig. 2.9c). If the single domain state is retained at $H_{ext} = 0$, the magnetization is aligned to the long axis of ellipse due to the lowest demagnetizing energy. This axis is called "easy axis." On the other hand, it is usually hard to direct the

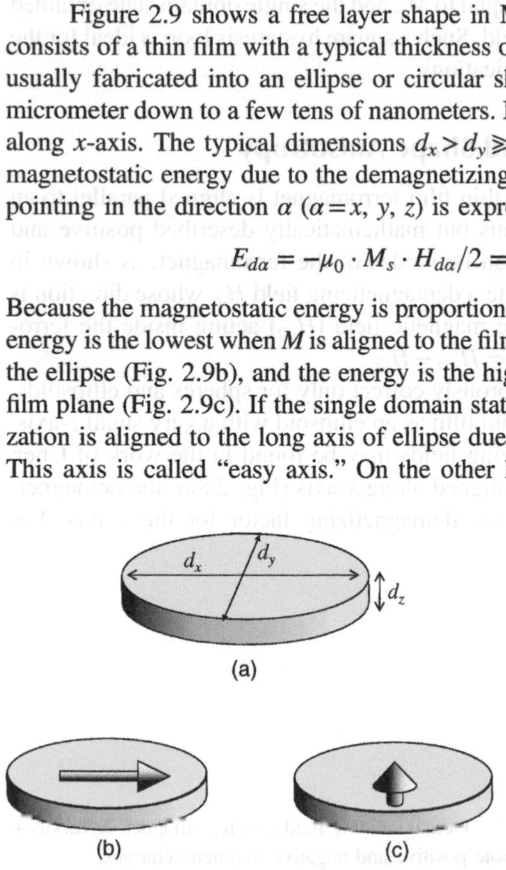

Figure 2.9 (a) Ellipse-shaped magnetic cell with the x-axis corresponding to the long axis and the z-axis corresponding to the out-of-plane axis. (b) In-plane magnetization along the long axis of ellipse (easy axis), (c) Perpendicular magnetization.

magnetization perpendicular to the plane because of the large demagnetizing energy. The perpendicular axis in this case is called "hard axis." The shape of ferromagnet yields a magnetic anisotropy energy called "shape magnetic anisotropy" and is expressed as $E_{shape} = \mu_0(N_y - N_x)M_s^2/2$.

2.2.3 Magnetocrystalline Anisotropy, Interface Magnetic Anisotropy, and Perpendicular Magnetic Anisotropy

When the effective magnetic field $H_{eff} = H_{ext} - H_d$ is used for the magnetization curve (M–H curve), we exclude the effects of shape and discuss the intrinsic magnetic properties of a ferromagnet. Typical examples of intrinsic magnetization curves are shown in Fig. 2.10. In isotropic "soft" ferromagnetic materials, the magnetization can be easily aligned parallel to the field direction and as a result can be saturated at a small magnetic field (Fig. 2.10a). A typical example of a soft ferromagnet is fcc $Ni_{0.81}Fe_{0.19}$ (Permalloy). When the magnetic energy depends on the magnetization direction, the directional dependence is called "magnetic anisotropy." In most common cases, the magnetic anisotropy energy (E_{MA}) can be expressed as

$$E_{MA} = K_u \sin^2 \theta, \tag{2.2}$$

where θ is the angle between the magnetization direction and a certain crystalline axis called the anisotropy axis. Such magnetic anisotropy is called "uniaxial magnetic anisotropy." When $K_u > 0$, the energy is the lowest for $\theta = 0$, and the magnetization tends to be directed along this axis, which corresponds to the easy axis of magnetization. Figure 2.10b schematically shows typical magnetization curves for the "easy" and "hard" axes in a ferromagnet with uniaxial anisotropy. The hard-axis magnetization curve shows a high saturation field (H_s), whereas the easy-axis magnetization curve exhibits a square hysteresis loop with a high coercive force (H_c).

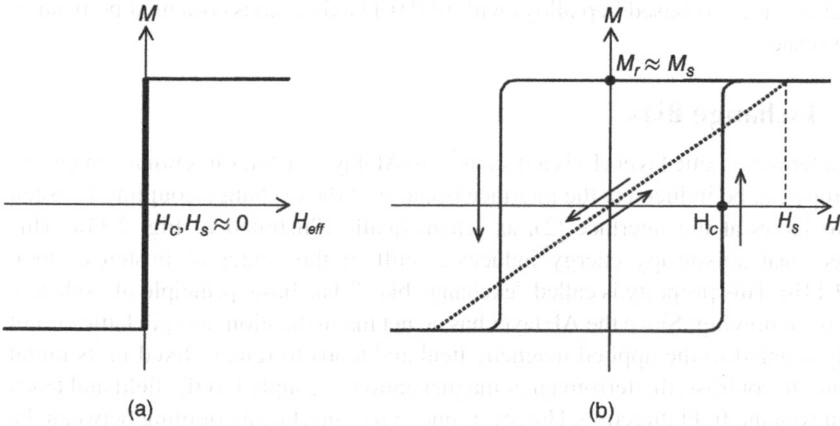

Figure 2.10 (a) Typical magnetization curve for soft, isotropic ferromagnet. (b) Typical magnetization curves for anisotropic ferromagnet. Solid and dotted lines represent magnetization curves for the easy and hard axes, respectively.

Magnetic anisotropy has several possible origins. Shape anisotropy is important in in-plane magnetized MRAM. In contrast, in out-of-plane magnetized MRAM, the two most important sources of anisotropy are (i) magnetocrystalline anisotropy and (ii) interfacial magnetic anisotropy.

(i) *Magnetocrystalline Anisotropy:* The anisotropy of crystalline lattice is usually reflected in the magnetic anisotropy. This is called magnetocrystalline anisotropy. When the lattice has a cubic crystallographic symmetry, the magnetocrystalline anisotropy also has cubic symmetry. When the lattice has a uniaxial symmetry (e.g., body-centered tetragonal (bct), face-centered tetragonal (fct), hexagonal close-packed (hcp), $L1_0$-type ordered structure), the magnetocrystalline anisotropy is often a uniaxial magnetic anisotropy. Some $L1_0$-ordered alloys such as $Fe_{50}Pt_{50}$ are known to have very high uniaxial magnetic anisotropy with the easy axis along [0 0 1] direction. When the easy axis of uniaxial anisotropy for a thin film ferromagnet is perpendicular to the film plane, the anisotropy is called "perpendicular magnetic anisotropy" (PMA).

(ii) *Interfacial Magnetic Anisotropy:* In a magnetic multilayer structure consisting of two or more materials, in which at least one of the materials is ferromagnetic, the interfaces can exhibit a magnetic anisotropy. This is called interfacial magnetic anisotropy. For some combinations of two materials, the interface yields a perpendicular magnetic anisotropy with the easy axis perpendicular to the plane. Co/Pt and Co/Pd interfaces are known to exhibit such strong perpendicular magnetic anisotropy.

The perpendicular magnetic anisotropy due to magnetocrystalline and/or interfacial anisotropy competes with the shape magnetic anisotropy of the thin film ferromagnet. If the perpendicular magnetic anisotropy is stronger than the shape magnetic anisotropy, the film can be magnetized out-of-plane at zero magnetic fields (Fig. 2.9c and Fig. 2.10b). Typical examples of such perpendicularly magnetized thin films are $L1_0$-type ordered FePt(0 0 1) and $Fe_{50}Pd_{50}$(0 0 1), Co/Pt and Co/Pd multilayers, and Co-based hcp alloys with [0 0 0 1] axis (*c*-axis) oriented perpendicular to plane.

2.2.4 Exchange Bias

When a ferromagnetic layer (F) is adjacent to an AF layer, a "unidirectional" magnetic anisotropy can be induced at the interface because of the exchange coupling between the two layers at the interface (2), as schematically illustrated in Fig. 2.11a. This unidirectional anisotropy energy induces a shift of the center of hysteresis loop (Fig. 2.11b). This property is called "exchange bias." The basic principle of exchange bias is the following: Since the AF layer has no net magnetization, its spin lattice is not directly coupled to the applied magnetic field and tends to remain fixed in its initial structure. In contrast, the ferromagnet magnetization is coupled to the field and tends to align with the field direction. However, due to the interfacial coupling between the AF spin lattice and the F magnetization, the last interfacial AF layer in contact with the F layer tends to exert a torque on its magnetization that hinders the switching of the F magnetization.

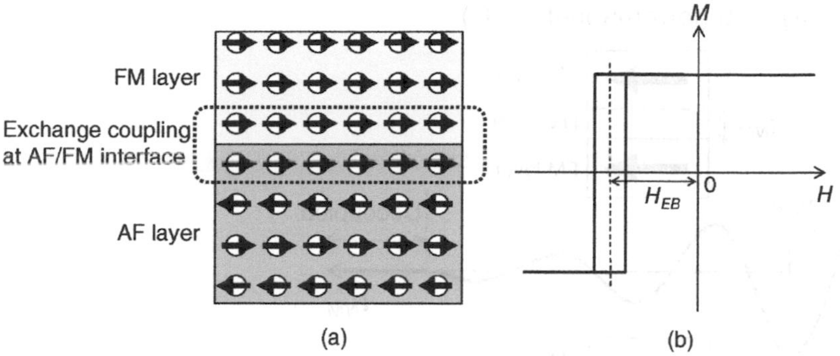

Figure 2.11 (a) Magnetic structure of a ferromagnet/antiferromagnet (FM/AF) bilayer. (b) Typical magnetization curve with exchange bias field, H_{EB}.

As a result, the magnetization of the ferromagnetic layer is pinned in a certain direction. To have a strong exchange bias, the bilayer AF/F structure must be deposited in a magnetic field and/or annealed and cooled down under a magnetic field from a temperature close to the ordering temperature of the AF material (T_N). Then, the magnetization is pinned along the cooling field direction. Typical AF materials used for exchange bias are $Ir_{20}Mn_{80}$ or $Pt_{50}Mn_{50}$ alloys.

2.2.5 Interlayer Exchange Coupling and Synthetic Antiferromagnetic Structure

When two ferromagnetic metal layers are separated by a nonmagnetic (NM) metal layer with a typical thickness of nanometer, an exchange coupling between the ferromagnetic layers is often observed. This coupling is called "interlayer exchange coupling". The coupling energy can be expressed as $E_{IEC} = -J_{IEC} \cdot \cos\theta$, where θ is the angle between the magnetizations and J_{IEC} is the exchange coupling constant. $J_{IEC} > 0$ ($J_{IEC} < 0$) corresponds to ferromagnetic (antiferromagnetic) IEC.

In the late 1980s, Grünberg et al. discovered the existence of an antiferromagnetic IEC in Fe/Cr/Fe structure and that this IEC could be used to change the relative orientation of the magnetization in the Fe layers from an antiparallel alignment at zero field to a parallel alignment under a saturating magnetic field (3). Fert (4) and Grünberg (3) then independently discovered a large change in the sheet resistance of the film associated with this change in the magnetization alignment. As explained in Chapter 1, this new phenomenon, known as the giant magnetoresistance (GMR) effect, triggered the development of this new field of electronics called spintronics. Because of the discovery of GMR, Fert and Grünberg were awarded the Nobel Prize in physics in 2007 (5). More details on GMR effect can be found in Chapter 1.

In 1990, Parkin et al. discovered an oscillatory IEC with respect to the thickness of nonmagnetic layer for various combinations of ferromagnetic and nonmagnetic materials (6). As schematically illustrated in Fig. 2.12, the IEC often shows a damped oscillatory character (7). The physical origin of IEC is interpreted as RKKY-type

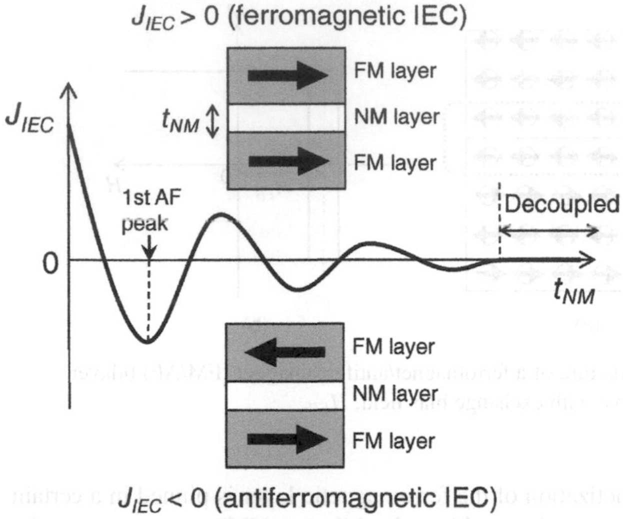

Figure 2.12 Schematic illustration of oscillatory interlayer exchange coupling (IEC).

interaction between two ferromagnetic layers or a formation of spin-polarized quantum well states (8,9). Nonmagnetic layer thickness (t_{NM}) > 5 nm is usually necessary to decouple the two FM layers. On the other hand, at the first peak of antiferromagnetic coupling (Fig. 2.12), the strongest AF coupling is attained. For example, a nonmagnetic Ru spacer layer combined with Co or Co–Fe alloys exhibit a very strong antiferromagnetic IEC at the first AF peak of about 0.4 nm. This type of strongly antiferromagnetically coupled FM/NM/FM structure is referred to as a "synthetic antiferromagnetic" structure. The SAF structure with Ru spacer is used as a building block in MTJs for device applications, as explained in the following section.

2.2.6 Spin-Valve Structure

An AF/FM/NM/FM structure, where NM denotes a "normal metal," meaning a "nonmagnetic" metal, or a tunnel barrier (Fig. 2.13a) is called "spin-valve" (10). Magnetization of the top FM layer (the "free layer") can be reversed by an external magnetic field or a current (via spin-transfer torque). Magnetization of the bottom FM layer (the "reference layer") is pinned via the exchange bias with the AF layer. This spin-valve-type magnetoresistive device can act as a magnetic sensor or as a memory cell for MRAM. The spin-valve structure shown in Fig. 2.13a, however, has a problem of unwanted stray magnetic field (H_{str}) from the reference layer. The stray field acting on the free layer induces an offset of the center of the free layer hysteresis loop, which causes various problems in magnetic sensor and memory applications. To minimize the unwanted stray field, a spin-valve device with a SAF type of reference layer (Fig. 2.13b) has been developed (11). Because the SAF trilayer forms a *flux closure* structure, most of the flux from the reference layer is absorbed in the pinned layer. This results in a suppression of the stray field and as a result induces a negligibly

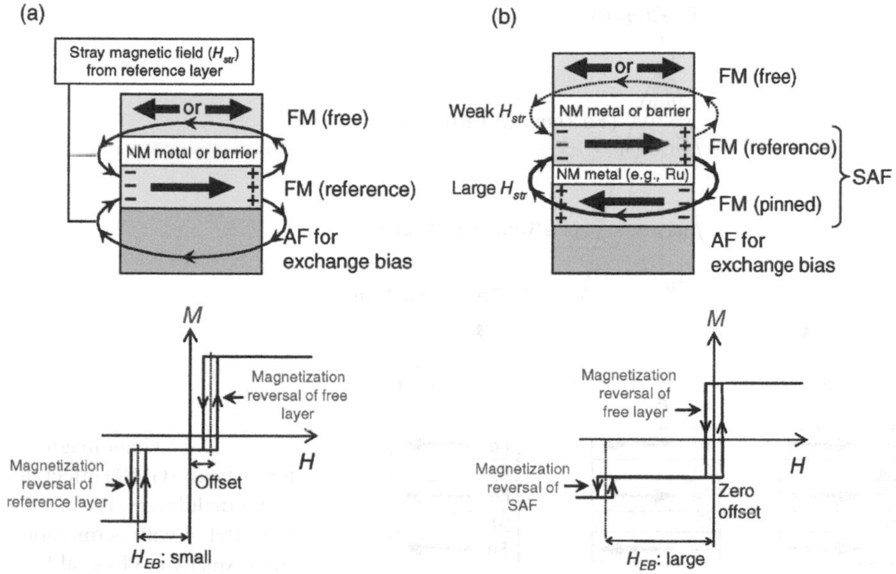

Figure 2.13 Typical spin-valve structures and magnetization curves. (a) Simple spin-valve structure with a single pinned layer. (b) Spin-valve structure with synthetic antiferromagnetic (SAF)-type pinned layer.

small offset of the free layer hysteresis loop. The SAF-type reference layer also enlarges the exchange bias field because of the smaller (or nearly zero) *net* magnetic moment in the SAF structure. The spin-valve-type MTJ with a SAF-type reference layer exhibits an ideal magnetization curve shown in Fig. 2.13b.

2.3 BASIC MATERIALS AND MAGNETOTRANSPORT PROPERTIES

2.3.1 Metallic Nonmagnetic Spacer for GMR Spin-Valve

In 1988, Baibich et al. fabricated a metallic magnetic multilayer consisting of Fe and Cr layers and obtained the GMR effect, in which electric resistance of the multilayer film changes by a few tens of percent due to the change in magnetization alignment (parallel or antiparallel), as shown in Fig. 2.14 (4). This new magnetoresistive effect was named *giant* magnetoresistance because the magnetoresistance (MR) ratio in GMR was more than 10 times than that in conventional anisotropic magnetoresistance (AMR), where the MR ratio is about 1% at ambient temperature.

The origin of the GMR effect is spin-dependent scattering of conduction electrons in the bulk of the FM layers and/or at FM/NM interfaces, as schematically illustrated in Fig. 2.15. Supposing that the conduction electrons with majority spin (spin parallel to magnetization of ferromagnetic layer) have lower probability to be scattered in the FM layers than the conduction electrons with minority spin (spin antiparallel to magnetization of ferromagnetic layer), the majority spin electrons can

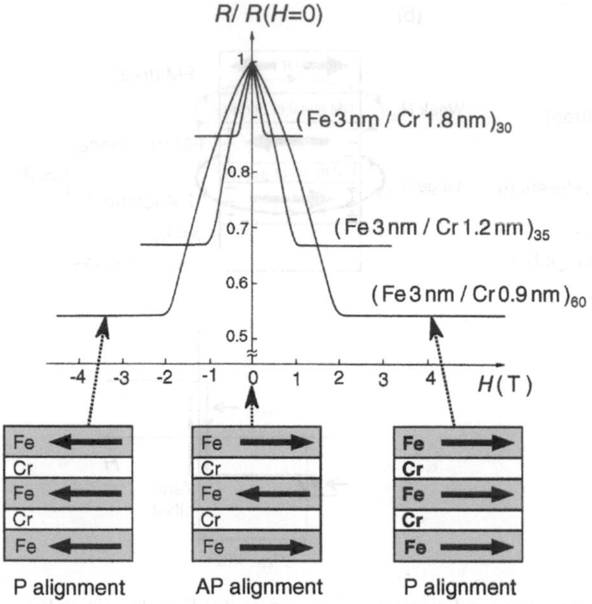

Figure 2.14 Giant magnetoresistance (GMR) effect in Fe/Cr multilayers. (Adapted from Ref. 4 with permission from American Physical Society.).

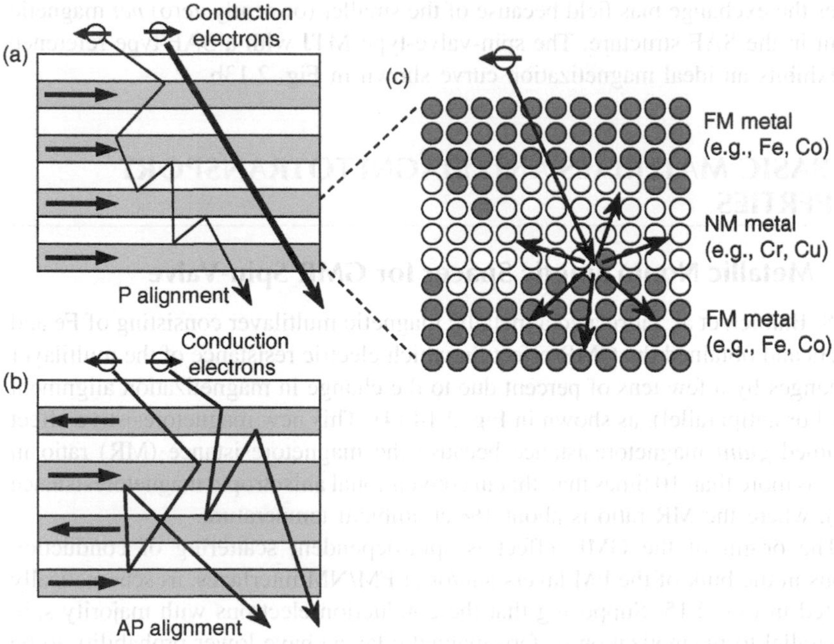

Figure 2.15 Schematic illustration of scattering of conduction electrons in magnetic multilayers. (a) Parallel (P) magnetization alignment. (b) Antiparallel (AP) magnetization alignment.

travel for a long distance without scattering in the parallel magnetization alignment, and, as a result, the total electric resistance of the film for the parallel alignment becomes smaller than that for the antiparallel alignment.

In principle, a finite GMR effect can occur for any combinations of ferromagnetic and nonmagnetic metals and alloys. High MR ratios, however, are obtained for certain material combinations such as Fe–Cr (3,4) and Co–Cu (12). This is because the electron scatterings at Fe–Cr and Co–Cu interfaces are largely spin dependent. At Fe–Cr interfaces, for example, scattering probability for minority-spin electrons is much smaller than that for majority-spin electrons because of the good matching between the Fe minority-spin band and the Cr band at the Fermi energy. At Co–Cu interfaces, on the other hand, the scattering probability is much smaller for majority-spin electrons.

For device applications such as magnetic sensors, the spin-valve structure (Fig. 2.13) with basically zero interlayer exchange coupling (the decoupled region in Fig. 2.12) is important. Such zero IEC is easily obtained in the Co/Cu/Co system with Cu thicker than 5 nm. (The Fe/Cr/Fe system usually exhibits much stronger IEC.) The Co/Cu/Co system is therefore useful in device applications such as magnetic sensors.

The GMR effect was originally measured with a current flowing parallel to the film plane (Fig. 2.16a). This geometry is called the "current-in-plane" (CIP) geometry. It is also possible to flow the current perpendicular to the layers by fabricating the multilayer film into a nanopillar junction structure. This geometry is called "current-perpendicular-to-plane" (CPP) geometry (13,14).

2.3.2 Magnetic Tunnel Junction with Amorphous AlO Tunnel Barrier

As explained in Section 2.1, a FM/NM/FM trilayer structure with an insulating NM spacer layer (tunnel barrier) is called a magnetic tunnel junction. The resistance of an MTJ depends on the relative magnetization alignment (parallel or antiparallel) of the electrodes. The tunnel resistance R of an MTJ is usually lower when the magnetizations are parallel than it is when the magnetizations are antiparallel, as shown in Fig. 2.3. That is, $R_P < R_{AP}$. This change in resistance with the relative orientation of

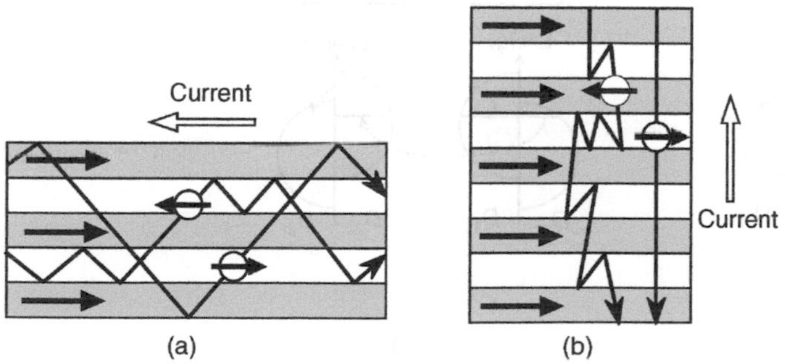

Figure 2.16 (a) Current-in-plane (CIP) GMR. (b) Current-perpendicular-to-plane (CPP) GMR.

the two magnetic layers, called the tunnel magnetoresistance effect, is one of the most important phenomena in spintronics. The size of this effect is measured by the fractional change in resistance $(R_{AP} - R_P)/R_P \times 100(\%)$, which is called the magnetoresistance ratio.

The TMR effect was first discovered in 1975 by Jullière (15), who found that a Fe/Ge-O/Co MTJ exhibited an MR ratio of 14% at 4.2 K. Although it received little attention for more than a decade because the TMR effect was not obtained at RT, it attracted renewed attention after the discovery of giant magnetoresistance in metallic magnetic multilayers in the late 1980s (3,4). Because researchers recognized that the GMR effect could be applied to magnetic sensor devices such as the read heads of hard disk drives (HDDs), extensive experimental and theoretical efforts were devoted to increasing the MR ratio at RT as well as to understanding the physics of spin-dependent transport. In 1995, Miyazaki and Tezuka (16) and Moodera et al. (17) made MTJs with amorphous aluminum oxide tunnel barriers and 3d transition metal ferromagnetic electrodes such as Fe and Co and obtained MR ratios above 10% at RT. Because these MR ratios were the highest then reported for a practical spin-valve structure, the TMR effect attracted a great deal of attention.

The origin of the TMR effect is tunneling of spin-polarized electrons. Jullière (15) proposed a simple phenomenological model (Fig. 2.17). According to this model, the MR ratio of an MTJ could be expressed in terms of the tunneling spin polarizations (TSP) P of the ferromagnetic electrodes:

$$MR = 2P_1P_2/(1 - P_1P_2), \tag{2.3}$$

where

$$P_\alpha \equiv [D_{\alpha\uparrow}(E_F) - D_{\alpha\downarrow}(E_F)]/[D_{\alpha\uparrow}(E_F) + D_{\alpha\downarrow}(E_F)]; \quad \alpha = 1, 2. \tag{2.4}$$

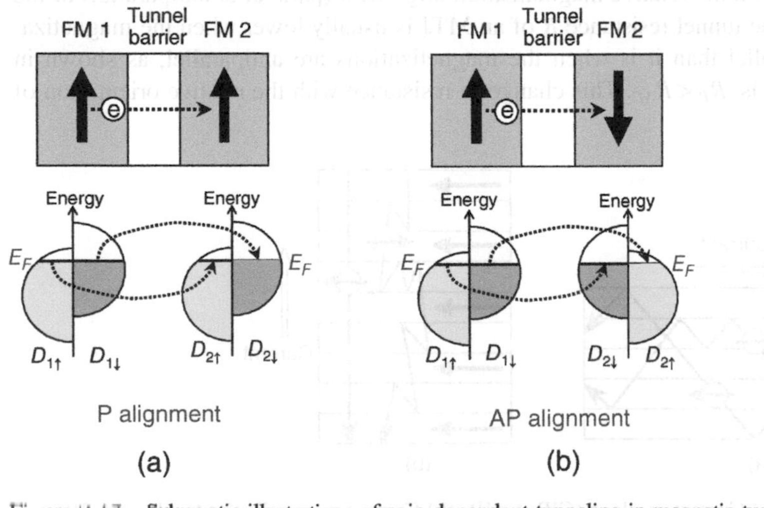

Figure 2.17 Schematic illustrations of spin-dependent tunneling in magnetic tunnel junction (MTJ). (a) Parallel (P) magnetization alignment. (b) Antiparallel (AP) magnetization alignment.

Here P_α is the spin polarization of a ferromagnetic electrode, and $D_{\alpha\uparrow}(E_F)$ and $D_{\alpha\downarrow}(E_F)$ correspond to the densities of states (DOS) of the electrode at the Fermi energy (E_F) for the majority-spin and minority-spin bands. In Jullière's model, spin polarization is an intrinsic property of an electrode material. When an electrode material is nonmagnetic, $P=0$. When the DOS of the electrode material is fully spin-polarized at E_F, $|P|=1$.

Since 1995, extensive experimental effort has been made on MTJs with amorphous AlO barriers and ferromagnetic electrode materials so that conditions for fabricating the AlO barrier could be optimized. As a result, RT MR ratios have been increased to about 70%, as shown in Fig. 2.18. These MR ratios of up to 70%, however, are still lower than those needed for many applications of spintronic devices. High-density spin-transfer torque MRAM (STT-MRAM, STT-RAM, or Spin-RAM), for example, will need to have MR ratios higher than 150% at RT, and the read head in next-generation ultrahigh-density HDDs will need to have both a high MR ratio and an ultralow tunneling resistance. The MR ratios of conventional AlO-based MTJs are simply not high enough for applications to next-generation devices.

One way to obtain an MR ratio significantly higher than 70% at RT is to use special kinds of ferromagnetic materials called *half-metals* as electrodes, which have full tunneling spin polarization ($|P|=1$) and are therefore theoretically expected to yield MTJs with huge MR ratios (up to infinity, according to Jullière's model). Some candidate half-metals are CrO_2, Heusler alloys such as Co_2MnSi, Fe_3O_4, and manganese perovskite oxides such as $La_{1-x}Sr_xMnO_3$. Very high MR ratios, above several hundred percent, have been obtained at low temperatures in $La_{1-x}Sr_xMnO_3$/$SrTiO_3$/$La_{1-x}Sr_xMnO_3$ MTJs (18) and Co_2MnSi/AlO/Co_2MnSi MTJs (19). However,

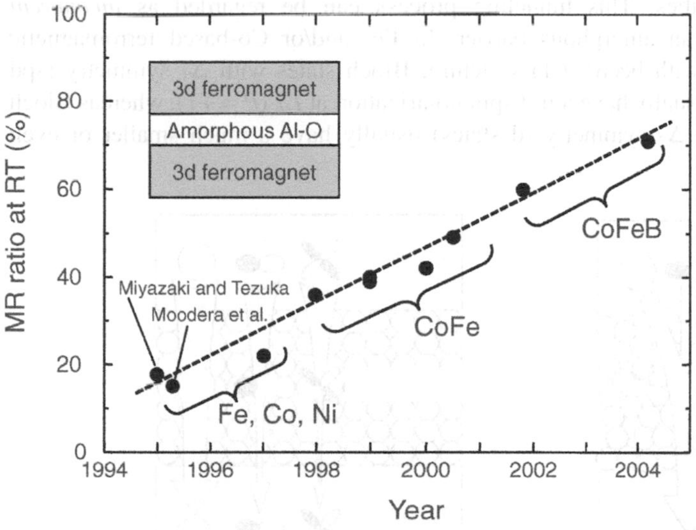

Figure 2.18 History of improvement in magnetoresistance (MR) ratio at room temperature (RT) for MTJs with amorphous AlO tunnel barriers. Early work by Miyazaki and Tezuka (16) and Moodera et al. (17) are indicated. (Reproduced from Ref. 17 with permission from American Physical Society.)

such high MR ratios have never been obtained at RT for half-metal electrodes when combined with amorphous tunnel barriers.

Another way to obtain a very high MR ratio is to use coherent spin-dependent tunneling in an epitaxial MTJ with a crystalline tunnel barrier such as MgO(0 0 1). This is explained in the following sections.

2.3.3 Magnetic Tunnel Junction with Crystalline MgO(0 0 1) Tunnel Barrier

In 2001, theoretical calculations predicted that epitaxial MTJs with a crystalline magnesium oxide tunnel barrier would have MR ratios of over 1000% (20,21). In 2004, huge MR ratios of about 200% were obtained at RT in MTJs with a crystalline MgO(0 0 1) barrier (22–24). The huge TMR effect in MgO-based MTJs is now called the *giant TMR effect* and is one of the most important technologies for device applications of spintronics.

2.3.3.1 Epitaxial MTJ with a Single-Crystal MgO(0 0 1) Barrier Before discussing the crystalline MgO(0 0 1) barrier, we will briefly review the tunneling process through an amorphous tunnel barrier (25). Tunneling in an MTJ with an amorphous AlO barrier is schematically illustrated in Fig. 2.19a, where the top electrode layer is Fe(0 0 1), being an example of a 3d ferromagnet. Various Bloch states with different symmetries of wave functions exist in the electrode. Because the tunnel barrier is amorphous, there is no crystallographic symmetry in the tunnel barrier. Because of this nonsymmetric structure, Bloch states with various orbital symmetries can couple with evanescent states in AlO and therefore have finite tunneling probabilities. This tunneling process can be regarded as *incoherent* tunneling through an amorphous barrier. In Fe- and/or Co-based ferromagnetic metals and alloys with bcc(0 0 1) structure, Bloch states with Δ_1 symmetry (spd hybridized states) usually have a full spin polarization at E_F ($P = +1$), whereas Bloch states with Δ_5 and Δ_2 symmetry (d states) usually have a much smaller or even

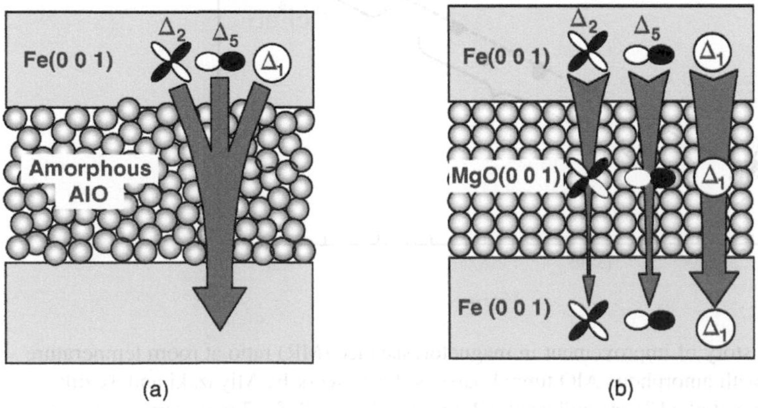

Figure 2.19 Schematic of electron tunneling through (a) an amorphous AlO barrier and (b) a crystalline MgO(0 0 1) barrier.

negative spin polarization at E_F. Because Bloch states with various spin polarizations contribute to the tunneling current, the tunneling spin polarization of the electrode is usually reduced below 0.6 for 3d ferromagnetic metals and alloys, which results in MR ratio below about 70% at RT (below about 100% at low temperature).

If only fully spin-polarized Δ_1 states coherently tunnel through a barrier, as illustrated in Fig. 2.19b, a very high spin polarization of tunneling current and thus a very high MR ratio are expected. Such ideal coherent tunneling is theoretically expected in an epitaxial Fe/MgO/Fe MTJ with a crystalline MgO(0 0 1) tunnel barrier (20,21). It should be noted that the Δ_1 Bloch states are highly spin-polarized not only in bcc Fe(0 0 1) but also in many other bcc ferromagnetic metals and alloys that are based on Fe and Co (e.g., bcc Fe–Co, bcc CoFeB, and some of the Heusler alloys). According to first-principles calculations, the TMR of a Co(0 0 1)/MgO(0 0 1)/Co(0 0 1) MTJ is even larger than that of a Fe(0 0 1)/MgO(0 0 1)/Fe(0 0 1) MTJ (26). A very large TMR should be characteristic of MTJs with 3d ferromagnetic alloy electrodes with a bcc (0 0 1) structure based on Fe and Co. Note also that very large TMR is theoretically expected not only for the MgO(0 0 1) barrier but also for other crystalline tunnel barriers such as ZnSe(0 0 1) (27) and $SrTiO_3$(0 0 1) (28). Large TMR has, however, never been obtained in MTJs with crystalline ZnSe(0 0 1) or $SrTiO_3$(0 0 1) barriers because of experimental difficulties in fabricating high-quality MTJs without pinholes in the barriers and interdiffusion at the interfaces.

Since the theoretical predictions of very large TMR effects in Fe/MgO/Fe MTJs, several experimental attempts have been made to fabricate fully epitaxial Fe(0 0 1)/MgO(0 0 1)/Fe(0 0 1) MTJs (29–31). Bowen et al. were the first to obtain a relatively high MR ratio in Fe(0 0 1)/MgO(0 0 1)/Fe(0 0 1) MTJs (30% at RT and 60% at 30 K), (30), but the RT MR ratios obtained in MgO-based MTJs did not exceed the highest of 70% obtained in AlO-based MTJs. In 2004, Yuasa et al. fabricated high-quality fully epitaxial Fe(0 0 1)/MgO(0 0 1)/Fe(0 0 1) MTJs with a single-crystal MgO (0 0 1) tunnel barrier by using molecular beam epitaxial (MBE) growth under an ultrahigh vacuum and attained RT MR ratios that were significantly higher than those of conventional AlO-based MTJs (Fig. 2.21) (23,24). A cross-sectional transmission electron microscope (TEM) image of an epitaxial MTJ is shown in Fig. 2.20 in which

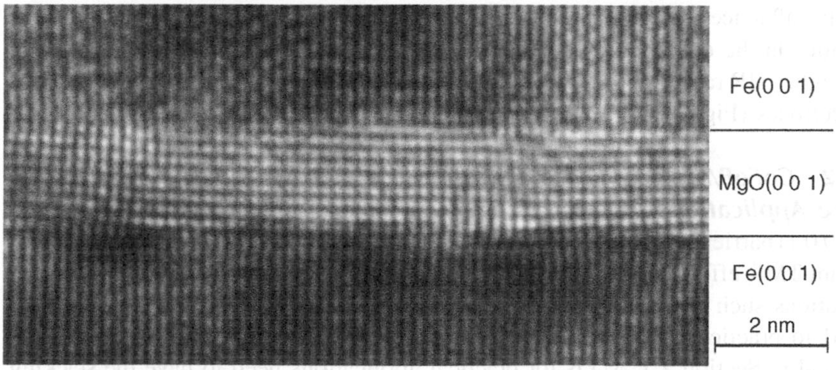

Figure 2.20 Cross-sectional transmission electron microscope (TEM) image of epitaxial Fe (0 0 1)/MgO(0 0 1) (1.8 nm)/Fe(0 0 1) MTJ. (Adapted from Ref. 24 with permission from Nature Publishing Group.)

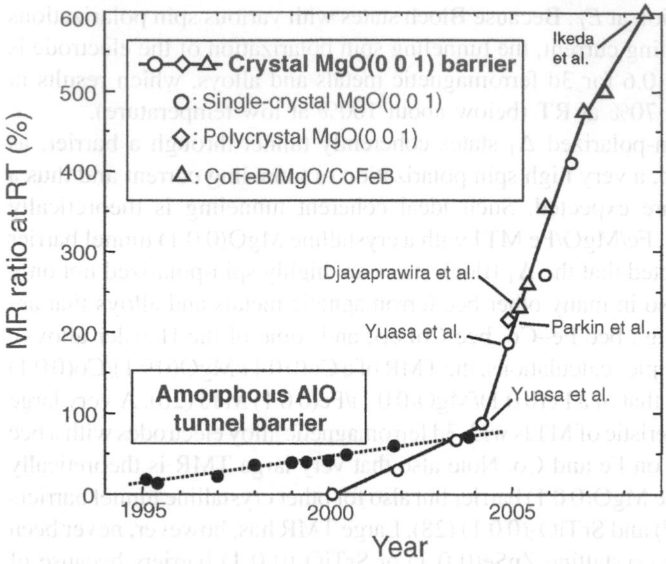

Figure 2.21 History of improvement in MR ratio at RT for MTJs with amorphous AlO barrier (solid circles) or crystalline MgO(0 0 1) barrier (open symbols). Data are shown for work by Yuasa et al. (22,23), Parkin et al. (24) Djayaprawira et al. (33), and Ikeda et al. (36).

single-crystal lattices of MgO(0 0 1) and Fe(0 0 1) are clearly evident. During the same period, Parkin et al. fabricated highly oriented polycrystalline (or textured) FeCo (0 0 1)/MgO(0 0 1)/FeCo(0 0 1) MTJs with a textured MgO(0 0 1) tunnel barrier by using sputtering deposition on an SiO_2 substrate. A tantalum nitrate seed layer was used to orient the entire MTJ stack in the (0 0 1) plane and obtained a giant MR ratio of up to 220% at RT (Fig. 2.21) (24). These were the first RT MR ratios that were higher than the highest MR ratios obtained with an AlO-based MTJ.

Fully epitaxial MTJs and textured MTJs are basically the same from a microscopic point of view if structural defects such as grain boundaries do not strongly influence the tunneling properties. It is therefore considered that the giant MR ratios in the epitaxial and textured MTJs originate from the same mechanism. Even higher MR ratios (above 400% at RT) have been attained by using FeCo and bcc Co electrodes (Fig. 2.21) (32).

2.3.3.2 CoFeB/MgO/CoFeB MTJ with a (0 0 1)-Textured MgO Barrier for Device Applications As explained above, epitaxial MTJs with a single-crystal MgO(0 0 1) barrier or textured MTJs with a polycrystalline MgO(0 0 1) barrier exhibit the giant TMR effect at RT. Although giant TMR is an attractive property for device applications such as spin-transfer torque MRAM, these MTJ structures cannot be applied to practical devices because of a problem concerning crystal growth. As explained in Section 2.1, MTJs for practical applications need to have the stacking structure shown in Fig. 2.2. Ir–Mn or Pt–Mn is used as the antiferromagnetic layer, and Co–Fe/Ru/Co–Fe trilayer is used as the SAF structure to attain an exchange bias

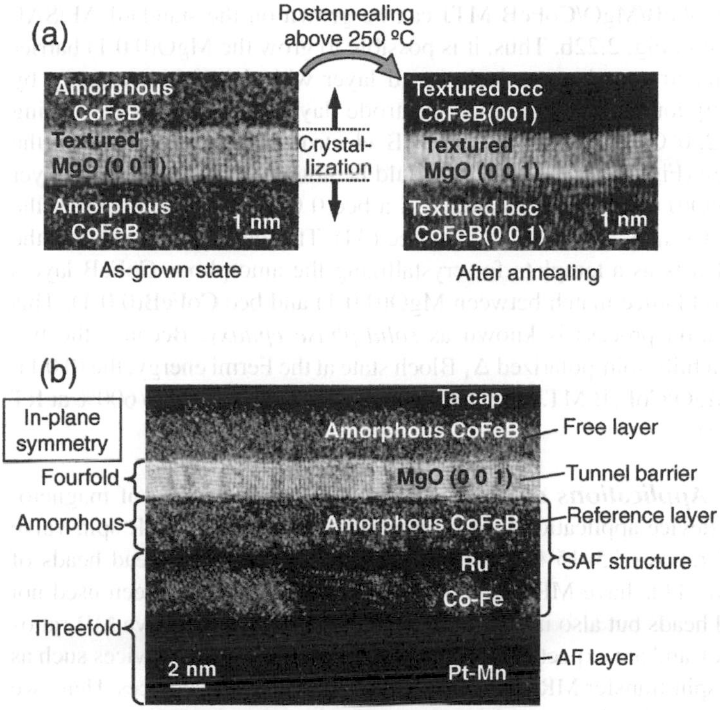

Figure 2.22 **Cross-sectional TEM image of CoFeB/MgO/CoFeB MTJ (a) in as-grown state and (b) after postannealing. Cross-sectional TEM image of CoFeB/MgO/CoFeB MTJ with practical AF/SAF structure underneath MTJ part. (Courtesy of Canon-Anelva Corporation.)**

that is strong enough for commercial devices. This type of reliable AF/SAF structure is based on an fcc structure with (1 1 1)-orientation.

A fundamental problem with this structure is that an NaCl-type MgO(0 0 1) tunnel barrier and a bcc(0 0 1) ferromagnetic electrode, both of which have fourfold in-plane crystallographic symmetry, cannot be grown on the fcc(1 1 1)-oriented AF/SAF structure, which has threefold in-plane crystallographic symmetry, because of the lattice mismatch. We could theoretically solve this problem by developing a new pinned layer structure with fourfold in-plane symmetry. This solution, however, is not generally acceptable to the electronics industry because it has spent more than a decade in developing the reliable pinned layer structure. The reliability of practical devices is determined by the reliability of the pinned layer.

To solve this problem concerning crystal growth, Djayaprawira et al. developed a new MTJ structure, CoFeB/MgO/CoFeB, by using a sputtering deposition technique (33). Figure 2.22a shows a cross-sectional TEM image of such a MTJ in an *as-grown* state. As can be seen in the figure, the bottom and top CoFeB electrode layers have an amorphous structure in the as-grown state. Surprisingly, the MgO barrier layer grown on the amorphous CoFeB is a (0 0 1)-oriented polycrystal (textured structure). Because the bottom CoFeB electrode is amorphous, this CoFeB/MgO/CoFeB MTJ can theoretically be grown on all kinds of underlayers. For practical applications, the CoFeB/MgO/CoFeB MTJ can be grown on the standard AF/SAF structure, as shown in Fig. 2.22b. Thus, it is possible to grow the MgO(0 0 1) barrier with fourfold symmetry on the practical pinned layer with threefold symmetry by inserting an amorphous CoFeB bottom electrode layer between them. During annealing above 250°C, the amorphous CoFeB electrode layers crystallize in the bcc(0 0 1) structure (Fig. 2.22b) (34,35). It should be noted that a $Co_{60}Fe_{20}B_{20}$ layer adjacent to an MgO(0 0 1) layer crystallizes in a bcc(0 0 1) structure, although the stable structure of $Co_{60}Fe_{20}B_{20}$ is not bcc but fcc (34). This clearly indicates that the MgO(0 0 1) layer acts as a template for crystallizing the amorphous CoFeB layers because of the good lattice match between MgO(0 0 1) and bcc CoFeB(0 0 1). This type of crystallization process is known as *solid-phase epitaxy*. Because the bcc CoFeB(0 0 1) has a fully spin-polarized Δ_1 Bloch state at the Fermi energy, the (0 0 1)-textured CoFeB/MgO/CoFeB MTJs exhibits a giant TMR effect of up to 600% at RT (Fig. 2.21) (33–36).

2.3.3.3 Device Applications of MgO-Based MTJs The history of magnetoresistance and its device applications are summarized in Fig. 2.23. GMR spin-valve devices have MR ratios of 5–15% at RT and have been used in the read heads of HDDs. AlO-based MTJs have MR ratios of 20–70% at RT and have been used not only in HDD read heads but also in MRAM cells. MgO-based MTJs have MR ratios of 150–600% at RT and are expected to be used in various spintronic devices such as HDD read heads, spin-transfer MRAM cells, and novel microwave devices. Here, we briefly describe recent developments in these applications.

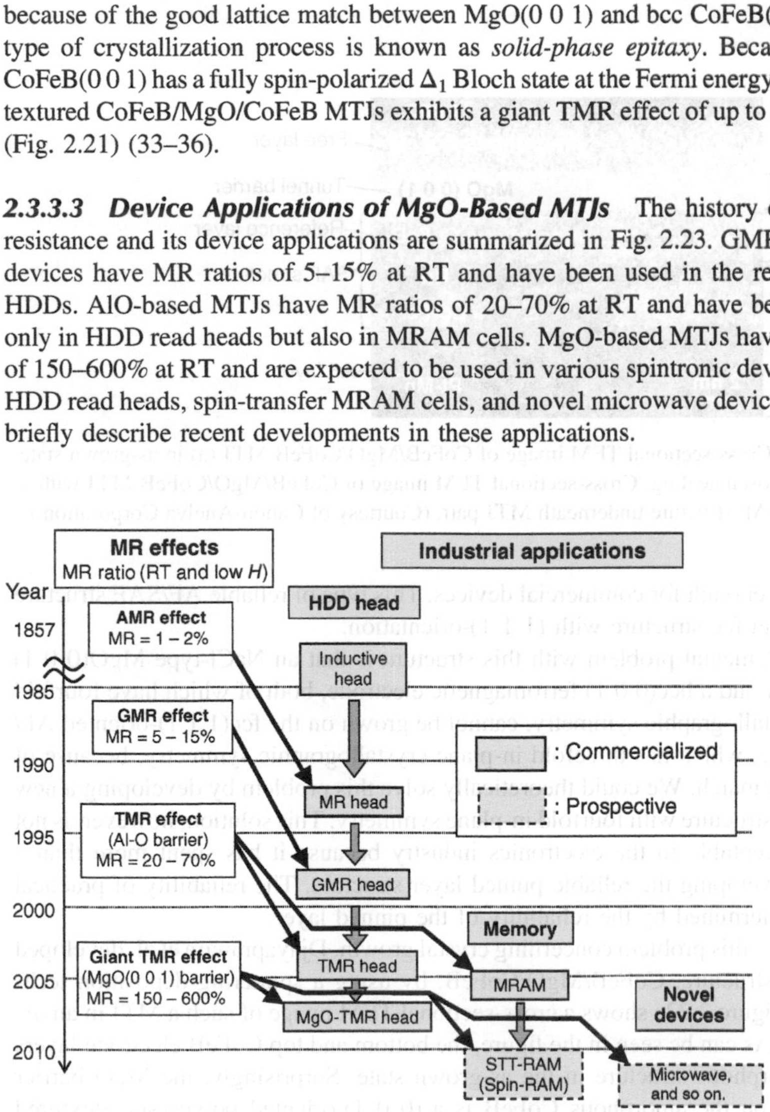

Figure 2.23 Historical and prospective magnetoresistive effects, MR ratios at RT and applications of MR effects in spintronic devices.

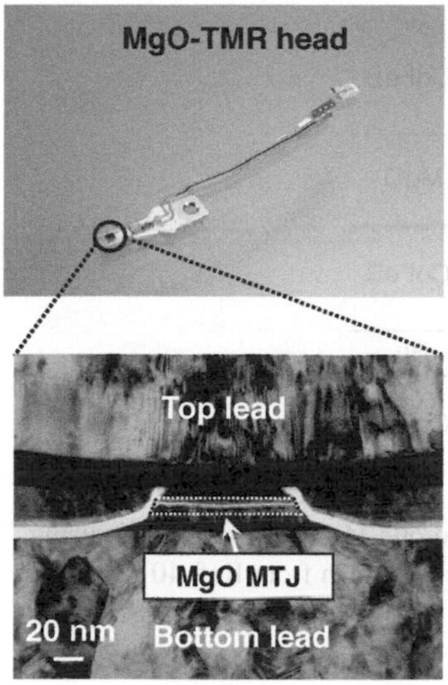

Figure 2.24 Photograph and cross-sectional TEM image of MgO-TMR read head for HDD with recording density of 250 Gbit/in^2. (Courtesy of Fujitsu Corp.)

(i) *Read Heads of High-Density HDDs*

MTJs with an amorphous AlO or TiO barrier have been used in the TMR read heads for HDDs with areal recording densities of 100–150 Gbit/in^2 since 2004 (37). These MTJs have low *RA* products (about 3 Ω μm^2) and MR ratios of 20–30% at RT. Although these properties are sufficient for recording densities of 100–150 Gbit/in^2, even lower *RA* products and much higher MR ratios are needed for recording densities above 200 Gbit/in^2.

By using ultralow-resistance CoFeB/MgO/CoFeB MTJs (38,39), HDD manufacturers were able to commercialize TMR read heads for ultrahigh-density HDDs in 2007 (35). For example, Fujitsu Corp. commercialized TMR read heads for ultrahigh-density HDDs in 2007 (Fig. 2.24). MgO-TMR heads have already been used for HDDs with recording densities of 250–800 Gbit/in^2. This means a factor of 5 increase in the recording density of HDDs, thanks to MgO-TMR heads combined with perpendicular magnetic recording media. MgO-TMR heads currently represent the mainstream technology for HDD read heads and are also expected to be applied to recording densities of up to about 1–2 Tbit/in^2.

(ii) *Spin-Transfer Torque MRAM*

The giant TMR effect in MgO-based MTJs is also useful in the development of spin-transfer torque MRAM (variously referred to as STT-MRAM, STT-RAM, or Spin-RAM). In 2005, Sony Corporation developed the 4-kbit prototype STT-MRAM based on CoFeB/MgO/CoFeB MTJs with in-plane magnetization (Fig. 2.25) and demonstrated reliable write and read operations (40). With the in-plane magnetized MTJs, however, high thermal stability, which is necessary to retain data for more than 10 years, is very difficult to achieve when the lateral size of MTJ cells is smaller than about 50 nm, which is a typical cell size in gigabit-scale STT-MRAM.

A solution to achieving gigabit-scale STT-MRAM is to use MTJs with perpendicularly magnetized electrodes (see Chapter 5). It is theoretically possible

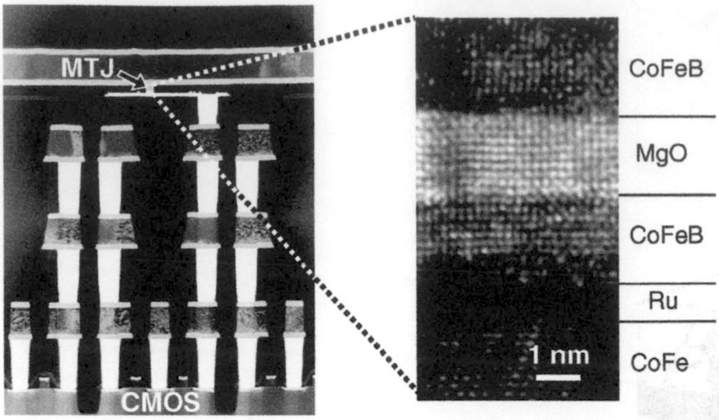

Figure 2.25 Cross-sectional TEM images of 4-kbit spin torque MRAM (Spin-RAM) using CoFeB/MgO/CoFeB MTJs. (Courtesy of Sony Corporation, redrawn from Ref. 40).

to simultaneously have high thermal stability in 50 nm-sized MTJ cells and a small write current density by using perpendicularly magnetized electrodes with a high perpendicular magnetic anisotropy, K_u. Since 2008, Toshiba Corporation and AIST have developed a prototype STT-MRAM using 30–50 nm-sized MgO-based MTJs with perpendicularly magnetized electrodes (Fig. 2.26) (41,42). The perpendicularly magnetized MgO-based MTJs are now the mainstream technology to achieve gigabit-scale STT-MRAM.

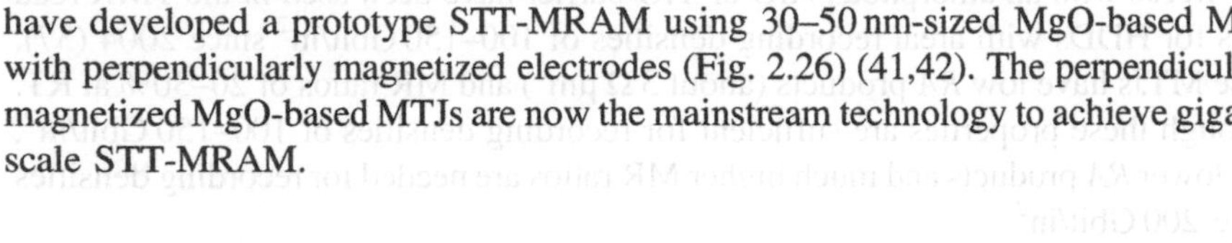

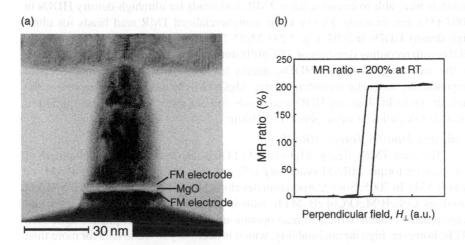

Figure 2.26 (a) Cross-sectional TEM image of MgO-based MTJ with perpendicularly magnetized electrodes with lateral diameter of 30 nm. (Courtesy of Toshiba Corporation.) (b) Typical magnetoresistance curve for perpendicular magnetic field. (Reproduced with permission from IEEE.)

REFERENCES

1. D.-X. Chen, J. A. Brug, and R. B. Goldfarb, "Demagnetizing factors for cylinders," *IEEE Trans. Magn.* **27**, pp. 3601–3619 (1991); doi: 10.1109/20.102932.
2. W. H. Meiklejohn and C. P. Bean, "A new magnetic anisotropy," *IEEE Trans. Magn.* **37**, pp. 3866–3876 (2002); doi: 10.1109/20.966120.
3. G. Binasch, P. Grünberg, F. Saurenbach, and W. Zinn, "Enhanced magnetoresistance in layered magnetic structures with antiferromagnetic interlayer exchange," *Phys. Rev. B* **39**, pp. 4828–4830 (1989); doi: http://dx.doi.org/10.1103/PhysRevB.39.4828.
4. M. N. Baibich, J. M. Broto, A. Fert, F. Nguyen Van Dau, F. Petroff, P. Etienne, G. Creuzet, A. Friederich, and J. Chazelas, "Giant magnetoresistance of (0 0 1)Fe/(0 0 1)Cr magnetic superlattices," *Phys. Rev. Lett.* **61**, pp. 2472–2475 (1988); doi: 10.1103/PhysRevLett.61.2472.
5. "The Nobel Prize in Physics 2007," www.nobelprize.org/nobel_prizes/physics/laureates/2007/.
6. S. S. P. Parkin, N. More, and K. P. Roche, "Oscillations in exchange coupling and magnetoresistance in metallic superlattice structures: Co/Ru, Co/Cr, and Fe/Cr," *Phys. Rev. Lett.* **64**, pp. 2304–2308 (1990); doi: 10.1103/PhysRevLett.64.2304.
7. J. Unguris, D. T. Pierce, R. J. Celotta, and J. A. Stroscio, "SEMPA studies of oscillatory exchange coupling," in *Magnetism and Structure in Systems of Reduced Dimension*, NATO ASI Series, Vol. **309**, eds. R. F. C. Farrow, B. Dieny, M. Donath, A. Fert, and B. D. Hermsmeier, Springer, 1993, pp. 101–112.
8. P. Bruno and C. Chappert, "Oscillatory coupling between ferromagnetic layers separated by a nonmagnetic metal spacer," *Phys. Rev. Lett.* **67**, pp. 1602–1605 (1991); doi: 10.1103/PhysRevLett.67.2592.
9. D. M. Edwards, J. Mathon, R. B. Muniz, and M. S. Phan, "Oscillations of the exchange in magnetic multilayers as an analog of de Haas–van Alphen effect," *Phys. Rev. Lett.* **67**, pp. 493–496 and 1467 (1991); doi: 10.1103/PhysRevLett.67.493 and 10.1103/PhysRevLett.67.1476.
10. B. Dieny, V. S. Speriosu, S. S. P. Parkin, B. A. Gurney, D. R. Wilhoit, and D. Mauri, "Giant magnetoresistance in soft ferromagnetic multilayers," *Phys. Rev. B* **43**, pp. 1297–1300 (1991); doi: 10.1103/PhysRevB.43.1297.
11. D. Heim, R. Fontana, C. Tsang, V. S. Speriosu, B. Gurney, and M. Williams, "Design and operation of spin valve sensors," *IEEE Trans. Magn.* **MAG-30**, pp. 316–321 (1994); doi: 10.1109/20.312279.
12. D. H. Mosca, F. Petroff, A. Fert, P. A. Schroeder, W. P. Pratt, and R. Loloee, "Oscillatory interlayer coupling and giant magnetoresistance in Co/Cu multilayers," *J. Magn. Magn. Mater.* **94**, pp. L1–L5 (1991); doi: 10.1016/0304-8853(91)90102-G.
13. W. P. Pratt, S.-F. Lee, J. M. Slaughter, R. Loloee, P. A. Schroeder, and J. Bass, "Perpendicular giant magnetoresistances of Ag/Co multilayers," *Phys. Rev. Lett.* **66**, pp. 3060–3063 (1991); doi: http://dx.doi.org/10.1103/PhysRevLett.66.3060
14. E. Tsymbal and D. G. Pettifor, "Perspectives of giant magnetoresistance," in *Solid State Physics*, Vol. **56**, ed. H. Ehrenreich and F. Spaepen, Academic Press, San Diego, CA, 2001, pp. 113–237.
15. M. Jullière, "Tunneling between ferromagnetic films," *Phys. Lett. A* **54**, pp. 225–226 (1975); doi: 10.1016/0375-9601(75)90174-7.
16. T. Miyazaki and N. Tezuka, "Giant magnetic tunneling effect in Fe/Al_2O_3/Fe junction," *J. Magn. Magn. Mater.* **139**, pp. L231–L234 (1995); doi: 10.1016/0304-8853(95)90001-2.
17. J. S. Moodera, L. R. Kinder, T. M. Wong, and R. Meservey, "Large magnetoresistance at room temperature in ferromagnetic thin film tunnel junctions," *Phys. Rev. Lett.* **74**, pp. 3273–3276 (1995); doi: 10.1103/PhysRevLett.74.3273.
18. M. Bowen, M. Bibes, A. Barthélémy, J.-P. Contour, A. Anane, Y. Lemaître, and A. Fert, "Nearly total spin polarization in $La_{2/3}$ $Sr_{1/3}$ MnO_3 from tunneling experiments," *Appl. Phys. Lett.* **82**, pp. 233–235 (2003); doi: 10.1063/1.1534619.
19. Y. Sakuraba, M. Hattori, M. Oogane, Y. Ando, H. Kato, A. Sakuma, T. Miyazaki, and H. Kubota, "Giant tunneling magnetoresistance in Co_2MnSi/Al–O/Co_2MnSi magnetic tunnel junctions," *Appl. Phys. Lett.* **88**, 192508 (2006); doi: 10.1063/1.2202724.
20. W. H. Butler, X.-G. Zhang, T. C. Schulthess, and J. M. MacLaren, "Spin-dependent tunneling conductance of Fe/MgO/Fe sandwiches," *Phys. Rev. B* **63**, 054416 (2001); doi: 10.1103/PhysRevB.63.054416.

21. J. Mathon and A. Umersky, "Theory of tunneling magnetoresistance of an epitaxial Fe/MgO/Fe(0 0 1) junction," *Phys. Rev. B* **63**, 220403 (2001); doi: 10.1103/PhysRevB.63.220403.
22. S. Yuasa, A. Fukushima, T. Nagahama, K. Ando, and Y. Suzuki, "High tunnel magnetoresistance at room temperature in fully epitaxial Fe/MgO/Fe tunnel junctions due to coherent spin-polarized tunneling," *Jpn. J. Appl. Phys.* **43**, pp. L588–L590 (2004); doi: 10.1143/JJAP.43.L588.
23. S. Yuasa, T. Nagahama, A. Fukushima, Y. Suzuki, and K. Ando, "Giant room-temperature magnetoresistance in single-crystal Fe/MgO/Fe magnetic tunnel junctions," *Nat. Mater.* **3**, pp. 868–871 (2004); doi: 10.1038/nmat1257.
24. S. S. P. Parkin, C. Kaiser, A. Panchula, P. M. Rice, B. Hughes, M. Samant, and S.-H. Yang, "Giant tunnelling magnetoresistance at room temperature with MgO (1 0 0) tunnel barriers," *Nat. Mater.* **3**, pp. 862–867 (2004); doi: 10.1038/nmat1256.
25. W. H. Butler, "Tunneling magnetoresistance from a symmetry filtering effect," *Sci. Technol. Adv. Mater.* **9**, pp. 014106 (2008); doi: 10.1088/1468-6996/9/1/014106.
26. X.-G. Zhang and W. H. Butler, "Large magnetoresistance in bcc Co/MgO/Co and FeCo/MgO/FeCo tunnel junctions," *Phys. Rev. B* **70**, 172407 (2004); doi: 10.1103/PhysRevB.70.172407.
27. Ph. Mavropoulos, N. Papanikolaou, and P. H. Dederichs, "Complex band structure and tunneling through ferromagnet/insulator/ferromagnet junctions," *Phys. Rev. Lett.* **85**, pp. 1088–1091 (2000); doi: 10.1103/PhysRevLett.85.1088.
28. J. P. Velev, K. D. Belashchenko, D. A. Stewart, M. Van Schilfgaarde, S. S. Jaswal, and E. Y. Tsymbal, "Negative spin polarization and large tunneling magnetoresistance in epitaxial Co/$SrTiO_3$/Co magnetic tunnel junctions," *Phys. Rev. Lett.* **95**, pp. 216601 (2005); doi: 10.1103/PhysRevLett.95.216601.
29. W. Wulfhekel, M. Klaua, D. Ullmann, F. Zavaliche, J. Kirschner, R. Urban, T. Monchesky, and B. Heinrich, "Single-crystal magnetotunnel junctions," *Appl. Phys. Lett.* **78**, pp. 509–511 (2001); doi: 10.1063/1.1342778.
30. M. Bowen, V. Cros, F. Petroff, A. Fert, C. Martinez Boubeta, J. L. Costa-Krämer, J. V. Anguita, A. Cebollada, F. Briones, J. M. de Teresa, L. Morellon, M. R. Ibarra Large, F. Güell, F. Peiró, and A. Cornet, "Large magnetoresistance in Fe/MgO/FeCo(0 0 1) epitaxial tunnel junctions on GaAs(0 0 1)," *Appl. Phys. Lett.* **79**, pp. 1655–1657 (2001); doi: 10.1063/1.1404125.
31. J. Faure-Vincent, C. Tiusan, E. Jouguelet, F. Canet, M. Sajieddine, C. Bellouard, E. Popova, M. Hehn, F. Montaigne, and A. Schuhl, "High tunnel magnetoresistance in epitaxial Fe/MgO/Fe tunnel junctions," *Appl. Phys. Lett.* **82**, pp. 4507–4509 (2003); doi: 10.1063/1.1586785.
32. S. Yuasa, A. Fukushima, H. Kubota, Y. Suzuki, and K. Ando, "Giant tunneling magnetoresistance up to 410% at room temperature in fully epitaxial Co/MgO/Co magnetic tunnel junctions with bcc Co(0 0 1) electrodes," *Appl. Phys. Lett.* **89**, 042505 (2006); doi: 10.1063/1.2236268.
33. D. D. Djayaprawira, K. Tsunekawa, M. Nagai, H. Maehara, S. Yamagata, N. Watanabe, S. Yuasa, Y. Suzuki, and K. Ando, "230% room-temperature magnetoresistance in CoFeB/MgO/CoFeB magnetic tunnel junctions," *Appl. Phys. Lett.* **86**, 092502 (2005); doi: 10.1063/1.1871344.
34. S. Yuasa, Y. Suzuki, T. Katayama, and K. Ando, "Characterization of growth and crystallization processes in CoFeB/MgO/CoFeB magnetic tunnel junction structure by reflective high-energy electron diffraction," *Appl. Phys. Lett.* **87**, pp. 242503 (2005); doi: 10.1063/1.2140612.
35. S. Yuasa and D. D. Djayaprawira, "Giant tunnel magnetoresistance in magnetic tunnel junctions with a crystalline MgO(0 0 1) barrier," *J. Phys. D Appl. Phys.* **40**, pp. R337–R354 (2007); doi: 10.1088/0022-3727/40/21/R01.
36. S. Ikeda, J. Hayakawa, Y. Ashizawa, Y. M. Lee, K. Miura, H. Hasegawa, M. Tsunoda, F. Matsukura, and H. Ohno, "Tunnel magnetoresistance of 604% at 300 K by suppression of Ta diffusion in CoFeB/MgO/CoFeB pseudo-spin-valves annealed at high temperature," *Appl. Phys. Lett.* **93**, pp. 082508 (2008); doi: 10.1063/1.2976435.
37. J.-G. Zhu and C. Park, "Magnetic tunnel junctions," *Mater. Today* **9**, pp. 36–45 (2006); doi: 10.1016/S1369-7021(06)71693-5.
38. K. Tsunekawa, D. D. Djayaprawira, M. Nagai, H. Maehara, S. Yamagata, N. Watanabe, S. Yuasa, Y. Suzuki, and K. Ando, "Giant tunneling magnetoresistance effect in low-resistance CoFeB/MgO(0 0 1)/CoFeB magnetic tunnel junctions for read-head applications," *Appl. Phys. Lett.* **87**, pp. 072503 (2005); doi: 10.1063/1.2012525.
39. Y. Nagamine, H. Maehara, K. Tsunekawa, D. D. Djayaprawira, N. Watanabe, S. Yuasac, and K. Ando, "Ultralow resistance-area product of 0.4 Ω $(\mu m)^2$ and high magnetoresistance above 50% in CoFeB/

MgO/CoFeB magnetic tunnel junctions," *Appl. Phys. Lett.* **89**, pp. 162507 (2006); doi: 10.1063/1.2352046.

40. M. Hosomi, H. Yamagishi, T. Yamamoto, K. Bessho, Y. Higo, K. Yamane, H. Yamada, M. Shoji, H. Hachino, C. Fukumoto, H. Nagao, and H. Kano,"A novel nonvolatile memory with spin torque transfer magnetization switching: Spin-RAM," *IEEE International Electron Devices Meeting, IEDM Technical Digest 19.1*, December 2005; doi: 10.1109/IEDM.2005.1609379.
41. T. Kishi, H. Yoda, T. Kai, T. Nagase, E. Kitagawa, M. Yoshikawa, K. Nishiyama, T. Daibou, M. Nagamine, M. Amano, S. Takahashi, M. Nakayama, N. Shimomura, H. Aikawa, S. Ikegawa, S. Yuasa, S. K. Yakushiji, H. Kubota, A. Fukushima, M. Oogane, T. Miyazaki, T. K. Ando,"Lower-current and fast switching of a perpendicular TMR for high speed and high density spin-transfer-torque MRAM," *IEEE International Electron Devices Meeting, IEDM Technical Digest 12.6*, December 2008; doi: 10.1109/IEDM.2008.4796680.
42. H. Yoda, T. Kishi, T. Nagase, M. Yoshikawa, K. Nishiyama, E. Kitagawa, T. Daibou, M. Amano, N. Shimomura, S. Takahashi, T. Kai, M. Nakayama, H. Aikawa, S. Ikegawa, M. Nagamine, J. Ozeki, S. Mizukami, M. Oogane, Y. Ando, S. Yuasa, K. Yakushiji, H. Kubota, Y. Suzuki, Y. Nakatani, T. Miyazaki, and K. Ando, "High efficient spin transfer torque writing on perpendicular magnetic tunnel junctions for high density MRAMs," *Curr. Appl. Phys.* **10**, pp. E87–E89 (2010); doi: 10.1016/j.cap.2009.12.021.

CHAPTER 3

MICROMAGNETISM APPLIED TO MAGNETIC NANOSTRUCTURES

Liliana D. Buda-Prejbeanu
SPINTEC, CEA-INAC, CNRS, Université Grenoble Alpes, Grenoble, France

3.1 MICROMAGNETIC THEORY: FROM BASIC CONCEPTS TOWARD THE EQUATIONS

Micromagnetism is a continuum description of ferromagnetic materials that exhibit a spontaneous magnetization M_s below a critical temperature T_C. A ferromagnetic system is rarely uniformly magnetized. In most of the cases, it consists of several regions with uniform magnetization vector **M** (magnetic domains) separated by transition regions (magnetic domain walls), inside of which the orientation of the magnetization changes with position. To describe such entities, several concepts were developed by Weiss (1) and Landau and Lifshitz (2), but it was Brown (3) who unified all of them in a unitary theory known as "micromagnetism", which is presented in this chapter.

Magnetic moments and magnetic order have quantum mechanical origin, but such an atomistic description is replaced in micromagnetism by continuous functions, which thus limits the smallest scale of applicability to few nanometers (see Fig. 3.1).

In a ferromagnet the individual magnetic moments are strongly coupled through exchange interactions that tend to align neighboring moments parallel, thus creating a local magnetization **M** (a net magnetic moment per unit volume) with a uniform magnitude M_S. The magnetization $\mathbf{M}(\mathbf{r}, t)$ is a continuous function that depends on space and time, related to the unit vector $\mathbf{M}(\mathbf{r}, t)$:

$$\begin{cases} \mathbf{M}(\mathbf{r}, t) &= M_s\mathbf{m}(\mathbf{r}, t), \\ |\mathbf{m}(\mathbf{r}, t)| &= 1. \end{cases} \tag{3.1}$$

The magnitude of the magnetization M_S is a temperature-dependent material parameter. However, the orientation of the magnetization $\mathbf{m}(\mathbf{r}, t)$ cannot be determined based only on the exchange coupling. The sources of nonuniform magnetization distribution are forces due to coupling with the crystalline structure, dipolar forces arising from magnetostatic charges, and forces due to external magnetic fields. These forces perturb the parallel alignment imposed by the exchange coupling, yielding

Introduction to Magnetic Random-Access Memory, First Edition. Edited by Bernard Dieny, Ronald B. Goldfarb, and Kyung-Jin Lee.

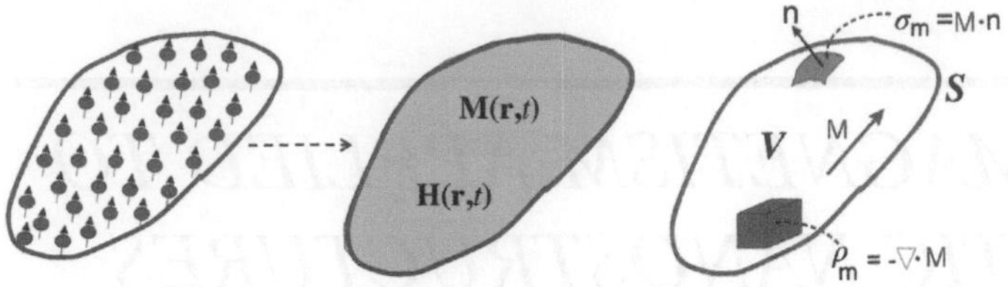

Figure 3.1 Continuous representation of a ferromagnetic system in micromagnetism. The sample has a volume V and a free surface S of normal vector **n**. The distribution of individual spins is replaced by a continuous magnetization and magnetic charges.

variations in the orientation of the magnetization vector. In micromagnetism, the magnetization is assumed to vary gradually in space. Particular cases such as magnetization discontinuities cannot be addressed properly.

Depending on the forces, external and internal, acting upon a magnetic system, different equilibrium magnetization configurations are foreseeable. Micromagnetic theory enables one to predict the spatial orientation of the magnetization under a set of conditions. A magnetic state corresponds to a local minimum of the total energy of the system. The temperature is assumed to be uniform in the material and for an isothermal process, the appropriate energy functional is the Gibbs free energy described below.

3.1.1 Free Energy of a Magnetic System

The free energy of a ferromagnetic system of volume V and under the influence of an external magnetic field contains several terms (4): exchange energy, magnetocrystalline anisotropy energy, demagnetizing energy, and Zeeman energy:

$$E_{tot} = E_{ex} + E_{anis} + E_{dem} + E_{app}. \tag{3.2}$$

3.1.1.1 Exchange Energy The exchange interaction is a short-range interaction coupling two neighboring spins. It arises from the quantum mechanical principle of exchange symmetry, which states that no observable physical quantity should change after exchanging two indistinguishable particles. The Hamiltonian of the exchange interaction can be written as (5)

$$\mathcal{H}_{ex} = -2\sum_{\langle i,j\rangle} J_{ij}\mathbf{S_i}\cdot\mathbf{S_j}, \tag{3.3}$$

where J_{ij} is the Heisenberg exchange integral. Since it is a short-range interaction, one can consider the sum being over the nearest neighbors only. The sign of J_{ij} determines whether the material is ferromagnetic (parallel alignment is favored) or antiferromagnetic (antiparallel alignment). For a constant modulus of the spins $|\mathbf{S_i}| = |\mathbf{S_j}| = S$, and small misalignment between neighboring spins, the scalar product $\mathbf{S_i}\cdot\mathbf{S_j}$ can be expressed as

$$\mathbf{S_i}\cdot\mathbf{S_j} = S^2\left\{1 - \frac{1}{2}\left[(\mathbf{r_{ij}}\cdot\nabla)\mathbf{m}(\mathbf{r_i})\right]^2\right\}, \tag{3.4}$$

where $\mathbf{r_{ij}}$ is the distance between the two neighboring spins. If we shift from a discrete to a continuous description and consider an isotropic exchange interaction ($J_{ij} = J$), the following Hamiltonian may be derived:

$$\mathcal{H}_{ex} = JS^2 \sum_i \left\{ \lfloor \Delta \mathbf{r_i} \cdot \nabla m_x(\mathbf{r_i}) \rfloor^2 + \lfloor \Delta \mathbf{r_i} \cdot \nabla m_y(\mathbf{r_i}) \rfloor^2 + \lfloor \Delta \mathbf{r_i} \cdot \nabla m_z(\mathbf{r_i}) \rfloor^2 \right\} + C. \quad (3.5)$$

Dropping the additive constant C and summing over all the spins leads to the expression of the exchange energy:

$$E_{ex} = \int_V A_{ex} \left\{ \lfloor \nabla m_x(\mathbf{r}) \rfloor^2 + \lfloor \nabla m_y(\mathbf{r}) \rfloor^2 + \lfloor \nabla m_z(\mathbf{r}) \rfloor^2 \right\} d\mathbf{r} \quad (3.6)$$

where A_{ex} is the exchange stiffness constant, having the dimensions of energy per unit length. Typical values of A_{ex} are on the order of 10^{-11} J/m. In the case of a simple cubic lattice with lattice parameter a, $A_{ex} = JS^2/a$.

3.1.1.2 *Magnetocrystalline Anisotropy Energy* The environment of a magnetic moment acts on its orientation to favor certain directions in space. The charge distribution of the ions that form the crystal lattice generates an anisotropic electrostatic field that influences the orbital angular momentum of the electrons. Due to the spin–orbit coupling (SOC), this in turns generates preferred orientations of the spins and therefore of the magnetization along particular directions in space (called the easy axis of magnetization). This coupling between magnetization and crystal proprieties is quantified by the magnetocrystalline anisotropy energy. Its expression depends on the symmetry of the crystalline structure (6). For instance, one can define uniaxial, cubic, or hexagonal magnetocrystalline anisotropy. The most common case is that of uniaxial anisotropy for which the corresponding energy has the expression,

$$E_{anis} = \int_V K_u \{1 - [\mathbf{u_k} \cdot \mathbf{m}(\mathbf{r})]^2\} d\mathbf{r}, \quad (3.7)$$

where $\mathbf{u_k}$ is a unit vector along the direction of the easy axis and K_u is the temperature-dependent anisotropy constant, expressed in joules per cubic meter (J/m^3). In thin films and multilayers, other anisotropy terms of surface or interfacial origin can have a major influence. This surface anisotropy arises from surface or interfacial phenomena such as electronic hybridization, stress, or symmetry breaking. An effective anisotropy constant K_{eff} is then expressed as the sum of the bulk and surface anisotropies: $K_{eff} = K_u + (K_{s1} + K_{s2})/t$, where K_{s1} and K_{s2} are the surface anisotropy constants that correspond to the top and bottom interfaces of the magnetic layer and t is the layer thickness.

3.1.1.3 *Demagnetizing Energy* A ferromagnetic material contains several magnetic domains pointing in different directions. The exchange interaction yields alignment of the magnetic moments inside a domain, but it does not explain why domain walls are formed. In fact, a long-range interaction, the magnetostatic interaction, accounts for the domain structure. Each magnetic moment in the ferromagnet is a dipole that produces a field experienced by other magnetic moments. Thus, a pair of dipoles driven only by the magnetostatic interaction will minimize its energy by pointing in opposite directions. Thus, in a ferromagnet, exchange and magnetostatic interactions are in competition, the former aligning the moments in the same direction, and the latter creating oppositely aligned domains over long distances. It follows that the typical size of the domains results from the relative strengths of these two interactions. Unlike the exchange interaction, which is local, the magnetostatic field at a given point is a sum over the contributions of all the magnetic moments in the whole magnetic volume. Subsequently, the numerical computation of this field is much more time-consuming than that of other fields.

In absence of any electrical current, the expression of the magnetostatic field $\mathbf{H_m}$ can be derived from three fundamental equations: the relationship between $\mathbf{H_m}$ and $\mathbf{M}$ given by

$$\mathbf{B} = \mu_0(\mathbf{H_m} + \mathbf{M}) \tag{3.8}$$

and the two Maxwell equations,

$$\begin{aligned} \nabla \cdot \mathbf{B} &= 0, \\ \nabla \times \mathbf{H_m} &= 0. \end{aligned} \tag{3.9}$$

The dipolar field is irrotational, which means that $\mathbf{H_m}$ is a conservative vector field. Therefore, there exists a scalar potential ϕ_m (magnetic scalar potential) such that $\mathbf{H_m} = -\nabla\phi_m$. The magnetostatic problem is reduced to Poisson's equation,

$$\nabla^2\phi_m = -\nabla \cdot \mathbf{M}, \tag{3.10}$$

complemented by the radiation condition at infinity, $\mathbf{H_m}(\mathbf{r} \to \infty) \to 0$. At the interface between two regions of the space (e.g., 1 and 2), the magnetic scalar potential ϕ_m is a continuous function, but its normal derivative is discontinuous:

$$\begin{cases} \phi_{m,1} = \phi_{m,2}, \\ \dfrac{\partial\phi_{m,1}}{\partial\mathbf{n}} - \dfrac{\partial\phi_{m,2}}{\partial\mathbf{n}} = -(\mathbf{M_1} - \mathbf{M_2}) \cdot \mathbf{n}, \end{cases} \tag{3.11}$$

where $\mathbf{n}$ is the normal vector pointing from region 1 to region 2. By analogy with electrostatics, $\rho_m = \nabla \cdot \mathbf{M}$ represents the volume magnetic charge density and $\sigma_m = -\mathbf{M} \cdot \mathbf{n}$ represents the surface magnetic charge density, respectively.

The magnetostatic Eqs. (3.10) and (3.11) may be solved by Green's method. The magnetic scalar potential for a three-dimensional sample is given by the integral expression,

$$\phi_m(\mathbf{r}) = \int_V G(\mathbf{r} - \mathbf{r}')\rho_m(\mathbf{r}')d\mathbf{r}' + \int_S G(\mathbf{r} - \mathbf{r}')\sigma_m(\mathbf{r}')dr', \tag{3.12}$$

where $G(\mathbf{r} - \mathbf{r}') = 1/(4\pi|\mathbf{r} - \mathbf{r}'|)$ is the associated Green's function. The magnetostatic field evaluation is straightforward:

$$\begin{aligned} \mathbf{H_m}(\mathbf{r}) &= -\textstyle\int_V \nabla G(\mathbf{r} - \mathbf{r}')\rho_m(\mathbf{r}')d\mathbf{r}' - \int_S G(\mathbf{r} - \mathbf{r}')\sigma_m(\mathbf{r}')dr', \\ &= -[\nabla G * \rho_m](\mathbf{r}) - [\nabla G * \sigma_m](\mathbf{r}), \end{aligned} \tag{3.13}$$

where $*$ is the convolution product.

Various names are given to the magnetostatic field. For the sake of clarity, it will be called stray field $\mathbf{H_{stray}}$ outside the material and demagnetizing field $\mathbf{H_d}$ inside.

The demagnetizing energy that quantifies the interaction of the magnetization with the magnetostatic field created by itself is given by the integral:

$$E_{dem} = -\frac{1}{2}\mu_0 \int_V M_s \mathbf{m}(\mathbf{r}) \cdot \mathbf{H_d}(\mathbf{r}) d\mathbf{r}. \tag{3.14}$$

Following the pole-avoidance principle (4), the demagnetizing energy is minimized if the magnetic charges are minimized. This occurs when the magnetization follows closed paths, as illustrated in Fig. 3.2. The formation of domains leads to a reduced magnetostatic energy at the expense of increased exchange energy. Since surface magnetic charges are located at the boundaries, the magnetization tends to align along the edge. The so-called shape anisotropy is therefore ascribed to the magnetostatic interaction.

Generally the demagnetizing field is nonuniform even if the magnetization is uniform. However, in the particular case of a body whose surface is of second degree, a uniform magnetization implies a uniform demagnetizing field. For Cartesian coordinates along the principal axes of the system, the equation of the surface boundary is $(x/a)^2 + (y/b)^2 + (z/c)^2 = 1$, where a, b, and c are the semi-axis lengths. If $c \rightarrow \infty$, the surface is an infinite cylinder of elliptical cross section. If a, b, and c all take finite values, the body is an ellipsoid. For a uniformly magnetized ellipsoid with only surface magnetic charges, the demagnetizing field is also uniform (7–9).

One can define a demagnetizing tensor $\mathbf{N}$ such that

$$\mathbf{H_d} = -\mathbf{N} \cdot \mathbf{M}. \tag{3.15}$$

The demagnetizing field does not require heavy computation for uniform magnetization, contrary to the general case of nonuniform magnetization, since the tensor $\mathbf{N}$ is diagonal if expressed on the basis of the principal axes:

$$\mathbf{N} = \begin{pmatrix} N_{xx} & 0 & 0 \\ 0 & N_{yy} & 0 \\ 0 & 0 & N_{zz} \end{pmatrix}. \tag{3.16}$$

The demagnetizing coefficients N_{xx}, N_{yy}, and N_{zz} are positive, since the field H_d "demagnetizes" the sample. Moreover, the trace of $\mathbf{N}$ is equal to 1, that is, $N_{xx}+$

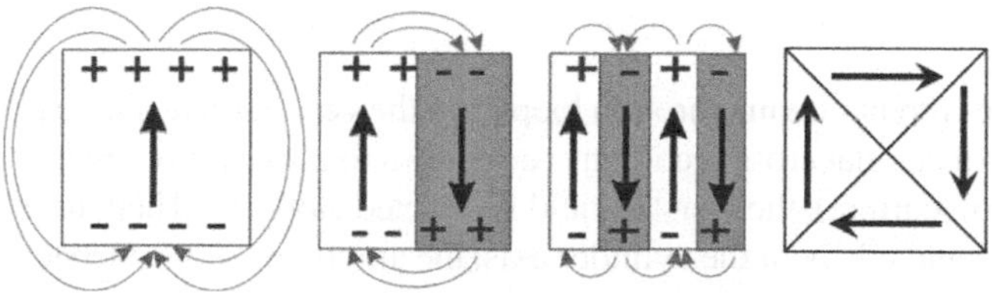

Figure 3.2 Magnetostatic field generated by a ferromagnetic sample. From left to right: single-domain configuration, two-domain configuration, four domain configuration, and flux-closure domains. The magnetostatic energy is decreased by dividing the sample into several magnetic domains. The surface magnetic charges (+, −) disappear for the flux-closure case, but volume magnetic charges form at the domain walls. The black arrows represent the magnetization within each domain.

$N_{yy}+N_{zz}=1$. The analytical expressions of the demagnetizing coefficients are well known for the three types of ellipsoids of revolution: (1) sphere $a=b=c$, (2) oblate spheroid $a=b>c$, (3) prolate spheroid $a=b<c$. The case of the sphere is the most simple, with $N_{xx}=N_{yy}=N_{zz}=1/3$, due to the symmetry. The formulas in the two other cases can be found in Ref. (10). In the limit of a, $b \gg c$, one obtains the case of a continuous thin film $N_{xx}=N_{yy}=0$ and $N_{zz}=1$.

3.1.1.4 Zeeman Energy If a magnetic field $\mathbf{H_{app}}$ is applied, thc magnetization M experiences a torque, which tends to align it parallel to the applied field direction. Due to the misalignment between $\mathbf{H_{app}}$ and $\mathbf{M}$, a supplementary contribution has to be included in the total energy,

$$E_{app} = -\mu_0 \int_V M_S\, \mathbf{m}(\mathbf{r},t) \cdot \mathbf{H_{app}}(\mathbf{r},t) d\mathbf{r}, \tag{3.17}$$

where $\mu_0=4\pi\times 10^{-7}$ H/m is the vacuum permeability. The externally applied field can be time-dependent and nonuniform in space. Most studies consider the basic case of an applied magnetic field regardless of its source. However, two cases are of particular interest: a magnetic field generated by an external magnet and a magnetic field generated by an electrical current.

1. The first case is typical in magnetic nanopillars composed of several ferromagnetic layers coupled through magnetostatic interactions. Using the formalism of the previous paragraph, one might evaluate the stray field $\mathbf{H_{stray}}(\mathbf{r})$ acting on a particular layer generated by the external magnets (e.g., the other magnetic layers of the stack).
2. In the second case, an electrical current distribution surrounding the magnetic sample, or even flowing through it, generates the commonly called Oersted or Ampère magnetic field. This situation is common in MRAM where the write field is generated by electric current pulses $\mathbf{J_{app}}(\mathbf{r},t)$ injected in the write lines. The Oersted field $\mathbf{H_{Oe}}$ is the solution of the Maxwell equation,

$$\nabla \times \mathbf{H_{Oe}}(\mathbf{r},t) = \mathbf{J_{app}}(\mathbf{r},t). \tag{3.18}$$

Integration over a contour C yields Ampère's circuital law:

$$\oint_C \mathbf{H_{Oe}} \cdot d\ell = I_{enclosed}, \tag{3.19}$$

where $I_{enclosed}$ is the current flowing through loop C. In the very particular case of an infinite cylinder of circular cross section traversed by a rotationally symmetric current (e.g., uniform current), the Oersted field is directed along $\mathbf{u_\theta}$. Then, for a circular contour C centered about the cylinder axis, the amplitude of the Oersted field is

$$H_{Oe}(r) = \frac{I_{enclosed}}{2\pi r}, \tag{3.20}$$

where r is the radius of C. If r is larger than the cylinder radius R, then the Oersted field decreases as $1/r$. For a uniform current density J_{app} inside the cylinder,

$H_{Oe}(r) = J_{app}r/2$ when $r < R$. The Oersted field is maximum at the edge of the cylinder, equal to $J_{app} R/2$. Therefore, for a given current density, the maximum intensity of the Oersted field inside a nanopillar depends on its lateral size (see Fig. 3.3).

For the general case (specific shape of the conductor, nonuniform current distribution) one has to use the formalism based on the magnetic vector potential **A** from which the Oersted field is derived: $\mathbf{H_{Oe}} = (1/\mu_0)\nabla \times \mathbf{A}$. This requires a solution of Poisson's equation for the magnetic vector potential **A**,

$$\nabla^2 \mathbf{A} = -\mu_0 \mathbf{J_{app}}, \tag{3.21}$$

complemented by the radiation condition at infinity $H_{Oe}(\mathbf{r} \to \infty) \to 0$. This equation must be solved numerically for nonuniform current distributions.

The list of the energies presented here is not exhaustive; other contributions might be included in the free energy functional depending on the additional interactions exhibited by the sample (e.g., coupling with an antiferromagnet, Ruderman–Kittel–Kasuya–Yosida (RKKY) exchange interactions).

3.1.2 Magnetically Stable State and Equilibrium Equations

The free energy functional might have several local minima; each of them is corresponding to a possible magnetically stable state. According to the variational principle (4,11), at equilibrium, the magnetization distribution inside the sample $\{\mathbf{m}(\mathbf{r}, t)|r \in V, \mathbf{m}(\mathbf{r}, t)| = 1\}$ satisfies simultaneously the following conditions:

$$\begin{cases} \delta E_{tot}(\mathbf{m}) = 0, \\ \delta^2 E_{tot}(\mathbf{m}) > 0, \end{cases} \tag{3.22}$$

where $\delta E_{tot}(\mathbf{m}) = E_{tot}(\mathbf{m} + \delta\mathbf{m}) - E_{tot}(\mathbf{m})$ is representing an infinitesimal variation of the energy induced by a small change in the magnetization $\delta\mathbf{m}$. One might show that the energy variation for a sample with uniaxial magnetocrystalline anisotropy takes the following relation:

$$\delta E_{tot}(\mathbf{m}) = \mu_0 \int_V M_s \left[\frac{2A_{ex}}{\mu_0 M_s} \nabla^2 \mathbf{m} + \frac{2K_u}{\mu_0 M_s} (\mathbf{u_k} \cdot \mathbf{m})\mathbf{u_k} + \mathbf{H_{app}} + \mathbf{H_d} \right] \cdot \delta\mathbf{m}\, dV + \oint_S 2A_{ex} \left(\mathbf{m} \times \frac{\partial \mathbf{m}}{\partial n} \right) dS. \tag{3.23}$$

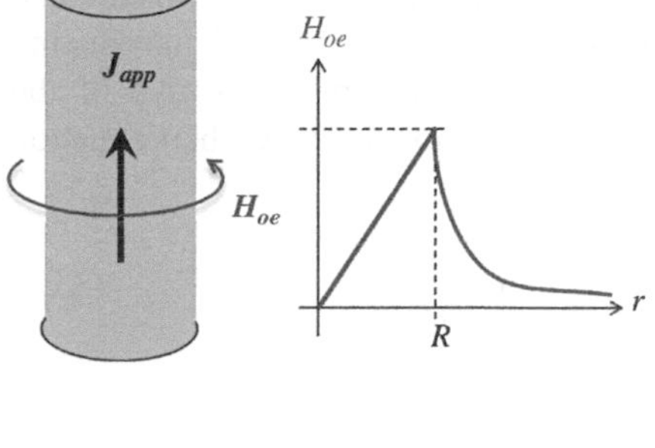

Figure 3.3 An electric current J_{app} flowing through an infinite cylindrical conductor generates the Oersted field H_{Oe} with characteristic space dependence. R is the radius of the cylinder, and r is the radial coordinate.

The quantity between the brackets from the first term represents an effective field,

$$\mathbf{H_{eff}} = \frac{2A_{ex}}{\mu_0 M_s}\nabla^2\mathbf{m} + \frac{2K_u}{\mu_0 M_s}(\mathbf{u_k}\cdot\mathbf{m})\mathbf{u_k} + \mathbf{H_{app}} + \mathbf{H_d}, \tag{3.24}$$

defined in general case as the variational derivative of the free energy with respect to the magnetization,

$$\mathbf{H_{eff}} = -\frac{1}{\mu_0 M_s}\frac{\delta E_{tot}}{\delta \mathbf{m}}. \tag{3.25}$$

To obtain the equilibrium condition, both the surface and volume integrals from Eq. (3.23) must vanish. This is possible only if simultaneously two conditions are fulfilled:

$$\begin{cases} \dfrac{\partial \mathbf{m}}{\partial n} = 0, \quad \forall \mathbf{r} \in S, \\ \mathbf{m}(\mathbf{r},t) \times \mathbf{H_{eff}}(\mathbf{r},t) = 0, \quad \forall \mathbf{r} \in V. \end{cases} \tag{3.26}$$

These conditions were deduced by Brown (4) and therefore are called the Brown equations. Their solution specifies the equilibrium state. The first one is a Neumann boundary condition, which forces the magnetization to be stationary near the sample surface S. The second equation states that for a magnetization distribution to be at equilibrium, the torque acting on **m** due to the effective field must be nil everywhere (the magnetization is aligned with the effective field).

3.1.3 Equations of Magnetization Motion

The Brown equations are completely defining the equilibrium state of a magnetic system, but they do not specify how the system reaches this state. The magnetization dynamics can be accessed through the Landau–Lifshitz–Gilbert equation. The starting point is represented by the well-known equation of Larmor:

$$\frac{\partial \mathbf{m}}{\partial t} = -\gamma(\mathbf{m} \times \mu_0 \mathbf{H}), \tag{3.27}$$

which describes the magnetization's gyrotropic reaction in the presence of the magnetic field **H**. Here γ is the gyromagnetic ratio of the free electron ($1.76 \times 10^{11}\,\mathrm{s^{-1}\,T^{-1}}$).

The Larmor equation is conservative. Its solution corresponds to a magnetization that precesses endlessly, with constant precession angle and energy. However, in real ferromagnetic materials, dissipation processes cause the system to minimize its energy and to reach, after certain time, a stable state ($\partial\mathbf{m}/\partial t = 0$). In order to take account of dissipation, a term has to be added to Eq. (3.27), allowing the magnetization to reach static equilibrium. Gilbert and Kelly (12,13) suggested introducing magnetic damping as a viscous force proportional to the time-derivative of the magnetization according to the Rayleigh dissipation functional. The Gilbert equation of motion includes precession and relaxation:

$$\frac{\partial \mathbf{m}}{\partial t} = -\gamma_0(\mathbf{m} \times \mathbf{H_{eff}}) + \alpha\left(\mathbf{m} \times \frac{\partial \mathbf{m}}{\partial t}\right). \tag{3.28}$$

The damping α is a dimensionless phenomenological constant, arising from all dissipation processes (e.g., magnon–magnon scattering, magnon–phonon scattering, and eddy currents) and $\gamma_0 = \mu_0|\gamma|$. The meaning and the measurement of α is an intricate matter, since its value depends not only on the material, but also on experimental conditions. For most common ferromagnetic materials, α is a scalar ranging from 5×10^{-3} to 0.1.

Since magnetization magnitude is conserved during the motion, if $\alpha \ll 1$, the Gilbert equation can be transformed to an equivalent form, previously introduced by Landau and Lifshitz (2):

$$\frac{\partial \mathbf{m}}{\partial t} = -\gamma_L(\mathbf{m} \times \mathbf{H}_{eff}) + \lambda \mathbf{m} \times (\mathbf{m} \times \mathbf{H}_{eff}). \tag{3.29}$$

Here, $\gamma_L = \gamma_0/(1+\alpha^2)$ and $\lambda = \gamma_0\alpha/(1+\alpha^2)$. The Landau–Lifshitz equation is of great interest for numerical resolution since the time derivative of $\mathbf{m}$ is directly expressed as function of $\mathbf{m}$ and $\mathbf{H_{eff}}$. The magnetization motion is illustrated in Fig. 3.4.

3.1.4 Length Scales in Micromagnetism

In micromagnetic modeling, particular attention has to be paid to length scales in order to always comply with the fundamental hypotheses of the micromagnetic model. The simulated system is divided into elementary cells and the micromagnetic equations are solved in each of these elements. The accuracy of the result depends strongly on the element size; the smaller, the more accurate, but also the longer the simulation. Therefore, an appropriate cell size has to be chosen. Since the exchange interaction has the shortest range, its strength with respect to other forces determines the typical scale over which the magnetization can vary. Two characteristic lengths can be defined: the exchange length λ_{ex} and the Bloch length λ_B. The competition between the exchange interaction and the demagnetizing energy is measured by the exchange length defined as

$$\lambda_{ex} = \sqrt{\frac{2A_{ex}}{\mu_0 M_s^2}}. \tag{3.30}$$

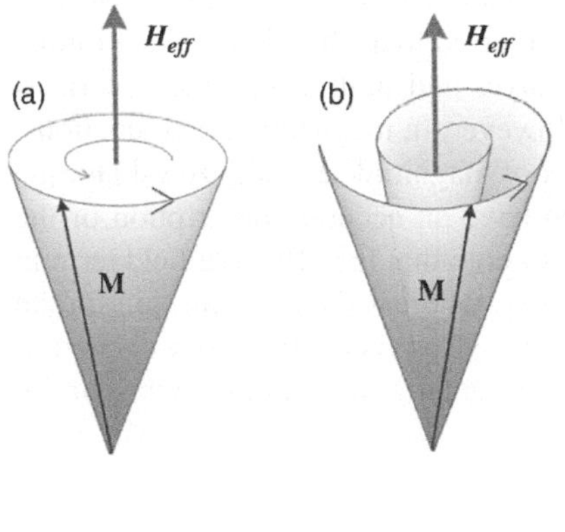

Figure 3.4 Precession of the magnetization vector $\mathbf{M}$ about the field $\mathbf{H_{eff}}$: (a) without damping and (b) with a damped motion ($\alpha > 0$).

Usually the exchange length is giving the width of magnetic vortex core. The Bloch length is the typical width of a Bloch wall (transition between two magnetic domains antiparallel orientated over which the magnetization rotates perpendicularly to the plane of the wall, unlike the Néel wall where the rotation is within the plane of the domain wall). It is given by the relative strength of the exchange interaction over the magnetocrystalline anisotropy:

$$\lambda_B = \sqrt{\frac{A_{ex}}{K_u}}. \tag{3.31}$$

For accurate micromagnetic simulations, the element size has to be smaller than these two lengths. For instance, in Permalloy ($Ni_{80}Fe_{20}$), the cell dimension is limited by the exchange length, which is about 5 nm. Therefore, an accurate simulation requires a typical cell size of about 2.5 nm.

3.1.5 Modification Related to Spin-Transfer Torque Phenomena and Spin–Orbit Coupling

The dynamic Eq. (3.28) or (3.29) is suitable for describing the evolution of a micromagnetic system excited by an external magnetic field. However, since the first evidence of the effect of an electric current on magnetization, this phenomenon, called spin-transfer torque (STT), has attracted more and more interest. In order to describe this new kind of interaction, the Landau–Lifshitz–Gilbert equation has to be adapted. The STT generated by a spin-polarized current on the magnetization is represented by an additional term that should be added to the magnetization equation of motion (Eq. (3.28)). The debate about the exact expression for the STT is still under consideration, but the most used models are briefly presented below.

1. For the case of nanopillars, spin-dependent transport theory predicts two terms (14):

$$\left(\frac{\partial \mathbf{m}}{\partial t}\right)_{STT} = -\gamma_0 a_J \mathbf{m} \times (\mathbf{m} \times \mathbf{p}) - \gamma_0 b_J (\mathbf{m} \times \mathbf{p}). \tag{3.32}$$

The vector **p** represents the direction of the pinned layer (of the polarizer), supposed to be fixed. Both terms are current dependent, since a_J and b_J are more or less complicated functions of the injected current. The term in b_J can be included in the expression of the effective field; it is therefore called field-like torque. In contrast, the term in a_J is often referred to as the Slonczewski torque or the in-plane torque. In metallic spin-valves and multilayers, the b_J term is usually much smaller than the a_J term. However, in magnetic tunnel junctions, the field-like term can be of the same order of magnitude as Slonczewski torque and should be also taken into account to have an accurate description of the magnetization dynamics. It is interesting to note that the effective field term in the LLG equation is deriving from an energy, whereas the in-plane spin-torque cannot be associated to any energy since the system is continuously excited. It actually behaves like a damping or antidamping term, depending on the current direction.

2. If the current flows through a magnetic nanowire containing a domain wall parallel to the wire cross section, the spin of the conduction electrons interacts with the magnetization of the magnetic domain wall. Two torque terms model this interaction (15): an adiabatic term and a nonadiabatic term,

$$\left(\frac{\partial \mathbf{m}}{\partial t}\right)_{STT} = -(\mathbf{u}\cdot\nabla)\mathbf{m} - \beta\mathbf{m}\times[(\mathbf{u}\cdot\nabla)\mathbf{m}]. \tag{3.33}$$

The vector $\mathbf{u}$ has units of velocity; it is parallel with the direction of the electron flow: $\mathbf{u} = \mathbf{J_{app}}(g\mu_B P/(2eM_s))$ (here, $g=2$ is Landé factor of the free electron, μ_B is the Bohr magneton, P is the current polarization fraction ($0<P<1$), β is the non-adiabatic coefficient, and e is the electron charge).

3. Novel out-of-equilibrium transport phenomena have also been demonstrated, such as current induced spin–orbit torques induced by the Rashba spin–orbit coupling and/or the spin Hall effect, leading to current induced magnetization reversal (16,17). This phenomenon might be included in the equation of motion by a SOC contribution proportional to the electric current:

$$\left(\frac{\partial \mathbf{m}}{\partial t}\right)_{SOC} = \gamma_0 C_R\left(\mathbf{m}\times\mathbf{J_{app}}\right) + \gamma_0 C_{SHE}\mathbf{m}\times\left(\mathbf{m}\times\mathbf{J_{app}}\right), \tag{3.34}$$

where C_R and C_{SHE} are sample-dependent coefficients.

Obviously, this list of physical phenomena that can be included in micromagnetic studies is not exhaustive. A very interesting and absolutely necessary step to understand the behavior of a magnetic body is the study of thermal effects.

3.1.6 Thermal Fluctuations

Magnetic properties are strongly dependent on temperature. In micromagnetism, the thermal fluctuations are taken into account, according to Brown's theory, by a random magnetic field $\mathbf{H_{th}}$ when the temperature of the sample is much less than the Curie temperature (the transition temperature between the ferromagnetic and the paramagnetic states). The mean value of the thermal field is zero $\langle \mathbf{H_{th}}(\mathbf{r}_i, t_k)\rangle = 0$ and its autocorrelation function is given by

$$\langle \mathbf{H_{th}}(\mathbf{r}_1, t_1)\cdot\mathbf{H_{th}}(\mathbf{r}_2, t_1)\rangle = D\delta(\mathbf{r}_1-\mathbf{r}_2)\delta(t_1-t_2), \tag{3.35}$$

where D is the variance expressed as $D = 2\alpha k_B T/(\mu_0 M_s V\gamma_0)$. Consequently, the thermal field has a Gaussian distribution centered about zero with a variance proportional to the temperature T and the damping factor α, and inversely proportional to the magnetic volume V. The thermal field is added to the effective field $\mathbf{H_{eff}}$ in the Landau–Lifshitz–Gilbert Eq. (3.28); in this case, Eq. (3.28) becomes a Langevin equation.

Since $\mathbf{H_{th}}$ is a random field, it accounts for nondeterministic (stochastic) processes. Therefore, running the same simulation several times at finite temperature leads to a distribution of magnetization trajectories; the larger the temperature, the broader the distribution.

3.1.7 Numerical Micromagnetism

The typical length of magnetization variation is the exchange length λ_{ex} (or the Bloch length λ_B, whichever is the smallest). Therefore, if the lateral size of a ferromagnetic sample is on the order of, or only a few times, the exchange length, the magnetization can be considered uniform: $\mathbf{m}(\mathbf{r}, t) = \mathbf{m}(t)$. This approximation is called macrospin approximation (or "single-domain" approximation, or "uniform-mode" approximation). Instead of using the complicated Eq. (3.13), the mean demagnetizing field and the magnetostatic energy can be easily calculated with an equivalent demagnetizing tensor (Eqs. (3.15 and 3.16)). The macrospin approximation is of great utility in building simple analytical models. Moreover, the LLG equation needs to be solved for only one spin (a "macrospin"), which makes macrospin simulations much faster than micromagnetic computations. The energy landscape can be plotted within the macrospin approximation since the state of the system is fully given by two independent parameters (the third one being deduced from the conservation of the norm, for example, the polar angles θ, ϕ). This is particularly useful to gain an insight into dynamic processes, such as magnetization reversal for example. Simple models can also be constructed and solved analytically, such as the Stoner–Wohlfarth model (18) to explain magnetic hysteresis during switching, and the Kittel law (19), which gives the ferromagnetic resonance (FMR) frequency of the uniform mode as a function of the applied field.

In most cases, the macrospin model is, however, a very rough approximation; three-dimensional micromagnetic simulations, for example, $\mathbf{m}(\mathbf{r}, t)$, are required for a description of nonuniform magnetization. There are two types of micromagnetic simulations: either the equilibrium state is sought or a complete time-varying computation is run to study the magnetization dynamics. Usually, the equilibrium state is needed to initialize the dynamic simulation. The most common method to find the micromagnetic equilibrium state is to relax the system with a large Gilbert damping (α) value. In this case, the path described by the magnetization (the intermediate states) has no physical meaning; only the final state corresponds to the physical state potentially reached by the system.

The resolution of the micromagnetic equations (nonlocal, integral equations) is demanding in terms of computational power. Time and space are discretized and the effective field is calculated in each element (or cell) at each time step. An approximate solution is given, but it has to converge to the exact result when the element size and the time step go to zero. A typical time step is about 0.1 ps (for a typical cell size of a few nanometers). The input parameters of any micromagnetic simulation are the saturation magnetization M_s, the exchange constant A_{ex}, and the anisotropy constant K_u. Moreover, the temperature T can be included, as well as a current density J_{app} and the spin polarization.

Many micromagnetic software programs are based on different approximations and implementation methods. The finite difference method (FDM) is the most common numerical method for micromagnetic simulations because it is easier and faster than the finite element method (FEM). The most widely used micromagnetic code utilizing the FDM is certainly the *Object-Oriented Micromagnetic Framework* (OOMMF) software (20), but many other software programs can be cited, such as the *LLG Micromagnetic Simulator* by Scheinfein (21), *MicroMagus* by Berkov and Gorn

(22), *Magsimus* by Oti (23), *MuMax* by Vansteenkiste and Van de Wiele (24), *GoParallel* by Torres and Martinez (25), and *MicroMagnum* by Drews (26). The last three use graphics cards (GPUs), whereas the others run on single CPU processors.

Among other micromagnetic solvers based on the finite element method, one can cite *Nmag* by H. Fangohr, M. Franchin, and T. Fischbacher; *Magpar* by W. Scholz; *FEMME* by D. Suess and T. Schrefl; *TetraMAG* by R. Hertel; *FastMag* by V. Lomakin; and *SallyMM* by O. Bottauscio and A. Manzin.

The predictions reported hereafter were obtained using the *Micro3D* solver (27), a finite difference code, applied to explore the magnetic behavior of magnetic nanosize samples such as thin films, nanodots, and nanowires.

3.2 MICROMAGNETIC CONFIGURATIONS IN MAGNETIC CIRCULAR DOTS

An expected trend in ferromagnetic systems is the increase of the number of the magnetically stable states upon increasing the size of a sample. Nanostructures with size <1 μm are of great interest from the point of view of applications since the number of possible equilibrium states is reduced (ideally, only two possible states for memory applications). The sample size is a key parameter for tuning the magnetic stable state. An illustration for the particular case of a cylindrical dot of diameter D and thickness t follows. The following material parameters were used: saturation magnetization $\mu_0 M_S = 1.76T$, uniaxial magnetocrystalline anisotropy (MCA) along the vertical axis (perpendicular to the plan of the dot, see Fig. 3.5a) with a constant $K_u = 5 \times 10^5$ J/m^3, and exchange stiffness $A_{ex} = 14p$ J/m. These parameters are typical for Co(0 0 0 1). The corresponding characteristic length scales are the exchange length $\lambda_{ex} = 3.4$ nm (Eq. 30) and the Bloch parameter $\lambda_B = 5.3$ nm (Eq. 31).

The magnetically stable state in zero applied field is studied as a function of the diameter of the dot varying between 40 and 200 nm for a given thickness of 5 nm. Two types of magnetization configurations are stable: a vortex-like state (V) and an in-plane quasi-single-domain state (SD) (Fig. 3.5b and c).

In the vortex state, the magnetization curls in a circular path along the dot edge, leading to a singularity (vortex) in the dot center where the magnetization points are out-of-plane. In order to properly describe such a magnetic singularity involving a rapid variation of the magnetization, the mesh size needs to be sufficiently small. The data reported here were obtained with a mesh size $<0.75\lambda_{ex}$, which provides an accurate description of the vortex internal structure. The vortex structure is a typical flux-closer state characterized by a considerable reduction of the demagnetizing energy. However, in the vortex core, the magnetization varies very sharply in space and this costs a lot of exchange energy. The vortex state has a degree of degeneracy of 4 (2 circulation senses and 2 pointing directions for the core).

In the quasi-single domain state (SD), the magnetization lies in the plane of the dot oriented almost everywhere in the same direction. However, nearby the dot edge, the magnetization tries to follow the curvature of the edge, avoiding the creation of surface magnetic charges. This configuration is almost a "macrospin". The SD state has no preferential direction in the plane of the dot since there is no in-plane

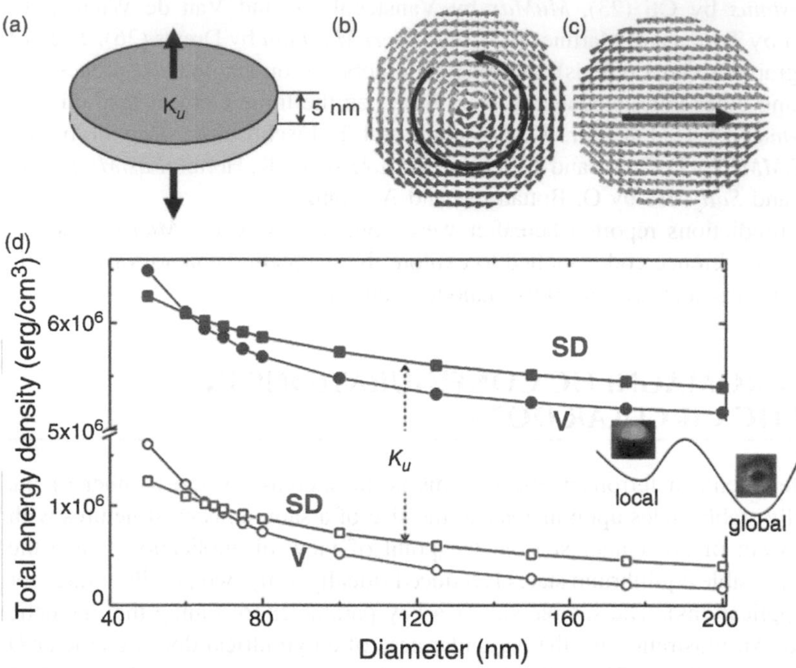

Figure 3.5 (a) Schematic view of a circular dot. Typical magnetization configuration for (b) a vortex state and (c) a quasi-in-plane single-domain state. (d) Total energy density evolution with the variation of the dot diameter for a fixed thickness of 5 nm. The open symbols correspond to the case without magnetocrystalline anisotropy (MCA) and the full dots to a strong MCA, $K_u = 5 \times 10^5\ \mathrm{J/m^3}$. The inset is a schematic representation of the energy with two MFM images (bright–dark contrast is indicative of a SD configuration and the dark spot is typical for a vortex).

anisotropy (shape or magnetocrystalline) to favor one particular direction. The degree of degeneracy of the SD state can be reduced by adjusting the shape of the dot from circular to elliptical cross section.

Both states (vortex and quasi-single domain) are actually stable but depend on the geometry of the dot (thickness and radius). One is associated with a global minimum of energy (fundamental state), whereas the other is only metastable (a local minimum of the free energy functional). The dependence of the free energy on the diameter of the dot is plotted in Fig. 3.5d for two kinds of dots: with and without perpendicular magnetocrystalline anisotropy. Very similar behavior is obtained in both cases. The total energy density of the single-domain and vortex dots with strong perpendicular magnetic anisotropy is just shifted upward by an amount equal to the magnetic anisotropy energy K_u. This shift is due to the fact that most spins are in-plane and hence point in a hard direction of magnetic anisotropy. The presence of strong perpendicular magnetic anisotropy does not influence the ground state configuration very much, nor the transition from the vortex-like state toward the single-domain state. It does, however, affect the onset of this transition: the critical diameter for Co dots with strong perpendicular magnetic anisotropy is 60 nm, whereas it is 67.5 nm for dots with no anisotropy.

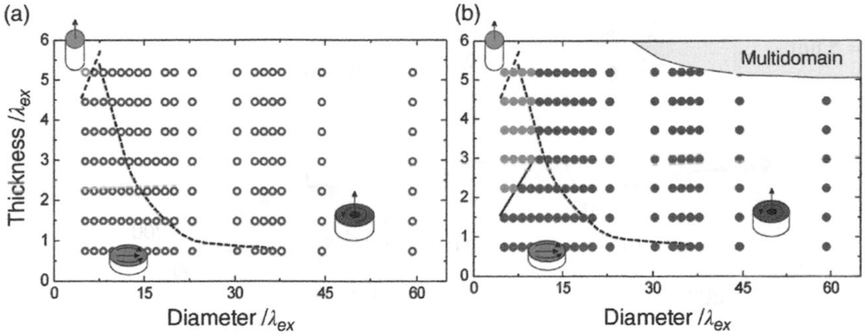

Figure 3.6 The ground state phase diagram in zero applied field for a Co dot (a) without and (b) with strong MCA. Green: out-of-plane single domain state; blue: quasi-single-domain state; red: vortex state. (Adapted from Ref. 27 with permission from Elsevier.)

The different ground state configurations in zero applied field of Co dots are summarized in Fig. 3.6a. The transition between SD and vortex-like states versus dot diameter decreases with increasing dot thickness. The total energy arises from the demagnetizing and exchange contributions. The energy of the SD state is dominated by the magnetostatic contribution because of the magnetic charges induced on the lateral surface of the dot. In contrast, for the vortex-like state, the magnetic flux being closed, the magnetostatic energy is drastically reduced. But the central region including the vortex core represents a large concentration of exchange energy that becomes comparable to the magnetostatic energy of the SD state if the dot diameter approaches the vortex core diameter. In addition to the in-plane single domain state and the vortex state, a third configuration corresponds to an out-of-plane single domain state when the thickness is comparable to, or larger than, the dot diameter. For this range of aspect ratios ($t \gg D$), the dot corresponds to an elongated wire and the shape anisotropy prefers alignment of the magnetization along the long wire axis. With a uniaxial magnetocrystalline anisotropy perpendicular to the dot plane, a similar ground-state phase diagram was computed for Co dots, as shown in Fig. 3.6b. While the presence of the anisotropy lowers the critical thickness of the boundary between the in-plane and the out-of-plane SD states (at constant diameter), it plays only a minor role for the in-plane SD to V transition. This difference is due to the relative volumes of perpendicular magnetization in the out-of-plane SD state and the vortex state. For the latter, this volume is small compared to the diameter and consequently the boundary between the in-plane SD and V ground states is only slightly shifted toward smaller values of t and D. A strong influence of the perpendicular magnetocrystalline anisotropy (MCA) is exhibited for a dot thickness >15–20 nm, where a transition from the vortex state to a weak circular stripe multidomain state takes place.

Figure 3.7 shows the evolution of a multidomain, circular stripe pattern state from a vortex state. The core of the vortex is extremely reduced in very flat circular elements. Upon increasing the thickness (increasing the aspect ratio t/D), the core of the vortex becomes wider and, for relatively thick elements, a circular stripe structure appears. The stripe structure is a clear signature of ferromagnetic samples dominated

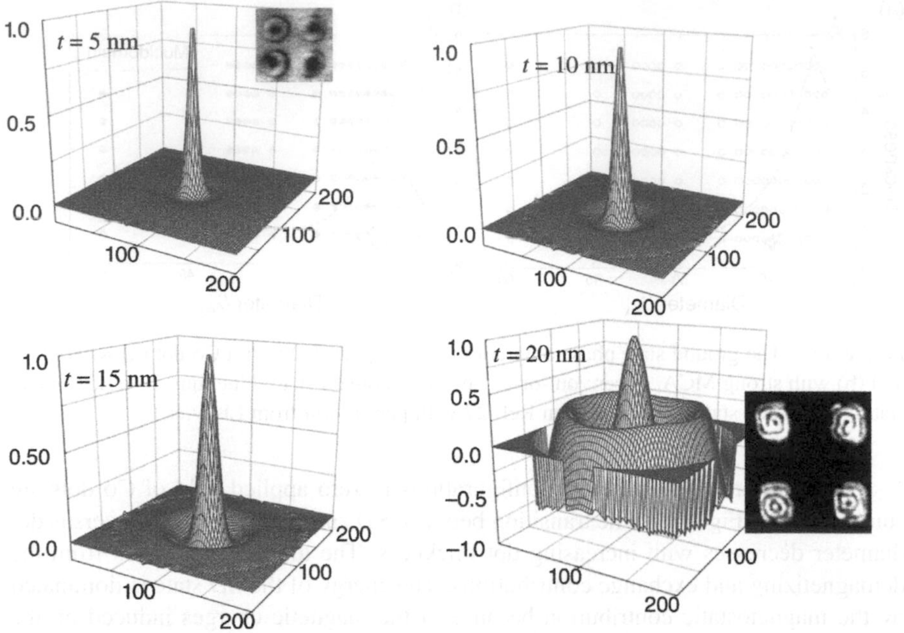

Figure 3.7 Evolution of the out-of-plane magnetization component with the thickness of the dot with a diameter of 200 nm and a MCA of $K_u = 5 \times 10^5$ J/m^3. The insets correspond to magnetic force microcopy images of real samples.

by magnetocrystalline anisotropy. One parameter quantifying the competition between the crystalline and shape anisotropy is the quality factor Q defined as $Q = 2K_u/(\mu_0 M_s^2)$. If $Q > 1$, the continuous thin films with out-of-plane magnetocrystalline uniaxial anisotropy are developing a state with the magnetization dominantly out-of-plane, in contrast if $Q < 1$, the magnetization is laying mostly in the plan of the thin film.

This static three-dimensional micromagnetic study shows that the magnetization configuration of a nanostructure can be selected by modifying the shape of the element and adjusting the material parameters. In the limit of very small elements, one should expect to find a behavior close to macrospin model.

3.3 STT-INDUCED MAGNETIZATION SWITCHING: COMPARISON OF MACROSPIN AND MICROMAGNETISM

The difference between macrospin behavior and a three-dimensional micromagnetic behavior has already been illustrated in a static study. The difference is even more obvious in terms of the magnetization dynamics. As an illustration, the case of the writing process in an MRAM cell is presented below. The free layer has a planar magnetization, the polarizer is also an in-plane magnetized layer, and both layers are placed in the middle of a nanopillar with an elliptical cross section 130 nm × 65 nm and total height of 125 nm.

The material parameters of the free layer are the following: $\mu_0 M_S = 1T$, magnetocrystalline anisotropy axis parallel with the long axis of the ellipse $K_u = 14.525\ \text{kJ/m}^3$, exchange stiffness $A_{ex} = 13p\ \text{J/m}$, damping parameter $\alpha = 0.01$, current polarization $P = 30\%$. The thickness of the free layer is 2.5 nm. We consider a very simple case: a uniform DC spin-polarized current is injected in the structure and only the magnetization of the free layer can evolve in time since the magnetization of the polarizer is fixed (+parallel to the long axis of the elliptical sample). As explained earlier, in zero applied field for an elliptical cross section dot, because of the shape anisotropy, only two possible stable states are possible in macrospin approximation: single domain with the magnetization pointing left or right. The magnetization can be switched from one direction to the other if a high-enough spin-polarized current is injected. The switching from parallel state (PP, free layer magnetization parallel with the polarizer) to the antiparallel state (AP, free layer magnetization antiparallel with the polarizer) is obtained for an injected current of 5 mA (7.5×10^7 A/cm^2).

The time traces of the mean magnetization component parallel with the long axis of the free layer are depicted in Fig. 3.8 for three different simulation assumptions: (1) The black curve was obtained with a macrospin model. After a precessional

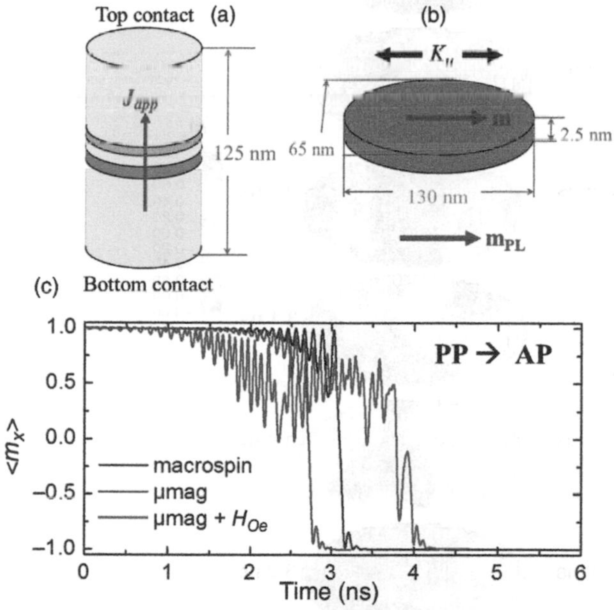

Figure 3.8 (a) Schematic view of a STT-MRAM cell with planar magnetization. The total thickness of the nanopillar is 125 nm. (b) Description of the free layer of the memory cell. (c) Time evolution of the longitudinal magnetization once the current is switched on (i.e., STT is active) as given by a three-dimensional macrospin model, with and without the Oersted field generated by the current itself.

motion of increasing amplitude, the switching occurs in 3.1 ns (i.e., the time for which $\langle m_x \rangle = 0$). (2) The red curve was computed by three-dimensional micromagnetic simulation (2400 discretization cells). The result is very similar to the macrospin prediction, but the estimated switching time is shorter, 2.7 ns. (3) If the Oersted field generated by the injected current itself is considered (see Section 1.1.4), the magnetization dynamics are quite different. Switching takes longer, 3.4 ns, and the time trace is quite complicated. The reason for this difference is intimately related to the magnetization dynamics, as one can see on the snapshots shown in Fig. 3.9. Even in the case considering only the spin-transfer torque action, the magnetization pattern is very nonuniform. This is counterintuitive, considering that the time traces of the magnetization, Fig. 3.8c, are rather similar to those for the macrospin case. Furthermore, when combining the influence of spin-transfer torque and Oersted field, the nonuniformity in magnetization distribution is enhanced. The central symmetry induced by the Oersted field fights against the uniaxial shape anisotropy. As a result, during the reversal process, a lot of spin waves are generated and their extinction takes more time (i.e., small damping value) with a direct impact on the switching time.

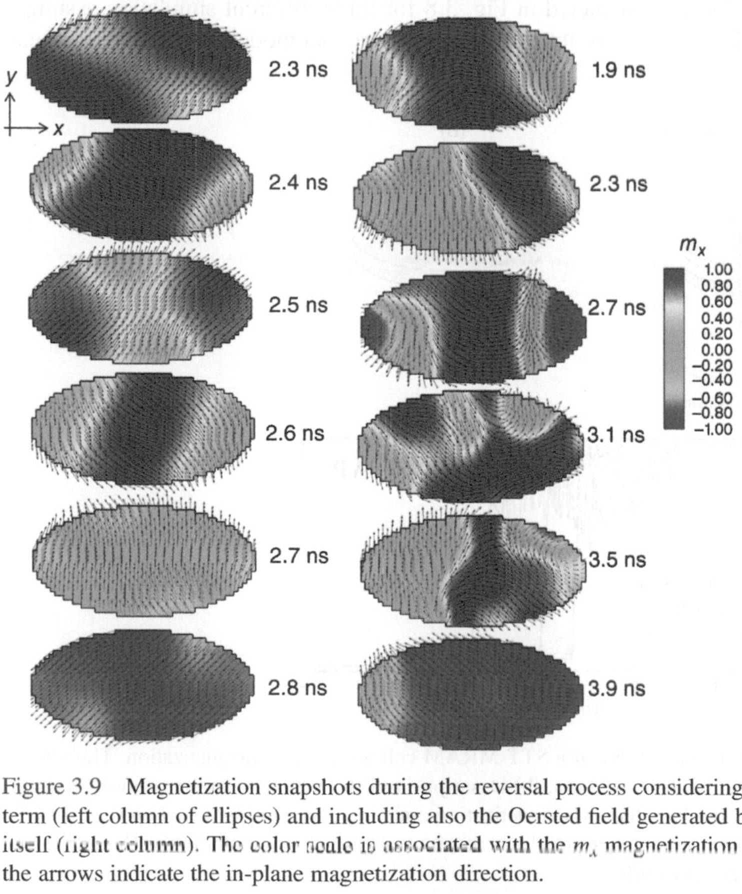

Figure 3.9 Magnetization snapshots during the reversal process considering only the STT term (left column of ellipses) and including also the Oersted field generated by the current itself (right column). The color scale is associated with the m_x magnetization component and the arrows indicate the in-plane magnetization direction.

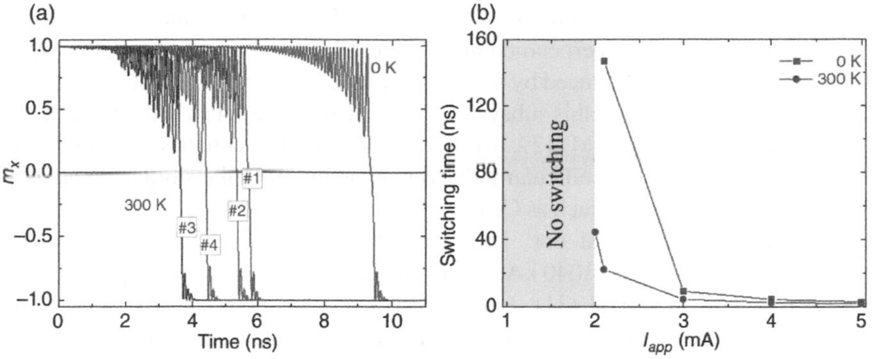

Figure 3.10 (a) Time traces of the longitudinal magnetization obtained by a macrospin model at 0 and 300 K (four events). (b) Average value of the switching time as a function of the injected current. The statistics were achieved using 4000 events.

The above results are for a temperature of 0 K (no thermal fluctuations). Insight about the impact of thermal fluctuations on the switching time is provided by carrying out simulations at 300 K as shown in Fig. 3.10. The macrospin model was used, 4000 events were simulated for each current value for an accurate statistics. Performing a similar analysis with a three-dimensional micromagnetic simulation would be too expensive in terms of computation time. The reversal is faster if the injected current is larger for both temperatures. But the impact of the thermal fluctuations is more evident close to the critical current line. This prediction is in agreement with experimental observations (28). The reversal depends strongly on thermal fluctuations. The latter are necessary to create an initial angle between the storage layers magnetization **M** and the spin current polarization **P** so that the STT, which is related to the cross product $\mathbf{M} \times \mathbf{P}$, becomes nonzero, which triggers the magnetization reversal. Therefore, if nothing is done to avoid it, there is a random incubation time preceding the magnetization switching, which can last several nanoseconds. This phenomenon must be carefully addressed in designing fast memories. Fortunately, especially in out-of-plane magnetized magnetic tunnel junctions (MTJs), the magnetization naturally bends at the edge of the pillar, which helps to nucleate the magnetization reversal, thus allowing fast magnetization switching.

3.4 EXAMPLE OF MAGNETIZATION PRECESSIONAL STT SWITCHING: ROLE OF DIPOLAR COUPLING

The use of micromagnetic modeling to explore new memory concepts is illustrated below for the case of precessional magnetization switching in a nanopillar. In general, the memory cell is composed of the interplay of several magnetic layers through the electronic transport but also through dipolar fields (stray fields). It is important to tailor the structure of the nanopillar to avoid undesirable effects due to the dipolar coupling.

The system under consideration is composed of one in-plane magnetized layer (free or storage layer) and a perpendicular polarizer. Since we are just interested here in the switching dynamics induced by the perpendicular polarizer on the storage layer, in this study we consider only this subsystem, which can be part of an orthogonal spin-transfer MRAM (OST-MRAM) (29,30). Such OST-MRAM comprises an in-plane MTJ with an additional perpendicular polarizer separated from the storage layer by a nonmagnetic spacer (a metal such as Cu or an insulator such as MgO). We assume that the free layer is an elliptical dot, 100 nm × 90 nm × 2.5 nm, with the following material parameters: $M_S = 1040\,\text{kA/m}$, exchange stiffness $A_{ex} = 13\,\text{pJ/m}$, and damping parameter $\alpha = 0.02$ (31). For the perpendicular polarizer, two cases are compared: single-layer polarizer and synthetic antiferromagnet polarizer (SAF) as depicted in Fig. 3.11a and b. One can see in Fig. 3.11c that the stray field generated by the single-layer polarizer on the free layer is highly nonuniform. Its mean value is ~80 kA/m, but the maximum in magnitude is as large as 130 kA/m. In comparison, the Oersted field goes up to 5.8 kA/m and its average is ~4 kA/m for a typical value of current density $J_{app} = 2.3 \times 10^{11}\,\text{A/m}^2$. In contrast, the SAF perpendicular polarizer is chosen so that the stray field generated on the free layer is negligible.

Current density and pulse length are swept and switching diagrams (Fig. 3.12) are plotted for the two current polarities and the two device structures. Due to a nonconstant a_J factor in the expression of the spin-torque (Eq. 32), the diagrams for two opposite polarities are not symmetric; in particular, the value of the critical current J_c is slightly different.

The addition of a stray field leads to slightly small critical current. At moderate current density ($<2 \times 10^{11}\,\text{A/m}^2$), the magnetization is still sufficiently uniform so that the final state is a single domain aligned along the easy axis (long axis of the elliptical dot). However, at higher currents, a vortex is formed, favored by the perpendicular polarizer stray field. The region where the final state is a vortex is shown in gray in Fig. 3.12c and d.

The evolution of the system under a pulse of 1210 ps and $2.5 \times 10^{11}\,\text{A/m}^2$ is shown in Fig. 3.13. For the SAF polarizer, two symmetrical, canted domains move along the edge of the free layer (Fig 3.13a) during the precession. When the current is switched off, the magnetization continues to rotate, achieving a whole turn before

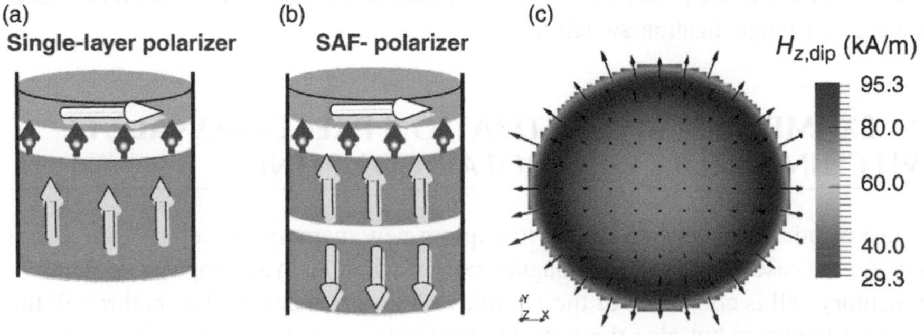

Figure 3.11 Schematic structure with (a) single-layer polarizer and (b) SAF polarizer. (c) Distribution of the stray field generated by the single-layer polarizer on the free layer.

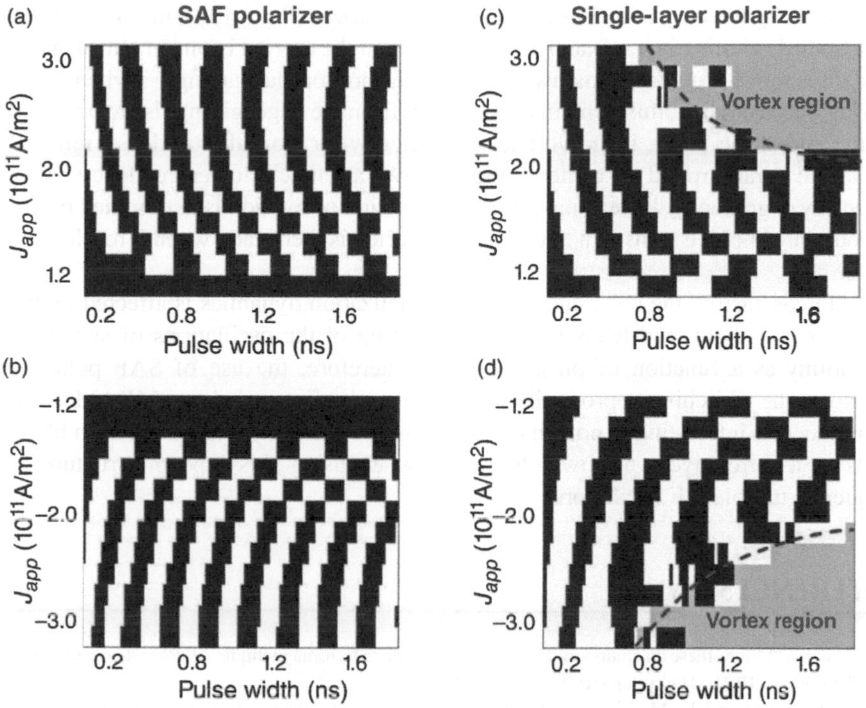

Figure 3.12 Switching diagrams. Black: no switching. Color scale: white is for switching, black is for nonswitching, and gray for a vortex in final state. SAF polarizer with (a) $J_{app} > 0$ and (b) $J_{app} < 0$. Single-layer polarizer (c) $J_{app} > 0$ and (d) $J_{app} < 0$. (Adapted from Ref. 31 with permission of AIP Publishing LLC.)

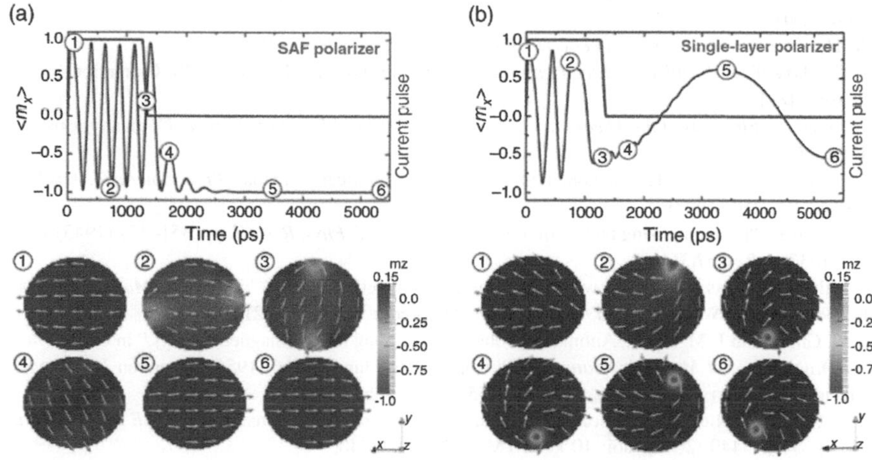

Figure 3.13 Simulations for a current density $J_{app} = 2.5 \times 10^{11}$ A/m^2 and pulse length $\Delta t = 1210$ ps with (a) a SAF perpendicular polarizer and (b) a single-layer perpendicular polarizer. The color scale represents the out-of-plane component of the magnetization. In part (b), when the current is off, the vortex describes a spiral around the free layer center. The system is at static equilibrium when the vortex reaches the center. (Adapted from Ref. 31.)

relaxing along the easy axis. For a single-layer polarizer (Fig 3.13b), the equilibrium state at the beginning is not exactly the same. When the current is on, an out-of-plane domain is formed; it points downward, since the spin-torque is along $-z$ when $J > 0$. During the precession, this domain moves away from the edge, giving rise to a vortex. At the end of the pulse, if the vortex is sufficiently far from the border, it remains stable and spirals around the center of the free layer at a much slower frequency with a trajectory of gradually decreasing radius. Its direction of rotation is determined by the surrounding in-plane spins that "drag" it. The vortex is then static when it reaches the center.

In conclusion, the coherence of the magnetization dynamics is affected by the polarizer stray field, and leads to a rapid damping of the oscillations of switching probability as a function of pulse duration. Therefore, the use of SAF polarizer improves the switching reproducibility in these ultrafast switching MRAM cells. Moreover, it is interesting to note that the improved coherence of the precession likely leads to a correlatively narrower linewidth when using this type of structure as frequency-tunable RF oscillators.

REFERENCES

1. P. Weiss, "L'hypothèse du champ moléculaire et la propriété ferromagnétique," *J. Phys.* **6**, pp. 661–690 (1907); doi: 10.1051/jphystap:019070060066100.
2. L. D. Landau and E. M. Lifshitz, "On the theory of the dispersion of magnetic permeability in ferromagnetic bodies," *Phys. Z. Sowjet.* **8**, pp. 153–169 (1935).
3. W. F. Brown, "Criterion for uniform micromagnetization," *Phys. Rev.* **105**, pp. 1479–1482 (1957); doi: 10.1103/PhysRev.105.1479.
4. W. F. Brown, "Thermal fluctuations of a single-domain particle," *Phys. Rev.* **130**, pp. 1677–1986 (1963); doi: 10.1103/PhysRev.130.1677.
5. K. H. J. Buschow and F. R. de Boer, *Physics of Magnetism and Magnetic Materials*, Kluwer, New York, 2003.
6. A. Hubert and R. Schafer, *Magnetic Domains*, Springer, New York, 1998.
7. J. C. Maxwell, *A Treatise on Electricity and Magnetism*, vol. **2**, §437 and §438, Clarendon Press, Oxford, 1881.
8. B. O. Peirce, *Elements of the Theory of the Newtonian Potential Function*, Ginn & Company, Boston, 1902, §69.
9. C. L. B. Shuddemagen, "The demagnetizing factors for cylindrical iron rods," *Proc. Am. Acad. Arts Sci.* **43**, pp. 185–256 (1907).
10. J. A. Osborn, "Demagnetizing factors of the general ellipsoid," *Phys. Rev.* **67**, pp. 351–357 (1945); doi: 10.1103/PhysRev.67.351.
11. J. Miltat, "Domains and domain walls in soft magnetic materials, mostly," in *Applied Magnetism*, eds. R. Gerber, C. D. Wright, and G. Asti, Kluwer, Dordrecht, 1994, pp. 221–308.
12. T. L. Gilbert and J. M. Kelly, "Anomalous rotational damping in ferromagnetic sheets," in *Conference on Magnetism and Magnetic Materials*, Pittsburgh, PA, June 14–16, 1955, American Institute of Electrical Engineers, New York, October 1955, pp. 253–263.
13. T. L. Gilbert, "A phenomenological theory of damping in ferromagnetic materials," *IEEE Trans. Magn.* **40**, pp. 3443–3449 (2004); doi: 10.1109/TMAG.2004.836740.
14. J. C. Slonczewski, "Current-driven excitation of magnetic multilayers," *J. Magn. Magn. Mater.* **159**, pp. L1–L7 (1996); doi: 10.1016/0304-8853(96)00062-5.
15. A. Thiaville, Y. Nakatani, J. Miltat, and Y. Suzuki, "Micromagnetic understanding of current-driven domain wall motion in patterned nanowires," *Europhys. Lett.* **69**, pp. 990–996 (2005); doi: 10.1209/epl/i2004-10452-6.

16. I. M. Miron, K. Garello, G. Gaudin, P-J. Zermatten, M.V. Costache, S. Auffret, S. Bandiera, B. Rodmacq, A. Schuhl, and P. Gambardella, "Perpendicular switching of a single ferromagnetic layer induced by in-plane current injection," *Nature* **476**, pp. 189–193 (2011); doi: 10.1038/nature10309.
17. L. Liu, C.-F. Pai, Y. Li, H. W. Tseng, D. C. Ralph, and R. A. Buhrman, "Spin-torque switching with the giant spin Hall effect of tantalum," *Science* **336**, pp. 555–558 (2012); doi: 10.1126/science.1218197.
18. E. C. Stoner and E. P. Wohlfarth, "A mechanism of magnetic hysteresis in heterogeneous alloys," *Philos. Trans. R. Soc. Lond. A* **240**, pp. 599–642 (1948); doi: 10.1098/rsta.1948.0007.
19. C. Kittel, *Introduction to Solid State Physics*, 4th ed., John Wiley & Sons, Inc., New York, 1971, pp. 596–601.
20. M. J. Donahue and D. G. Porter, *OOMMF User's Guide,* Version 1.0, NISTIR 6376, National Institute of Standards and Technology, Gaithersburg, MD, http://math.nist.gov/oommf. September (1999).
21. M. R. Scheinfein, *LLG Micromagnetics Simulator*, Portland, OR, http://llgmicro.home.mindspring.com.
22. D. V. Berkov and N. L. Gorn, *MicroMagus*, Jena, Germany, http://www.micromagus.de.
23. J. Oti, *MagOasis*, http://www.magoasis.com/aboutus.php.
24. A. Vansteenkiste and B. Van de Wiele, "MuMax: a new high-performance micromagnetic simulation tool," *J. Magn. Magn. Mater.* **323**, pp. 2585–2591 (2011); doi: 10.1016/j.jmmm.2011.05.037.
25. L. Torres and E. Martinez, *GoParallel*, Salamanca, Spain, http://www.goparallel.net.
26. A. Drews, *MicroMagnum*, University of Hamburg, Germany, http://micromagnum-tis.informatik.uni-hamburg.de.
27. L. D. Buda, I. L. Prejbeanu, U. Ebels, and K. Ounadjela, "Micromagnetic simulations of magnetisation in circular cobalt dots." *Comput. Mater. Sci.* **24**, pp. 181–185 (2002); doi: 10.1016/S0927-0256(02)00184-2.
28. T. Devolder, J. Hayakawa, K. Ito, H. Takahashi, S. Ikeda, P. Crozat, N. Zerounian, J.-V. Kim, C. Chappert, and H. Ohno, "Single-shot time-resolved measurements of nanosecond-scale spin-transfer induced switching: stochastic versus deterministic aspects," *Phys. Rev. Lett.* **100**, pp. 2–5 (2008); doi: 10.1103/PhysRevLett.100.057206.
29. O. Redon, B. Dieny, and B. Rodmacq, U.S. Patent No. 6,532,164 B2 (2000).
30. A. D. Kent, B. Ozyilmaz, and E. del Barco, "Spin-transfer induced precessional magnetization reversal," *Appl. Phys. Lett.* **84**, pp. 3897–3899 (2004); doi: 10.1063/1.1739271.
31. A. Vaysset, C. Papusoi, L. D. Buda-Prejbeanu, S. Bandiera, M. Marins de Castro, Y. Dahmane, J.-C. Toussaint, U. Ebels, S. Auffret, R. Sousa, L. Vila, and B. Dieny, "Improved coherence of ultrafast spin-transfer-driven precessional switching with synthetic antiferromagnet perpendicular polarizer," *Appl. Phys. Lett.* **98**, 242511 (2011); doi: 10.1063/1.3597797.

CHAPTER 4

MAGNETIZATION DYNAMICS

William E. Bailey
Materials Science and Engineering, Department of Applied Physics and Applied Mathematics, Columbia University, New York, USA

4.1 LANDAU–LIFSHITZ–GILBERT EQUATION

The Landau–Lifshitz–Gilbert (LLG) equation for magnetization dynamics includes two types of motion: *precession* and *relaxation* (or *damping*). Historical context for the equation is given in Section 4.1.3. Section 4.1.3.1 shows how the LLG equation can be motivated intuitively through magnetomechanical experiments and how additional terms (such as those resulting from spin-torque) can be added to the equation of motion. In Section 4.1.3.2, we show how the Landau–Lifshitz (LL) and Landau–Lifshitz–Gilbert damping terms are equivalent in terms of the resulting motion and can be used interchangeably.

4.1.1 Introduction

Our direct experience with magnets tells us about the final state of magnetization dynamics. The magnetization $\mathbf{M}$ of a soft ferromagnetic material tends to align with applied magnetic fields $\mathbf{H}$, minimizing the Zeeman energy density,

$$U = -\mu_0 \mathbf{M} \cdot \mathbf{H}, \tag{4.1}$$

where μ_0 is the permeability of free space.

One might then guess that the *full motion* of $\mathbf{M}$, or $\dot{\mathbf{M}}$ (where the dot denotes a derivative with respect to time), describes rotation along a direct, energy-minimizing path. However, the dynamics of magnetization $\dot{\mathbf{M}}$ have two terms rather than only one, and the energy-minimizing term, *relaxation*, is usually one or two orders of magnitude smaller than an energy-preserving term, the *precession*. For precessional motion only, no energy is lost; $U = -\mu_0 \mathbf{M} \cdot \mathbf{H}$ remains constant as the magnetization $\mathbf{M}$ simply rotates around the field $\mathbf{H}$. The precession and relaxation terms were written first by Landau and Lifshitz in 1935 (1).

Introduction to Magnetic Random-Access Memory, First Edition. Edited by Bernard Dieny, Ronald B. Goldfarb, and Kyung-Jin Lee.

4.1.2 Variables in the Equation

Before describing the LL equation, we define the variables involved, the choice of which contains important assumptions about the physics of magnetization motion.

Reduced Magnetization Ferromagnets are distinguished by their spontaneous magnetic order. For any magnetic material, the magnetization can be written as $M = N_v\langle\mu\rangle$, where N_v is the volume density of dipole moments (in units of m^{-3}) and $\langle\mu\rangle$ is an ensemble average dipole moment (A m^2), taken along the direction of the applied field. The magnitude $|\mu|$ is constant in a ferromagnet, independent of applied field **H** for constant temperature. Within a single *domain*, the magnetization is equal to the saturation magnetization $M = M_s = N_v|\mu|$. The direction of **M** can vary; the influence of **H** is to change the direction. It becomes sensible to define the *reduced magnetization* **m**,

$$\mathbf{m} \equiv \mathbf{M}/M_s, \tag{4.2}$$

as a unit vector in the magnetization direction. Magnetization dynamics are thus *rotational*, at right angles to **m**, describing trajectories on the unit sphere,

$$\mathbf{m} \cdot \dot{\mathbf{m}} = 0. \tag{4.3}$$

Clearly, in ferromagnetic materials the demagnetized state $\mathbf{M}=0$ is possible, but only as a large-scale average over domains, averaging to zero magnetization. Micromagnetic calculations make the similar assumption that in each finite element, **m** is a unit vector, M_s does not change, and the task is to calculate **m(r)** for all spatial coordinates **r**.

Effective Field The magnetization direction **m** influences the energy of a ferromagnet in several ways. The Zeeman energy, Eq. (4.1), created by an applied external field $\mathbf{H_b}$, is one energy term, but the variation of **m(r)** can create other terms as well. Anisotropies due to shape (dipolar fields), interaction with crystal fields (magnetocrystalline anisotropy, induced anisotropy in alloys), stress, and exchange, all have different possible energy terms. The most convenient way to treat these additional energies is through the definition of an *effective field* acting on particle *i*:

$$\mathbf{H_{eff}^i} = -\frac{1}{\mu_0 M_s}\frac{\partial U_i}{\partial \mathbf{m}}. \tag{4.4}$$

Since magnetic fields directed along **m** cannot exert a torque to rotate the magnetization, they are not effective. A Cartesian basis for $\mathbf{H_{eff}}$ in three variables is thus less compact than a spherical basis:

$$\mathbf{H_{eff}^i} = -\frac{1}{\mu_0 M_s}\left(\frac{\partial U_i}{\partial m_x}\hat{\mathbf{x}} + \frac{\partial U_i}{\partial m_y}\hat{\mathbf{y}} + \frac{\partial U_i}{\partial m_z}\hat{\mathbf{z}}\right) \tag{4.5}$$

can be written more compactly as

$$\mathbf{H_{eff}^i} = -\frac{1}{\mu_0 M_s}\left(\frac{\partial U_i}{\partial \varphi}\hat{\boldsymbol{\varphi}} + \frac{1}{\sin\theta}\frac{\partial U_i}{\partial \theta}\hat{\boldsymbol{\theta}}\right), \tag{4.6}$$

where the derivative in r is not allowed. In the usual convention, the magnetization direction is $\hat{\mathbf{r}} = \cos\varphi \sin\theta\, \hat{\mathbf{x}} + \sin\varphi \sin\theta\, \hat{\mathbf{y}} + \cos\theta\, \hat{\mathbf{z}}$.

4.1.3 The Equation

In Landau and Lifshitz's original paper (1), they proposed the equation

$$\dot{\mathbf{s}}/\mu_0 = [\mathbf{fs}] + \lambda\{\mathbf{f} - (\mathbf{fs})\mathbf{s}/s^2\} \quad \text{(cgs)}. \tag{4.7a}$$

In modern notation, with $\mathbf{s} \to \mathbf{M}$ (spin magnetization), $\mu_0 \to |\gamma|$ (explicitly for g_{eff}= 2), $\mathbf{f} \to \mathbf{H_{eff}}$ (effective field; see Section 4.1.2), square brackets to cross products, and parentheses to dot products, we get

$$\frac{\dot{\mathbf{M}}}{|\gamma|} = \mathbf{H_{eff}} \times \mathbf{M} + \lambda \left[\mathbf{H_{eff}} - \left(\frac{\mathbf{H_{eff}} \cdot \mathbf{M}}{M_s^2} \right) \mathbf{M} \right] \quad \text{(cgs)}. \tag{4.7b}$$

In the 80 years since their paper was published, the equation stands uncorrected. Alternative forms proposed for the second (relaxation) term (see Section 4.1.3.2) turn out to be equivalent. The gyromagnetic ratio can be expressed in SI units as

$$\gamma = \frac{e}{2m_e} g_{eff}, \tag{4.8}$$

where e is the electronic charge and m_e is the electronic rest mass. Numerically, in SI, γ is

$$\gamma = \gamma_0 \left(\frac{g_{eff}}{2} \right), \quad \gamma_0 = -2\pi \times 27.99\ \text{GHz/T}, \tag{4.9}$$

with $g_{eff} = 2$, because LL considered pure spin-moments only. The cgs form is numerically identical, but more conveniently expressed as $\gamma_0 = -2\pi \times 2.799$ MHz/Oe. The assumption $g_{eff} = 2$ has been relaxed in subsequent years and values of g_{eff} up to 2.2 have been established experimentally given the mostly quenched, but nevertheless finite amounts of orbital moment in magnetic materials (2). See Section 4.2.3 for details.

Modern treatments of the LL equation typically use an alternative form, although the original form is no less valid and is still used occasionally. Dividing Eq. (4.7b) by M_s, making use of the vector identity $\mathbf{a} \times \mathbf{b} \times \mathbf{c} = \mathbf{b}(\mathbf{a}\cdot\mathbf{c}) - \mathbf{c}(\mathbf{a} \cdot \mathbf{b})$, thus $-\mathbf{m} \times \mathbf{m} \times \mathbf{H_{eff}} = \mathbf{H} - \mathbf{m}(\mathbf{m} \cdot \mathbf{H_{eff}})$, and introducing the *dimensionless damping parameter* α, we can write

$$\dot{\mathbf{m}} = -|\gamma| \mathbf{m} \times \mathbf{H_{eff}} - \alpha|\gamma|(\mathbf{m} \times \mathbf{m} \times \mathbf{H_{eff}}) \quad \text{(cgs)}, \tag{4.10}$$

$$\dot{\mathbf{m}} = -\mu_0|\gamma| \mathbf{m} \times \mathbf{H_{eff}} - \alpha\mu_0|\gamma|(\mathbf{m} \times \mathbf{m} \times \mathbf{H_{eff}}) \quad \text{(SI)}. \tag{4.11}$$

The dimensionless damping α (the same in both unit systems) can also be expressed as a rate λ:

$$\lambda \equiv |\gamma| M_s \alpha \quad \text{(cgs)}, \tag{4.12}$$

$$\lambda \equiv \frac{1}{4\pi} |\gamma| \mu_0 M_s \alpha \quad \text{(SI)}, \tag{4.13}$$

also defined here to have the same value in s^{-1} for both cgs and SI units. Alternative definitions of λ in SI units can differ by a factor of 4π; see Appendix of Ref. 3.

4.1.3.1 Precessional Term The first term of the LL equation is conservative. All energy stored in the magnetization system and its interactions with the lattice, expressed through the energy in the effective field, $U = -\mu_0 M_s \mathbf{m} \cdot \mathbf{H}_{eff}$, remains stored under operation of this term. Because the precessional term is dominant, constant-energy lines for $\mathbf{m}(\theta,\varphi)$ can be a good approximation for trajectories.

The form for the precessional term can be derived from Ehrenfest's theorem in quantum mechanics, as shown in elementary texts such as Ref. 4. We will instead motivate the precessional term semiclassically, through the Einstein-de Haas experiments (5).

Einstein-de Haas Experiments Einstein showed experimentally that there is a real connection between magnetization **M** and angular momentum **L**. The magnetization of a cylinder of soft iron was suspended by a thin wire. An external solenoid coil reversed the magnetization, changing it by $\Delta M = 2M_s$ along its axis. The reversal caused the cylinder to rotate around its axis. The ratio between the mechanical angular momentum and magnetic angular momentum was identified as γ, where $\Delta\mathbf{M} = \gamma\Delta\mathbf{L}$. This relation can be generalized to a dynamical equation. The time derivative of both sides yields

$$\dot{\mathbf{M}} = \gamma\boldsymbol{\tau}, \quad \dot{\mathbf{M}} = \gamma\mu_0 \mathbf{M} \times \mathbf{H}, \tag{4.14}$$

where $\boldsymbol{\tau}$ is a mechanical torque on the body. Here, we substitute the torque on a point dipole per unit volume (magnetization) in a uniform field, $\boldsymbol{\tau} = \mu_0 \mathbf{M} \times \mathbf{H}$, appropriate under the single-domain approximation detailed earlier. Equation (4.14) provides an alternative and surprising derivation of the first term in the LL equation not mentioned in Ref. 1.

Magnetomechanical values of γ were in reasonable agreement with the free electron result *e*/*m* (*e*/*mc* in cgs). It is remarkable that Einstein was able to carry out these experiments in his spare time while he developed the theory of relativity, although students with similar ambitions should be aware that his measurements were found to be in error by a factor of 2 (6). Magnetomechanical and microwave resonance measurements of γ were later found by Barnett to agree within 10% (7); his measurements continued into the early 1950s (8).

Additional Torque Terms Novel terms to magnetization dynamics, such as the influence of spin-torque, can be added into the LLG. Both the torque and the magnetic moment are total, volume-summed quantities, so to convert an additional torque term to dynamics of reduced magnetization **m**, the torque term needs to be expressed as

$$\dot{\mathbf{m}} = \cdots - \frac{|\gamma|}{StM_s}\boldsymbol{\tau} \quad \text{(cgs)}, \quad \dot{\mathbf{m}} = \cdots - \frac{|\gamma|}{St\mu_0 M_s}\boldsymbol{\tau} \quad \text{(SI)}. \tag{4.15}$$

Here, the volume of a thin film is taken as $S{\cdot}t$, where S is the area and t is the thickness

4.1.3.2 Relaxation Term

The relaxation term has been more controversial and more widely discussed in the years since LL's paper. LL added the term completely ad hoc in order to have the magnetization decay toward $\mathbf{H}_{\text{eff}}$. In the equivalent expression, Eq. (4.10), it is clear how relaxation operates geometrically: $\mathbf{m}\times\mathbf{H}$ gives the axis about which $\mathbf{m}$ needs to rotate to bring the $\mathbf{m}$ and $\mathbf{H}$ into alignment, and $\mathbf{m}$ moves at right angles to it.

Gilbert Form An alternative form for the damping term was proposed by Gilbert and Kelly in 1955 (9,10), allowing the new LL equation to be written as

$$\dot{\mathbf{m}} = -\mu_0|\gamma|\mathbf{m}\times\mathbf{H}_{\text{eff}} + \alpha(\mathbf{m}\times\dot{\mathbf{m}}). \qquad (4.16)$$

This form is known as the *Landau–Lifshitz–Gilbert* equation. Gilbert justified the new damping term by claiming that the "higher" damping observed for $Ni_{81}Fe_{19}$ platelets would be better represented by a "viscous" term proportional to $\dot{\mathbf{m}}$. For low damping, $\alpha \ll 1$ (true for almost all known cases), it can be shown easily that the LL and LLG forms are equivalent. If the relaxation term to the motion is small compared with the precessional term, one can obtain Eq. (4.16) from Eq. (4.10) by substituting $\dot{\mathbf{m}} \approx |\gamma|\mathbf{m}\times\mathbf{H}_{\text{eff}}$.

Functional Equivalence of the LL and Gilbert Damping The identity between the LL and LLG forms is more exact than a small-damping approximation. Magnetization trajectories are equivalent for all angles subject to a small redefinition of γ. Decomposition of Eqs. (4.11 and 4.16) into spherical coordinates yields the directly integrable expressions:

$$\begin{bmatrix}\dot{\theta}\\ \dot{\varphi}\end{bmatrix} = \mu_0|\gamma|\begin{bmatrix}\alpha & 1\\ \dfrac{-1}{\sin\theta} & \dfrac{\alpha}{\sin\theta}\end{bmatrix}\begin{bmatrix}H_\theta^{eff}\\ H_\varphi^{eff}\end{bmatrix} \quad \text{(SI, LL)}, \qquad (4.17)$$

$$\begin{bmatrix}\dot{\theta}\\ \dot{\varphi}\end{bmatrix} = \frac{\mu_0|\gamma|}{1+\alpha^2}\begin{bmatrix}\alpha & 1\\ \dfrac{-1}{\sin\theta} & \dfrac{\alpha}{\sin\theta}\end{bmatrix}\begin{bmatrix}H_\theta^{eff}\\ H_\varphi^{eff}\end{bmatrix} \quad \text{(SI, LLG)}, \qquad (4.18)$$

so that LL and LLG trajectories are equivalent if we take $\gamma_{LL} = \gamma_{LLG}/\left(1+\alpha^2\right)$. Gilbert recognized the equivalence in an expanded version of his paper (10).

No physical behavior can be explained by the LLG that could not be explained by the LL, as long γ and α are both free parameters. The gyromagnetic ratio would need to be resolved to better than one part in α^{-2} to separate the expressions even in principle. Table 4.1 shows a best precision in measurement of γ to about 0.007, on the order of α for $Ni_{81}Fe_{19}$. While there may be formal reasons to prefer one form over another, a debate that continues (11), the difference between these forms would be quite challenging to resolve in an experiment. The LL form is more convenient for algebraic manipulations since the right side of the equation contains no terms in $\dot{\mathbf{m}}$.

Finally we note that λ, the relaxation rate (in units of s^{-1}) in the LL damping scheme, has an analogous rate G in the Gilbert damping scheme, with $\lambda = G$. These values can be used interchangeably. See Table 4.1 for values.

TABLE 4.1 Room-Temperature Values of Materials Parameters for Several Typical Ferromagnets and Ferromagnetic Alloys.

Material	g_{eff} (12,13)	$\mu_0 M_s$ (T) (14)	α_0 ($\times 10^{-3}$)	λ or G (MHz)
Fe	2.094 ± 0.02	2.1545	1.8	57 (15–17)
Co	2.170 ± 0.02	1.8173	8.5	170–280 (16)
Ni	2.185 ± 0.02	0.6165	23.3	220 (18)
$Ni_{81}Fe_{19}$ (Py)	2.129 ± 0.02	0.95 (19) – 1.07 (20)	7.0 (21)	98–111
$Co_{60}Fe_{20}B_{20}$ (20)	2.07	1.14	6.5 (20)	107
$Co_{72}Fe_{18}B_{10}$ (22)	2.07	1.76	6.0 (20)	153

Precision is no better than 1% for g_{eff} and no better than 5% for α and G.

4.2 SMALL-ANGLE MAGNETIZATION DYNAMICS

In this section, we consider small-angle rotations of the magnetization for in-plane magnetization. First, in Section 4.2.1, we see how the LLG can be linearized for small angles θ with respect to an in-plane equilibrium magnetization. Solutions for this equation under periodic driving are shown in Section 4.2.2. The corresponding experimental technique is ferromagnetic resonance (FMR), which measures the absorption of microwaves through the frequency-dependent susceptibility. FMR is the source of most data on the dynamical response of ferromagnetic materials and heterostructures. Experimental parameters, both in the "bulk" and possible modification through film surfaces (finite-size effects), are reviewed in Section 4.2.3. Solutions for pulsed magnetization motion, relevant for the beginning or end of switching processes, are shown in Section 4.2.4.

4.2.1 LLG for Thin-Film, Magnetized in Plane, Small Angles

We will treat the magnetization dynamics of an ultrathin film, relevant for magnetic random access memory (MRAM). Even for elements patterned to a 100 nm pillar, a 5 nm thick ferromagnetic film can be approximated reasonably well as a semi-infinite film. The magnetization is assumed to have its zero-field orientation in the y–z plane. We take the film normal to be $\hat{\mathbf{x}}$. The magnetization is assumed to be saturated along $\hat{\mathbf{z}}$ in the presence of external applied fields along $\hat{\mathbf{z}}$, an approximation strictly true in limiting cases (Note that full alignment is strictly true only in the case of pure hard axis or pure-easy axis alignment) (see Section 4.3).

Much of the interesting behavior in magnetization dynamics concerns only small excursions of the magnetization from its equilibrium direction. We introduce the following reduced variables (23):

$$\omega_M \equiv \mu_0 |\gamma| M_s, \quad \omega_H \equiv \mu_0 |\gamma| H_z', \tag{4.19}$$

where $H_z' = H_b + H_K + \cdots$ is defined as an *effective field* including any uniaxial (or unidirectional) anisotropy fields maintaining the magnetization along the biased direction. (Surface anisotropy terms will be discussed later.) We can express

Eq. (4.11) in the form

$$\begin{bmatrix} \dot{m}_x \\ \dot{m}_y \end{bmatrix} = \begin{bmatrix} -\alpha(\omega_M + \omega_H) & -\omega_H \\ \omega_M + \omega_H & -\alpha\omega_H \end{bmatrix} \begin{bmatrix} m_x \\ m_y \end{bmatrix} + \begin{bmatrix} \alpha & 1 \\ -1 & \alpha \end{bmatrix} \begin{bmatrix} \gamma H_x^{eff} \\ \gamma H_y^{eff} \end{bmatrix}$$

under the assumption $m_x, m_y \ll 1$ and $m_z \simeq 1$.

This is the most compact form of the LL equation, valid for small angles. This system of two coupled ordinary differential equations in m_x, m_y can be converted to a single second-order ordinary differential equation in m_y by differentiating the first and substituting the second into the first. After some algebra, we can convert these equations into the classical equation for a driven harmonic oscillator,

$$\ddot{m}_y + \eta \dot{m}_y + \omega_0^2 m_y = f_0^2 \cos \omega t, \tag{4.20}$$

$$\eta = \alpha(\omega_M + 2\omega_H), \quad \omega_0 = \sqrt{\omega_H(\omega_H + \omega_M)}, \; f_0^2 \approx (\omega_M + \omega_H)\gamma H_{y,0}, \tag{4.21}$$

where the α term in the drive is neglected. A similar equation could be written for m_x. Note that the relaxation rate η, resonant frequency ω_0, and driving amplitude f_0 are all functions of the applied field through ω_H.

Re-expanding the variables in SI units, we can write the *Kittel equation* for the resonant frequency of precession as

$$\upsilon_0 = \frac{\omega_0}{2\pi} = \frac{|\gamma_0|}{2\pi}\left(\frac{g_{eff}}{2}\right)\mu_0\sqrt{H_z'(H_z' + M_s)}. \tag{4.22}$$

4.2.2 Ferromagnetic Resonance

In their 1935 paper, Landau and Lifshitz worked through Eq. (4.7a), derived a correct expression for the susceptibility $\chi(\omega)$ for a sphere, and predicted a large enhancement of microwave power absorption on resonance—all far before any experimental observation of magnetic resonance. Rabi and coworkers at Columbia reported the first experimental observation of nuclear magnetic resonance (NMR) using time-of-flight techniques in 1938 (24). The observation of resonant absorption at microwave frequencies, relying on advances in radar technology (25), did not appear until 1945; the first observation of electron spin resonance (ESR) (26) was followed rapidly by the first observation of NMR in 1946 (27), and then by FMR later that year (28). The role of dipolar fields in increasing the in-plane resonance frequency for FMR (as $\gamma\sqrt{BH}$) was first understood by Kittel (29).

Microwave Susceptibility The absorption of microwaves in an insulating ferromagnet is found through the complex susceptibility $\chi''(\omega)$; eddy currents bring complications in conductive ferromagnets. From the equation for susceptibility in a forced harmonic oscillator, we can find $\chi(\omega)$ for the ferromagnetic thin film; neglecting terms in α (2),

$$\chi_{M,//} = \frac{M_y}{H_y} = \frac{M_s}{H_y}\frac{f_0^2}{(\omega_0^2 - \omega^2) + i\eta\omega} = \frac{\omega_M(\omega_M + \omega_H)}{(\omega_0^2 - \omega^2) + i\eta\omega}, \tag{4.23}$$

which can be expressed in terms of reduced variables,

$$h \equiv \frac{\omega_H}{\omega_M}, \quad h \equiv \frac{H'_z}{M_s}, \quad \Omega \equiv \frac{\omega}{\omega_M}, \quad \chi_M = \frac{h+1}{h(h+1) - \Omega^2 + i(1+2h)\alpha\Omega}. \tag{4.24}$$

On and near resonance, we can express $h = h_r + \Delta h$, where h_r is the field for resonance at fixed frequency Ω. We then have for the imaginary part of the susceptibility χ''_M, neglecting terms Ä$h^2 \ll 1$ and Ä$h \ll h_r$,

$$\chi''_M \approx \alpha\Omega \frac{h_r + 1}{(1+2h_r)} \frac{1}{(\Delta h)^2 + \alpha^2\Omega^2}.$$

We see that there is a Lorenzian (peaked) enhancement of imaginary susceptibility (power absorption) at the resonant frequency $\omega = \omega_0$, by a maximum of α^{-1}. Expressing the full-width in field for which χ'' reaches half its peak value, we have

$$\Delta H(\omega) = \Delta H_0 + \frac{2\alpha}{|\gamma|}\omega \quad (cgs), \quad \mu_0 \Delta H(\omega) = \mu_0 \Delta H_0 + \frac{2\alpha}{|\gamma|}\omega \quad \text{(SI)} \tag{4.25}$$

to which we have added an inhomogeneous broadening term ΔH_0 that might arise due to large-scale magnetic disorder.

Extraction of Materials Parameters Equation (4.25) describes how to measure the damping α in a swept-field, variable frequency FMR experiment. The damping α is found through a linear fit to frequency-dependent linewidth $\Delta H(\omega)$, where α is given by the slope and ΔH_0 is given by the zero-frequency offset. (An alternative view holds that disorder combined with exchange coupling between grains can also produce a frequency-dependent linewidth (30), but these complications are generally exhausted at relatively low frequencies (10–20 GHz).) The range up to 70 GHz is best for materials such as Fe. Measurements in the range of 0–24 GHz are most reliable where ferromagnetic materials are very soft (easily magnetized) and there is little contribution from magnetocrystalline anisotropy (e.g., Permalloy (Py), $Ni_{81}Fe_{19}$).

The resonance position provides a good estimate of other materials parameters influencing the effective field. From Eq. (4.20) we have $\omega_0 = \mu_0\gamma\sqrt{H'_z(H'_z + M_s)}$. In the absence of surface anisotropy, the slope of $\omega(H_b)$ is determined primarily by the film magnetization M_s. In the presence of surface anisotropy (see Section 4.2.3.2), the slope of $\omega(H_b)$ contains a contribution from surface anisotropy. In either case, the offset of $\omega(H_b)$ provides a measure of H_K. For in-plane measurements, γ can be taken from the literature (12,13); see Table 4.1.

Eddy Currents Equation (4.25) cannot be used for the damping of thick metallic ferromagnetic films. Ament and Rado (31) have developed an expression for the equivalent permeability $\mu_{eff}(\omega)$ of semi-infinite films. The finite conductivity of the metal film implies a finite skin depth for microwaves; the skin depth is reduced compared with the nonmagnetic case by a factor $\mu_r^{-1/2} = \sqrt{1+\chi_M}$. The inhomogeneous magnetization profile in the ferromagnet gives rise to inhomogeneous exchange fields, broadening the resonance. Experimental extraction of λ often failed in bulk whiskers (18) and oft-cited (32,33) low-temperature values of λ for Co and Fe are in fact poorly known. Accurate measurements of α in ferromagnetic metals were enabled only through the production of ultrathin ferromagnetic films, where the Rado-Ament analysis was not necessary. For films thinner than those hat support a spin wave, Lock (34) has written an estimate for the damping λ_e due to eddy currents, converted here to a SI value for α:

$$\alpha_{eddy} = \frac{\mu_0 \omega_M t_F^2}{12\rho} \tag{4.26}$$

where t_F designates the ferromagnetic layer thickness. For Fe, with large M_s and small ρ, the eddy current damping is relatively more important than that of Py, forming a sizable fraction of its bulk damping at $t_F = 50$ nm. Py is much less sensitive (35). For films much thinner than the (magnetic) skin depth, eddy-current damping became insignificant, and it became possible to extract α directly from Eq. (4.25) starting in the late 1960s (36). The experimental program of Heinrich has centered on extraction of α in ultrathin epitaxial films (37,38).

4.2.3 Tabulated Materials Parameters

The materials parameters that should be used in the LLG are reasonably well known in the bulk for elemental solids and typical soft alloys at room temperature. Finite-size effects (thickness dependence) remain under investigation, particularly for the damping α. We show parameters for ferromagnetic metals in the "bulk" in Table 4.1. Tabulated dynamics parameters have been measured by FMR, which remains the most quantitative of magnetization dynamics techniques. Values of $\mu_0 M_s$ and g_{eff} come from magnetometry and are confirmed by magnetomechanical experiments, respectively.

4.2.3.1 Bulk Values *Magnetic Moments* Saturation magnetizations M_s were tabulated by Stearns in the Landolt–Börnstein tables (14), reproduced here for 300 K (for the elemental solids Fe, Co, and Ni, the values given here for 300 K have been extrapolated from the 286.5 K values cited using factors of $1 - c\Delta T$, where $c_{Fe} = 1.2 \times 10^{-4} K^{-1}$, $c_{Co} = 5.9 \times 10^{-5} K^{-1}$, and $c_{Ni} = -4.2 \times 10^{-4} K^{-1}$. Values are taken for Fe[1 1 0], Co[0 0 0 1], Ni[1 1 1]; nonnegligible anisotropy in $\mu_0 M_s$ exists, to 0.5% in Co at zero temperature, on the order of 5×10^{-4} in the cubic solids). The alloy values are less certain. For nominal $Ni_{81}Fe_{19}$ alloys, fits to FMR data have indicated a range of 0.95–1.07 T. Interpolation of the Ni, Fe values in Table 4.1 would yield an estimate of 0.91 T with a sensitivity to composition of 0.015 T/%. For constant alloy sputtering target composition (nominal film composition), even though the exiting flux from the target is thought to converge to the internal composition, deposited compositions can vary by several percent, as shown, for example, in the Ni–Ti system (39), due to differences in thermalization of the ejected atoms.

Gyromagnetic Ratios Gyromagnetic ratios are known from a combination of magnetomechanical measurements (g') and FMR measurements (g), taken on bulk samples, in some cases on the same bulk samples (12); in careful experiments, good agreement (to better than 0.7% error) is found between the measurements satisfying the relationship $1/g' + 1/g = 1$ (40). The error cited for each material is the error for the composition, typically <1%; fits to binary alloy series should improve the error. Later investigations confirmed these values (13). Taking g_{eff} as a free parameter, allowed to range much outside of the band of values listed, or allowed to vary with thickness, can be error-prone.

Damping The most structure-sensitive of the parameters is the damping α, but only in a particular sense. The values given here are α_0, the lowest values found in the literature for variable-frequency FMR spanning linewidth variations significantly greater than the inhomogeneous broadening; see Section 2.2. The Fe value of $\lambda/G = 57$ MHz ($\alpha = 0.0018$) is equal in sputtered films (16) and single-crystal whiskers (15) at room temperature in measurements up to 90 GHz. Ni also shows $\lambda/G = 220$ MHz ($\alpha = 0.023$) equally in the two measurements (18). Co has not been shown to fit the Rado–Ament analysis for whiskers. Measurements of nominally FCC films on MgO have shown different Gilbert damping parameters α for two different crystal directions, not expected for a cubic material.

A higher value of α_{eff} is often obtained for imperfect materials. However, α_{eff} is not expected to be constant with frequency: where damping is increased due to magnetic inhomogeneities (ΔH_0), effective $\alpha_{eff}(\omega)$ is a decreasing function of frequency, converging toward α_0; see Ref. 35 ($\alpha \neq \alpha_0$ thus reflects some departure from a single-domain model). Complicating the matter further is that the lowest values of α have been observed in sputtered films; crystallographically optimized Fe films made by molecular-beam epitaxy (MBE) exhibit significantly higher values of α (41), attributed to the presence of vacancies (38).

4.2.3.2 *Finite-Size Effects* *Effective Magnetization* There is some question whether the bulk magnetization applies for ultrathin films. Experiments on epitaxial Co/Cu report, in roughly equal numbers, enhanced, reduced, or unchanged magnetic moment μ_B per Co interface *atom*; see Table 21 in Ref. 42. In magnetization dynamics experiments, possible finite-size effects on magnetization are overwhelmed by *surface anisotropy*, proposed by Néel in the 1950s (43). Surface anisotropy introduces free energy terms per unit area of the top and bottom ferromagnetic interfaces as $U_{i,area} = -\sigma_i m_x^2$, where the interfaces favor in-plane magnetization for $\sigma < 0$. Thus,

$$\mathbf{H}_{\mathbf{eff}} = 2\frac{\sigma_{top} + \sigma_{bottom}}{\mu_0 M_s}\frac{1}{t_{FM}} m_x \hat{\mathbf{x}}, \quad \mu_0 M_s^{eff} = \mu_0 M_s - \frac{4\sigma}{M_s}\frac{1}{t_{FM}}, \tag{4.27}$$

which, as can be seen, has the sole effect of changing the effective $\mu_0 M_s$ resulting from a fit of the form of Eq. (4.22). First identifications of surface anisotropy in FMR experiments date from the 1970s, although film preparation is in question for experiments from this era. The Permalloy–air interface was found to have a surface anisotropy of $\sigma = 0.08$ mJ/m^2 (44). More recently, Rantschler et al. have estimated σ in Permalloy–Cu, Permalloy–Ag, Permalloy–Ta, finding 0.100, 0.100, and 0.070 mJ/m^2, respectively (45). The positive surface anisotropy, which favors perpendicular magnetization, has the effect of *reducing* the resonance frequencies as if $\mu_0 M_s$ were reduced. In a Cu/Permalloy (5 nm)/Cu film, $\mu_0 M_s^{eff}$ becomes reduced by roughly 0.1 T.

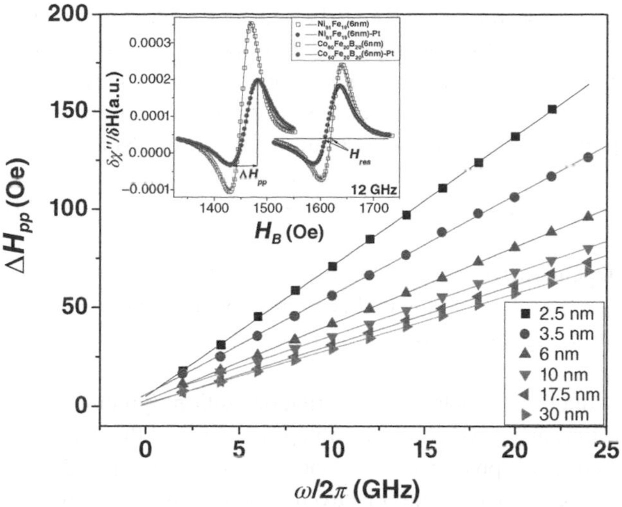

Figure 4.1 Illustration of damping size effect: FMR linewidth ΔH_{pp} as a function of frequency $\omega/2\pi$ for $Ni_{81}Fe_{19}$ (t_{FM}) with Cu (3 nm)/Pt (3 nm) overlayers. There is an inverse dependence of α on the t_{FM} as shown in the slope of the curve. *Inset:* Experimental trace of absorption derivative $\partial\chi''/\partial H$.

Gyromagnetic Ratio It is not clear whether there is a size effect in the gyromagnetic ratio. Some have proposed that the ratio of orbital to spin moment μ_L/μ_S could be enhanced in structures of finite dimension (46), making larger values of g_{eff} plausible in ultrathin films. Variable frequency, perpendicular FMR (normal condition) is best suited to determine g_{eff} since it enables its isolation from the slope $\Delta H_{res}/\Delta\omega$. Identifications of a size effect in g_{eff} have relied on in-plane FMR (47) or pulsed inductive microwave magnetometry (PIMM) (48), where (at low frequency) $\nu^2 \propto g_{eff}^2 M_s^{eff}$; values of g_{eff} as high as 2.6 were extracted for the Co (3 nm)–Cu system, but with $\mu_0 M_s$ correspondingly lower than that given in Table 4.1 (see Fig. 4.1).

Damping α There is good evidence that the damping α itself exhibits a size effect, where $\alpha = \alpha_0 + Kt^n$. In structures with "spin sinks" in proximity to the FM layer, either heavy elements such as Pt or other FM layers, *spin pumping* can increase α by a factor of roughly 2 over bulk values at thicknesses between 2–5 nm, with inverse dependence on FM thickness ($n = -1$). A detailed review of experiments and theory to 2004 was given by Tserkovnyak et al. (49). However, even in structures without a spin-sink layer, a similar size effect in damping is present. The origin of this size effect is not known.

4.2.4 Pulsed Magnetization Dynamics

Equation (4.20), derived under the assumption of small amplitudes, $\alpha \ll 1$, and sinusoidal driving fields, can be rewritten for arbitrary transverse external fields $H_y(t)$:

$$\ddot{m}_y + \eta \dot{m}_y + \omega_0^2 m_y = (\omega_M + \omega_H)\mu_0 \gamma H_y(t), \tag{4.28}$$

where the longitudinal field $\mu_0 H'_{b,z} = \omega_H/|\gamma|$ is taken to be constant. For a step function in transverse magnetic field, the magnetization trajectory describes a damped sinusoid with initial conditions dependent on the initial state. The simplest case is where a finite rotation, in-plane angle $\theta_0 \approx m_{y0} = m_y{}^y(t = 0)$, relaxes to zero after removal of a finite-width, fast fall-time pulse (falling step). The solution for Eq. (4.28) is

$$m_y(t) = m_{y0}e^{-t/\tau}\cos\omega_0 t, \quad \tau = \frac{2}{\eta}, \tag{4.29}$$

where η and ω (both field-dependent) are given as in Eq. (4.21). It is less straightforward to incorporate inhomogeneous broadening in this expression; one can calculate a frequency linewidth by multiplying $\Delta H(\omega)$ by $\partial\omega_0/\partial H$, which can be substituted in the expression for η.

The first high-speed inductive measurements of magnetization dynamics were carried out at IBM-Zurich in the early 1960s (50); the 350 ps rise time of the pulse was sufficient to look at low-frequency precessional dynamics (~1 GHz). The NIST-Boulder group updated the technique using electronics with higher frequency capability (51), and have investigated precessional dynamics up to 3 GHz (52). Pump-probe magneto-optical Kerr effect measurements of small-angle magnetization dynamics, demonstrated first on EuS films (53), were extended to metallic ferromagnetic systems (3), also by the NIST-Boulder group. Similar experiments have been carried out on lithographically defined micrometer-size structures, monitored through time-dependent magnetoresistance (54).

4.3 LARGE-ANGLE DYNAMICS: SWITCHING

In this section, we consider the limit of motions of up to 180°. In Section 4.3.1, we develop a model for the quasistatic switching of a single domain with arbitrary in-plane anisotropy axis. We show that a critical field exists for spontaneous switching at any temperature, with an energy barrier for lower fields. In Section 4.3.2, we develop a simple model for thermal reversal rates over the barrier. Thermally activated switching is relevant for the small patterned elements used in MRAM and is essential in some switching schemes. Finally, in Section 4.3.3, we show integrated switching trajectories for single-domain thin film elements and illustrate the effect of damping α.

4.3.1 Quasistatic Limit: Stoner–Wohlfarth Model

In this section, we consider the quasistatic switching behavior of a single-domain magnetic particle with well-defined anisotropy. Note that any behavior discussed here can be described, in greater detail, through direct integration of the LLG equation. The LLG equation reaches equilibrium $\dot{\mathbf{m}} = 0$, where the precessional term is zero. Thus, the torque

$$\tau = \mathbf{m} \times \mathbf{H_{eff}} = 0 \tag{4.30}$$

or, equivalently, magnetization $\mathbf{m}$ and applied effective field $\mathbf{H_{eff}}$ are collinear. This is a useful property in simulation: micromagnetic ground states (for N particles coupled through magnetostatics or otherwise) can be found by integrating the LLG forward to convergence, accelerated by taking unphysically large values of α.

Energy Expression We assume that the particle has some built-in structure, or *anisotropy*, which favors magnetization in a particular direction. This model was first considered by Stoner and Wohlfarth in 1948 (55). The anisotropy can be expressed as a contribution to the energy U_A, which depends on the magnetization orientation ϕ,

$$U_A(\phi) = K_u \sin^2(\phi - \phi_u), \tag{4.31}$$

where ϕ_u is the angle of the preferred direction of the magnetization. The energy is minimized where $\phi = \phi_u$ and maximized where $\phi = \phi_u \pm (\pi/2)$ (hard axis), but is equivalent for $\phi = \phi_u$ and $\phi = \phi_u \pm \pi$ (easy axis). The anisotropy axis thus has no direction, only an orientation. K_u is an anisotropy constant and takes units of erg/cm^3 in cgs units or J/m^3 in SI units.

With the definitions

$$H_K \equiv \frac{2K_u}{M_s}, \quad h_b = \frac{H_b}{H_K}, \quad u = \frac{U}{M_s H_K}, \tag{4.32}$$

the normalized total energy including Zeeman and anisotropy terms is

$$u = -m_z h_b + \frac{1}{4}\left[1 - \cos 2\phi \cos 2\phi_u + \sin 2\phi \sin 2\phi_u\right]. \tag{4.33}$$

Hard and Easy Axes Magnetization In the hard axis case, $\phi_u = (\pi/2)$; in the easy axis case, $\phi_u = 0$. For these two cases, Eq. (4.33) becomes

$$u_{HA} = -m_z h_b + \frac{1}{2} m_z^2, \quad u_{EA} = -m_z h_b + \frac{1}{2}\left(1 - m_z^2\right), \tag{4.34}$$

since $m_z = \cos\phi$. We can find the equilibrium magnetization by taking the first derivative of u with respect to m_z, and find

$$m_z^{HA} = h_b, \quad m_z^{EA} = -h_b, \tag{4.35}$$

which is a stable minimum only for the hard axis case. The energy minimum for the easy axis case must lie on either extremum, $\phi = 0$ or $\phi = \pi$. Clearly, the lowest energy is found for $\phi = 0$ for $h_Z > 0$ and $\phi = \pi$ for $h_Z < 0$.

Switching does not respond simply to the energy minimum. For an initial state $\phi = \pi$ under the application of $0 < h_z < 1$, or $H_b < H_K$, the $\phi = 0$ state becomes lower and lower in energy compared with the $\phi = \pi$ state, but the single domain cannot rotate into it. As illustrated in Fig. 4.2, an energy barrier develops between the two states. Small excursions away from $\varphi = \pi$ only raise the energy. Only when the condition $du/d\phi < 0$ is fulfilled over the whole domain $-1 < h_b < 1$ will the particle switch. This condition becomes true for $h_b \geq 1$ or $H_b > H_K$. For lower fields, thermal activation is required to climb the energy barrier.

General Case: The Switching Astroid Richer behavior emerges if we consider the problem more generally. We would like to consider the critical field

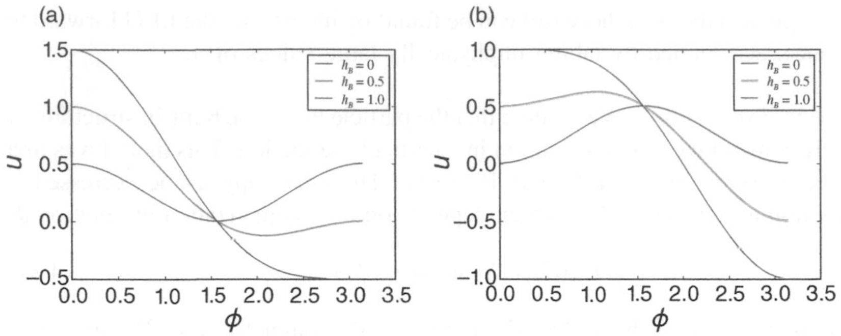

Figure 4.2 Energy landscapes: hard axis (a) and easy axis (b). Note the metastable equilibria at $\phi = \pi$ for the easy axis case if $H_b < 1$. Angles in radians.

for switching h_0. We now allow both the reduced magnetization $\mathbf{m} = \mathbf{M}/M_s$ and the reduced field $\mathbf{h} = \mathbf{H}/H_K$ to take any direction in the x–y plane. We will see that the critical field is a function of angle as well as magnitude, so $\mathbf{h_0}$ is a dividing line in the control plane. Here we follow Thiaville's notation (56). A convenient aspect of this formulation is that it can be used for any form of the in-plane anisotropy energy, given here as $G(\phi)$. The magnetization takes angle ϕ with respect to the x-axis, giving us magnetization vector $\mathbf{m}$ and its orthogonal in-plane vector $\hat{\mathbf{e}} \equiv \hat{\phi}$; this is a different coordinate system than the y–z film plane considered up until now.

$$\mathbf{m}(\phi) = \cos\phi\,\hat{\mathbf{x}} + \sin\phi\,\hat{\mathbf{y}}, \qquad \hat{\phi} = \hat{\mathbf{e}} = -\sin\phi\,\hat{\mathbf{x}} + \cos\phi\,\hat{\mathbf{y}}, \tag{4.36}$$

$$\frac{\partial \mathbf{m}(\phi)}{\partial \phi} = \hat{\mathbf{e}}, \qquad \frac{\partial^2 \mathbf{m}(\phi)}{\partial \phi^2} = -\mathbf{m}. \tag{4.37}$$

To find equilibria, we already know to take

$$\frac{\partial[-\mathbf{m}(\phi)\cdot\mathbf{h_0} + u_A(\phi)]}{\partial \phi} = -\hat{\mathbf{e}}\cdot\mathbf{h_0} + \frac{\partial u_A(\phi)}{\partial \phi} = 0 \tag{4.38}$$

and for stability, $\frac{\partial^2 u}{\partial \phi^2} = 0$.

$$\frac{\partial\left[-\hat{\mathbf{e}}\cdot\mathbf{h}_0 + (\partial u_A(\phi))/\partial\phi\right]}{\partial \phi} = \mathbf{m}\cdot\mathbf{h}_0 + \frac{\partial^2 u_A(\phi)}{\partial \phi^2} = 0. \tag{4.39}$$

After decomposing $\mathbf{h_0} = h_0^m\mathbf{m} + h_0^e\mathbf{e}$, these two criteria give

$$\mathbf{h_0} = -\frac{\partial^2 u_A(\phi)}{\partial \phi^2}\mathbf{m} + \frac{\partial u_A(\phi)}{\partial \phi}\hat{\mathbf{e}} = -G''\mathbf{m} + G'\hat{\mathbf{e}}. \tag{4.40}$$

From here, we can transfer back into the Cartesian system using Eqs. (4.36 and 4.37,

$$\mathbf{h_0} = -(G''\cos\phi + G'\sin\phi)\hat{\mathbf{x}} + (-G''\sin\phi + G'\cos\phi)\hat{\mathbf{y}}. \tag{4.41}$$

If we fix $\hat{\mathbf{x}}$ for the uniaxial anisotropy easy axis, after some algebra, this yields

$$\mathbf{h_0} = -\cos^3\phi\,\hat{\mathbf{x}} + \sin^3\phi\,\hat{\mathbf{y}}. \tag{4.42}$$

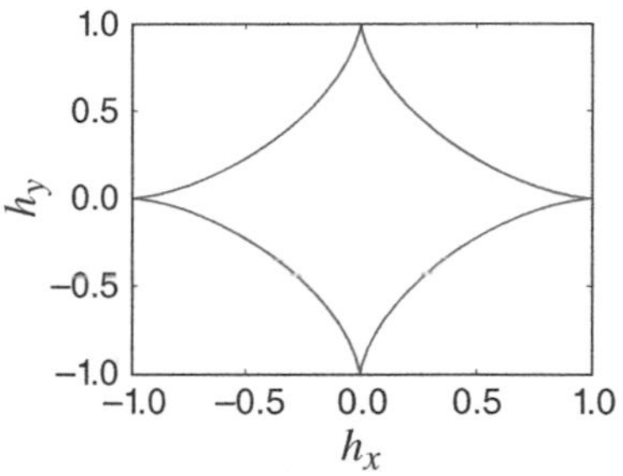

Figure 4.3 The switching astroid described by Eq. (4.42). Magnetization does not become destabilized until the field |h| lies outside the boundary.

For $\phi = 0$ ($m = 1$) and $\phi = \pi$ ($m = -1$), the magnetization lies along the easy axis. In order to destabilize the positive (negative) magnetization $m = \pm 1$, it is necessary to apply an opposite, negative (positive) field $H = \mp H_K$, or $h = \mp 1$. The boundary between switching and no switching, known as the *switching astroid*, is shown in Fig. 4.3.

4.3.2 Thermally Activated Switching

For easy axis switching, we saw that for applied fields large enough, $h > -m$, there is no energy barrier for reversal. Magnetization reversal is continuously energetically favored for rotation of **m** away from $\varphi = 0$. For smaller, subcritical fields $|h| < 1$ directed along the easy axis, an energy barrier exists for switching, as shown in Fig. 4.2. If no thermal energy were available, a particle would never switch for $|h| < 1$. Thermal fluctuations help drive the magnetization over the barrier, and yield a finite, thermally activated switching rate that depends on the applied field and volume of the particle.

In Eq. (4.34), if we take our starting point as $m_z = 1$ ($\phi = 0$) and apply a field h along $-\hat{\mathbf{z}}$ (opposite to the magnetization direction), we find that the energy difference between the initial state and the unstable equilibrium at $m_z = -h$ is

$$\Delta u = u_{eq} - u(\varphi = 0), \quad \Delta u = \frac{1}{2}(1 - h_b)^2. \tag{4.43}$$

The Arrhenius–Néel equation for kinetics (57) states that the frequency ν of overcoming the barrier (and switching) is

$$\nu(\Delta U, V, T) = \nu_0 e^{-\Delta E_b/k_B T}, \quad \Delta E_b = \frac{1}{2} M_s H_K V \left(1 - \frac{H_b}{H_K}\right)^n, \quad n = 2 \tag{4.44}$$

for an energy barrier ΔE_b. Here k_B is the Boltzmann constant, $k_B(300\text{ K}) = 25.86$ meV, and ν_0 is a temperature-independent attempt frequency. The value $n = 2$ obtains for easy axis switching; for the full Stoner–Wohlfarth model with arbitrary orientation of the anisotropy axes (not shown here), $1.5 \leq n \leq 2$. With a characteristic time $\tau = \nu^{-1}$, if a switching experiment is carried out many times, the number of times N the particle does *not* reverse in time t is given by

$$\frac{\partial N}{\partial t} = -\frac{N}{\tau}, \quad N(t) = N_0 e^{-t/\tau} \tag{4.45}$$

for N_0 attempts. The probability $P_s(t)$ for switching after time t is thus

$$P_s(t) = 1 - P_{ns}(t) = 1 - e^{-t/\tau}, \quad \tau = \nu_0^{-1} e^{\Delta E_b/k_B T}, \tag{4.46}$$

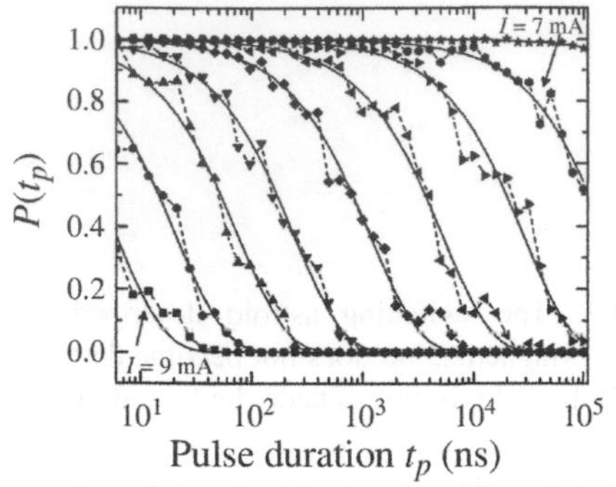

Figure 4.4 Probability $P(t_p)$ of not-switching a magnetic tunnel junction in response to field pulses of width t_p and variable pulse amplitude I. (Reproduced from Ref. 59 with permission from AIP Publishing LLC.)

where ΔE_b was given in Eq. (4.44). Wernsdorfer et al. (58) applied Eq. (4.46) to the low-temperature (0.2–6 K) and low-speed switching of a single 25 nm diameter Co nanoparticle, measured using a SQUID microbridge with temporal resolution of ~250 μs.

Thermal Switching Experiments: MRAM In the context of MRAM, Rizzo and coworkers at Motorola applied Eq. (4.46), $n=2$, to the thermal switching in magnetic tunnel junctions (59) for pulse durations t_p of 5 ns to 100 μs at room temperature. The magnetic state was characterized through the magnetoresistance. Data are shown in Fig. 4.4 for the probability of not-switching after pulse duration t_p $P_{ns}(t_p)$ under the application of field pulses with different values (currents) of H/H_K (given there as i/i_{sw}). Excellent agreement is shown with the single-domain/single-energy-barrier model. Other contemporaneous experiments showed more complicated behavior, with multiple energy barriers indicating thermal activation of domain walls (60). Further experiments characterized the full thermal switching astroid, with arbitrary pulse direction, at constant sweep rate (61).

4.3.3 Switching Trajectory

In this section we will examine a calculated trajectory for switching of a thin-film element. A convenient dimensionless form can be used for integration if we take out a factor $T = \mu_0 \gamma M_s$. We define a dimensionless $t' = T \cdot t$,

$$T^{-1}\begin{bmatrix}\dot{\theta}\\ \dot{\phi}\end{bmatrix} = \begin{bmatrix}\partial\theta/\partial t'\\ \partial\phi/\partial t'\end{bmatrix} = \frac{1}{1+\alpha^2}\begin{bmatrix}\alpha & 1\\ -1/\sin\theta & \alpha/\sin\theta\end{bmatrix}\begin{bmatrix}h_\theta^{eff}\\ h_\phi^{eff}\end{bmatrix}. \tag{4.47}$$

Note that for Permalloy, for example, $t' = 1$ is reached for $t = 64.5$ ps. Fields can be transferred from the rotating frame to the Cartesian frame through

$$\begin{bmatrix}h_\theta\\ h_\phi\end{bmatrix} = \begin{bmatrix}\cos\theta\cos\phi & \cos\theta\sin\phi & -\sin\theta\\ -\sin\phi & \cos\phi & 0\end{bmatrix}\cdot\begin{bmatrix}h_x\\ h_y\\ h_z\end{bmatrix}. \tag{4.48}$$

Example: Influence of α For a useful example, we can take a thin film magnetized in plane, evolving under the influence of an in-plane switching field. To

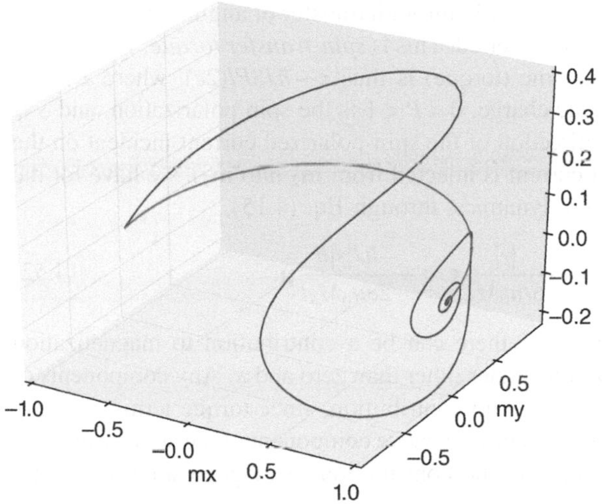

Figure 4.5 Switching on the unit sphere (phase plot, time is implicit) under the influence of $h = +1$, initial condition $m_x = -1$, thin film magnetized in the x–y film plane. The z-axis is taken to be the film normal. Trajectories for $\alpha = 1$ and $\alpha = 0.1$ are shown. Total integration time is $t' = 100$, or ~6 ns

be consistent with the form where the polar axis is $\hat{\mathbf{z}}$, the film plane is x–y, and we have effective fields as follows:

$$h_x = h_b. \quad h_z = -\cos\theta \tag{4.49}$$

$$\begin{bmatrix} h_\theta \\ h_\phi \end{bmatrix} = \begin{bmatrix} h_b \cos\theta\cos\phi + \sin\theta\cos\theta \\ -h_b \sin\phi \end{bmatrix}. \tag{4.50}$$

The expression for direct integration is then given as

$$\begin{bmatrix} \partial\theta/\partial t' \\ \partial\phi/\partial t' \end{bmatrix} = \frac{1}{1+\alpha^2} \begin{bmatrix} \alpha & 1 \\ -1/\sin\theta & \alpha/\sin\theta \end{bmatrix} \begin{bmatrix} h_b \cos\theta\cos\phi + \sin\theta\cos\theta \\ -h_b \sin\phi \end{bmatrix}. \tag{4.51}$$

We show integrated trajectories from $m_x = -1$ to $m_x = +1$ in Fig. 4.5. Notice the influence of α on the convergence toward a new equilibrium direction, very rapid for $\alpha = 1$, and less direct for $\alpha = 0.1$. The effective field from demagnetization keeps the magnetization trajectory more in-plane than out-of-plane.

4.4 MAGNETIZATION SWITCHING BY SPIN-TRANSFER

In this final section, we show how spin-torque terms are added to the LLG equation (Section 4.4.1), write the full LLG equation with spin-torque terms (Section 4.4.2), and show a simple estimate for the spin-torque switching current. Sun has presented a thorough set of calculations of spin-torque switching for single domains, with anisotropy and thermal activation (62).

4.4.1 Additional Terms to the LLG

Spin-polarized currents can act as an additional source for torque on the magnetization (63,64). A spin-polarized electron carries an angular momentum $\hbar$. If the electron is scattered in the FM layer, reversing its spin through transfer of angular momentum to **m**, **m** rotates according to the torque exerted. This is *spin-transfer torque*. The angular momentum transferred per unit time (torque) is thus $\tau \sim \hbar JSP/(2e)$, where J is the current density, e is the electronic charge, $0 < P < 1$ is the spin polarization, and S is the area. Taking $\hat{\mathbf{p}}$ as the *spin* direction of the spin-polarized current incident on the FM (i.e., the direction of $\mathbf{m_2}$ if current is injected from $\mathbf{m_2}$ into $\mathbf{m_1}$), we have for the additional term to magnetization dynamics, through Eq. (4.15).

$$\dot{\mathbf{m}}_{STT} = -\frac{|\gamma|}{St\mu_0 M_s}\boldsymbol{\tau} = -\frac{\hbar J|\gamma|P}{2e\mu_0 M_s t}\hat{\mathbf{p}}. \tag{4.52}$$

Most interesting to note here is that there can be a contribution to magnetization dynamics only where **m** and **p** form angles other than zero and π. Any components of $\dot{\mathbf{m}}_{STT}$ that are parallel to **m** will make no contribution, since torque terms can only rotate the magnetization, $\dot{\mathbf{m}} \cdot \mathbf{m} = 0$. The transverse component of the spin current can be isolated by taking $\mathbf{m} \times \mathbf{m} \times \mathbf{p}$ and the contribution to magnetization dynamics given by spin-torque is

$$\dot{\mathbf{m}} = \cdots + \gamma A_J \,\mathbf{m} \times \mathbf{m} \times \mathbf{p} + \gamma B_J \,\mathbf{m} \times \mathbf{p}, \quad A_J \equiv A \approx \frac{\hbar P}{2e\mu_0 M_s t} J, \tag{4.53}$$

where A_J is the *Slonczewski* spin-torque. A_J is an experimental parameter, proportional to (an odd function of) current density J, with estimate as given. B_J is the *field-like* spin-torque, introduced separately; its dependence on current and voltage is under investigation (65,66). Both terms have units of field (T).

4.4.2 Full-Angle LLG with Spin-Torque

In terms of the magnetization coordinates, Eq. (4.52) can be expressed as

$$\dot{\mathbf{m}} = \cdots + \gamma A_J \,\hat{\mathbf{m}} \times \hat{\mathbf{m}} \times \left(p_\theta \hat{\boldsymbol{\theta}} + p_\phi \hat{\phi}\right) + \gamma B_J \,\hat{\mathbf{m}} \times \left(p_\theta \hat{\boldsymbol{\theta}} + p_\phi \hat{\phi}\right). \tag{4.54}$$

Defining $T = \gamma \mu_0 M_s$, normalized time $t' = T \cdot t$, and normalized spin-torque parameters $A'_J = A_J/M_s$, $B'_J = B_J/M_s$, the single-particle, full-angle LLG with spin-torque can be written as

$$T^{-1}\begin{bmatrix}\dot{\theta}\\ \dot{\phi}\end{bmatrix} = \frac{1}{1+\alpha^2}\begin{bmatrix}\alpha & 1\\ -1/\sin\theta & \alpha/\sin\theta\end{bmatrix}\begin{bmatrix}h_\theta^{eff}\\ h_\phi^{eff}\end{bmatrix}$$
$$+\frac{1}{1+\alpha^2}\begin{bmatrix}-A'_J - \alpha B'_J & \alpha A'_J - B'_J\\ (B'_J - \alpha A'_J)/\sin\theta & (-A'_J - \alpha B'_J)/\sin\theta\end{bmatrix}\begin{bmatrix}p_\theta\\ p_\phi\end{bmatrix}.$$

Nulling of Damping If the effective field and the injected spin polarization, **H** and $\hat{\mathbf{p}}$, are collinear, the effective damping can be made zero through the spin-torque. Damping is expressed in the on-diagonal terms of the right-hand-side of the equation. These cancel in the case of

$$A_J \gtrsim \alpha H. \tag{4.56}$$

Since it is known that A_J is proportional to current, we can express

$$A_J = AJ \geq \alpha H, \quad J \geq \frac{\alpha H}{A}. \tag{4.57}$$

Zero damping implies that equilibrium magnetization becomes unstable to any perturbation. A remarkable result of the nonlinear aspects of the LLG is that a limit cycle, oscillating around the equilibrium, can replace the stable state. The stable precession is the basis of the spin-torque oscillator, a microwave power source driven by dc currents.

For higher values of current density J, the limit cycle becomes destabilized, replaced by stable equilibrium π away. This is *spin-torque switching*. Equation (4.5) suggests that lower current (lower-power) spin-torque switching is favored by materials with lower values of damping α.

Interested readers may refer to Ref. 38 for more details on the physical basis of ferromagnetic relaxation, Ref. 67 on magnetic anisotropies in ultrathin films, Ref. 68 on electrical measurements of magnetization dynamics in submicrometer structures, Ref. 69 on spin-transfer torque, and Ref. 70 on nonlinear magnetization dynamics in nanosystems.

ACKNOWLEDGMENTS

I thank the National Science Foundation (NSF-ECCS 0925829) and *Fondation Nanosciences* for support as well as SPINTEC for hospitality during the preparation of this chapter.

REFERENCES

1. L. Landau and E. Lifshitz, "On the theory of the dispersion of magnetic permeability in ferromagnetic bodies," *Phys. Zeitsch. Sowjet.* **8**, pp. 153–169 (1935). Reprinted and translated as *Ukr. J. Phys.* **53**, pp. 14–22 (2008).
2. H. Brooks, "Ferromagnetic anisotropy and the itinerant electron model," *Phys. Rev.* **58**, pp. 909–918 (1940); doi: 10.1103/PhysRev.58.909.
3. T. Crawford, T. Silva, C. Teplin, and C. Rogers, "Subnanosecond magnetization dynamics measured by the second-harmonic magneto-optic Kerr effect," *Appl. Phys. Lett.* **74**, pp. 3386–3388 (1999); doi: 10.1063/1.123353.
4. R. Liboff, *Introduction to Quantum Mechanics*, vol. 4, Addison-Wesley, Boston, MA, 2002.
5. A. Einstein and W. De Haas, "Experimental proof of Ampere's molecular currents," *Verh. Dtsch. Phys. Ges.* **17**, pp. 152–170 (1915).
6. V. Y. Frenkel, "On the history of the Einstein–de Haas effect," *Sov. Phys. Usp.* **22**, pp. 580–584 (1979); doi: 10.1070/PU1979v022n07ABEH005587.
7. S. J. Barnett, "Magnetization by rotation," *Phys. Rev.* **6**, pp. 239–270 (1915); doi: 10.1103/PhysRev.6.239.
8. S. Barnett and G. Kenny, "Gyromagnetic ratios of iron, cobalt and many binary alloys of iron, cobalt and nickel," *Phys. Rev.* **87**, pp. 723–734 (1952); doi: 10.1103/PhysRev.87.723.
9. T. L. Gilbert and J. M. Kelly,"Anomalous rotational damping in ferromagnetic sheets," in *Conference on Magnetism and Magnetic Materials*, Pittsburgh, PA, June 14–16, 1955, American Institute of Electrical Engineers, New York, October 1955, pp. 253–263.

10. T. Gilbert, "A phenomenological theory of damping in ferromagnetic materials," *IEEE Trans. Magn.* **40**, pp. 3443–3449 (2004); doi: 10.1109/TMAG.2004.836740.
11. W. Saslow, "Landau–Lifshitz or Gilbert damping? That is the question," *J. Appl. Phys.* **105**, 07D315 (2009); doi: 10.1063/1.3077204.
12. A. Meyer and G. Asch, "Experimental g' and g values of Fe, Co, Ni and their alloys," *J. Appl. Phys.* **32**, pp. 330–333 (1961); doi: 10.1063/1.2000457.
13. R. A. Reck and D. L. Fry, "Orbital and spin magnetization in Fe-Co, Fe-Ni, and Ni-Co," *Phys. Rev.* **184**, pp. 492–495, (1969); doi: 10.1103/PhysRev.184.492.
14. Landolt–Börnstein Tables, Chapter III-13: Metals: Phonon States, Electron States, and Fermi Surfaces, Springer, Berlin, 1990, pp. 100–101.
15. Z. Frait and D. Fraitova, "Ferromagnetic resonance and surface anisotropy in iron single crystals," *J. Magn. Magn. Mater.* **15–18**, (Part 2), pp. 1081–1082 (1980); doi: 10.1016/0304-8853(80)90895-1.
16. F. Schreiber, J. Pflaum, Z. Frait, Th. Mühge, and J. Pelzl, "Gilbert damping and g-factor in Fe_xCo_{1-x} alloy films," *Solid State Commun.* **93**, pp. 965–968 (1995); doi: 10.1016/0038-1098(94)00906-6.
17. C. Scheck, L. Cheng, I. Barsukov, Z. Frait, and W. Bailey, "Low relaxation rate in epitaxial vanadium-doped ultrathin iron films," *Phys. Rev. Lett.* **98**, 117601 (2007); doi: 10.1103/PhysRevLett.98.117601.
18. S. Bhagat and P. Lubitz, "Temperature dependence of ferromagnetic relaxation in the 3d transition metals," *Phys. Rev. B* **10**, pp. 179–185 (1974); doi: 10.1103/PhysRevB.10.179.
19. Y. Guan and W. E. Bailey, "Ferromagnetic relaxation in $(Ni_{81}Fe_{19})_{1-x}Cu_x$ thin films: band filling at high Z," *J. Appl. Phys.* **101**, 09D104 (2007); doi: 10.1063/1.2709750.
20. A. Ghosh, J. F. Sierra, S. Auffret, U. Ebels, and W. E. Bailey, "Dependence of nonlocal Gilbert damping on the ferromagnetic layer type in ferromagnet/Cu/Pt heterostructures," *Appl. Phys. Lett.* **98**, 052508 (2011); doi: 10.1063/1.3551729.
21. Y. Li and W. E. Bailey, "Wave-number-dependent Gilbert damping in metallic ferromagnets," *Phys. Rev. Lett.* **116**, 117602 (2014); doi: 10.1103/PhysRevLett.116.117602.
22. C. Bilzer, T. Devolder, J.-V. Kim, G. Counil, C. Chappert, S. Cardoso, and P. P. Freitas, "Study of the dynamic magnetic properties of soft CoFeB films," *J. Appl. Phys.* **100**, 053903 (2006); doi: 10.1063/1.2337165.
23. N. Chan, V. Kambersky, and D. Fraitova, "Impedance matrix of thin metallic ferromagnetic films and SSWR in parallel configuration," *J. Magn. Magn. Mater.* **214**, pp. 93–98 (2000); doi: 10.1016/S0304-8853(99)00776-3.
24. I. I. Rabi, J. R. Zacharias, S. Millman, and P. Kusch, "A new method of measuring nuclear magnetic moment," *Phys. Rev.* **53**, pp. 318–327 (1938); doi: 10.1103/PhysRev.53.318.
25. P. Forman, "Swords into ploughshares: breaking new ground with radar hardware and technique in physical research after World War II," *Rev. Mod. Phys.* **67**, pp. 397–455 (1995); doi: 10.1103/RevModPhys.67.397.
26. E. Zavoisky, "Paramagnetic relaxation of liquid solutions for perpendicular fields," *J. Phys. USSR*, **9**, pp. 211–216 (1945).
27. E. M. Purcell, H. C. Torrey, and R. V. Pound, "Resonance absorption by nuclear magnetic moments in a solid," *Phys. Rev.* **69**, pp. 37–38 (1946); doi: 10.1103/PhysRev.69.37.
28. J. Griffiths, "Anomalous high-frequency resistance in ferromagnetic metals," *Nature* **158**, pp. 670–671 (1946); doi: 10.1038/158670a0.
29. C. Kittel, "Interpretation of anomalous Larmor frequencies in ferromagnetic resonance experiment," *Phys. Rev.* **71**, pp. 270–271 (1947); doi: 10.1103/PhysRev.71.270.2.
30. R. McMichael and P. Krivosik, "Classical model of extrinsic ferromagnetic resonance linewidth in ultrathin films," *IEEE Trans. Magn.* **40**, pp. 2–11 (2004); doi: 10.1109/TMAG.2003.821564.
31. W. Ament and G. Rado, "Electromagnetic effects of spin wave resonance in ferromagnetic metals," *Phys. Rev.* **97**, pp. 1558–1566 (1955); doi: 10.1103/PhysRev.97.1558.
32. K. Gilmore, Y. U. Idzerda, and M. D. Stiles, "Identification of the dominant precession-damping mechanism in Fe, Co, and Ni by first-principles calculations," *Phys. Rev. Lett.* **99**, 027204 (2007); doi: 10.1103/PhysRevLett.99.027204.
33. M. Fähnle, J. Seib, and C. Illg, "Relating Gilbert damping and ultrafast laser-induced demagnetization," *Phys. Rev. B* **82** 144405 (2010); doi: 10.1103/PhysRevB.82.144405.
34. J. Lock, "Eddy current damping in thin metallic ferromagnetic films," *Br. J. Appl. Phys.* **17**, pp. 1645–1647 (1966); doi: 10.1088/0508-3443/17/12/415.

35. C. Scheck, L. Cheng, and W. Bailey, "Low damping in epitaxial sputtered Fe films," *Appl. Phys. Lett.* **88**, 252510 (2006); doi: 10.1063/1.2216031.
36. C. E. Patton, "Linewidth and relaxation processes for the main resonance in the spin-wave spectra of Ni–Fe alloy films," *J. Appl. Phys.* **39**, pp. 3060–3068 (1968); doi: 10.1063/1.1656733.
37. B. Heinrich, K. Urquhart, A. Arrott, J. Cochran, K. Myrtle, and S. Purcell, "Ferromagnetic-resonance study of ultrathin BCC Fe(1 0 0) films grown epitaxially on FCC Ag(1 0 0) substrates," *Phys. Rev. Lett.* **59**, pp. 1756–1759 (1987); doi: 10.1103/PhysRevLett.59.1756.
38. B. Heinrich, "Spin relaxation in magnetic metallic layers and multilayers," in *Ultrathin Magnetic Structures III: Fundamentals of Nanomagnetism*, eds. J. A. C. Bland and B. Heinrich, Chapter 5, Springer, Berlin, 2005, pp. 143–210; doi: 10.1007/3-540-27163-5_5.
39. M. Bendahan, P. Canet, J.-L. Seguin, and H. Carchano, "Control composition study of sputtered Ni-Ti shape memory alloy film," *Mater. Sci. Eng. B* **34**, pp. 112–115 (1995); doi: 10.1016/0921-5107(95)01237-0.
40. C. Kittel, "On the gyromagnetic ratio and spectroscopic splitting factor of ferromagnetic substances," *Phys. Rev.* **76**, pp. 743–748 (1949); doi: 10.1103/PhysRev.76.743.
41. R. Urban, G. Woltersdorf, and B. Heinrich, "Gilbert damping in single and multilayer ultrathin films: role of interfaces in nonlocal spin dynamics," *Phys. Rev. Lett.* **87**, 217204 (2001); doi: 10.1103/PhysRevLett.87.217204.
42. C. A. F. Vaz, J. A. C. Bland, and G. Lauhoff, "Magnetism in ultrathin film structures," *Rep. Prog. Phys.* **71**, 056501 (2008); doi: 10.1088/0034-4885/71/5/056501.
43. L. Néel, "Anisotropie magnétique superficielle et surstructures d'orientation," *J. Phys. Radium*, **15**, pp. 225–239 (1954); doi: 10.1051/jphysrad:01954001504022500.
44. G. C. Bailey and C. Vittoria, "Presence of magnetic surface anisotropy in Permalloy films," *Phys. Rev. B* **8**, pp. 3247–3251 (1973); doi: 10.1103/PhysRevB.8.3247.
45. J. O. Rantschler, P. J. Chen, A. S. Arrott, R. D. McMichael, J. W. F. Egelhoff, and B. B. Maranville, "Surface anisotropy of Permalloy in NM/NiFe/NM multilayers," *J. Appl. Phys.* **97**, 10J113 (2005); doi: 10.1063/1.1853711.
46. P. Gambardella, S. Rusponi, M. Veronese, S. S. Dhesi, C. Grazioli, A. Dallmeyer, I. Cabria, R. Zeller, P. H. Dederichs, K. Kern, C. Carbone, and H. Brune, "Giant magnetic anisotropy of single cobalt atoms and nanoparticles," *Science* **300**, pp. 1130–1133 (2003); doi: 10.1126/science.1082857.
47. J.-M. Beaujour, J. Lee, A. Kent, K. Krycka, and C.-C. Kao, "Magnetization damping in ultrathin polycrystalline Co films: evidence for nonlocal effects," *Phys. Rev. B* **74**, 214405 (2006); doi: 10.1103/PhysRevB.74.214405.
48. J. P. Nibarger, R. Lopusnik, Z. Celinski, and T. J. Silva, "Variation of magnetization and the Landé *g* factor with thickness in Ni-Fe films," *Appl. Phys. Lett.* **83**, pp. 93–95 (2003); doi: 10.1063/1.1588734.
49. Y. Tserkovnyak, A. Brataas, G. Bauer, and B. Halperin, "Nonlocal magnetization dynamics in ferromagnetic heterostructures," *Rev. Mod. Phys.* **77**, pp. 1375–1421 (2005); doi: 10.1103/RevModPhys.77.1375.
50. P. Wolf, "Free oscillations of the magnetization in Permalloy films," *J. Appl. Phys.* **32**, pp. S95–S96 (1961); doi: 10.1063/1.2000514.
51. T. Silva, C. Lee, T. Crawford, and C. Rogers, "Inductive measurement of ultrafast magnetization dynamics in thin-film Permalloy," *J. Appl. Phys.* **85**, pp. 7849–7862 (1999); doi: 10.1063/1.370596.
52. A. B. Kos, T. J. Silva, and P. Kabos, "Pulsed inductive microwave magnetometer," *Rev. Sci. Instrum.* **73**, pp. 3563–3569 (2002); doi: 10.1063/1.1505657.
53. M. R. Freeman, "Picosecond pulsed-field probes of magnetic systems," *J. Appl. Phys.* **75**, pp. 6194–6198 (1994); doi: 10.1063/1.355454.
54. S. E. Russek, S. Kaka, and M. Donahue, "High-speed dynamics, damping, and relaxation times in submicrometer spin-valve devices," *J. Appl. Phys.* **87**, pp. 7070–7073 (2000); doi: 10.1063/1.372934.
55. E. C. Stoner and E. P. Wohlfarth, "A mechanism of magnetic hysteresis in heterogeneous alloys," *Philos. Trans. R. Soc. A* **240**, pp. 599–642 (1948); doi: 10.1098/rsta.1948.0007.
56. A. Thiaville, "Extensions of the geometric solution of the two-dimensional coherent magnetization rotation model," *J. Magn. Magn. Mater.* **182**, pp. 5–18 (1998); doi: 10.1016/S0304-8853(97)01014-7.
57. L. Néel, "Some theoretical aspects of rock-magnetism," *Adv. Phys.* **4**, pp. 191–243 (1955); doi: 10.1080/00018735500101204.

58. W. Wernsdorfer, E. B. Orozco, K. Hasselbach, A. Benoit, B. Barbara, N. Demoncy, A. Loiseau, H. Pascard, and D. Mailly, "Experimental evidence of the Néel–Brown model of magnetization reversal," *Phys. Rev. Lett.* **78**, pp. 1791–1794 (1997); doi: 10.1103/PhysRevLett.78.1791.
59. N. D. Rizzo, M. DeHerrera, J. Janesky, B. Engel, J. Slaughter, and S. Tehrani, "Thermally activated magnetization reversal in submicron magnetic tunnel junctions for magnetoresistive random access memory," *Appl. Phys. Lett.* **80**, pp. 2335–2337 (2002); doi: 10.1063/1.1462872.
60. R. H. Koch, G. Grinstein, G. A. Keefe, Y. Lu, P. L. Trouilloud, W. J. Gallagher, and S. S. P. Parkin, "Thermally assisted magnetization reversal in submicron-sized magnetic thin films," *Phys. Rev. Lett.* **84**, pp. 5419–5422 (2000); doi: 10.1103/PhysRcvLctt.84.5419.
61. J. Z. Sun, J. C. Slonczewski, P. L. Trouilloud, D. Abraham, I. Bacchus, W. J. Gallagher, J. Hummel, Y. Lu, G. Wright, S. S. P. Parkin, and R. H. Koch, "Thermal activation-induced sweep-rate dependence of magnetic switching astroid," *Appl. Phys. Lett.* **78**, pp. 4004–4006 (2001); doi: 10.1063/1.1379596.
62. J. Z. Sun, "Spin-current interaction with a monodomain magnetic body: a model study," *Phys. Rev. B* **62**, pp. 570–578 (2000); doi: 10.1103/PhysRevB.62.570.
63. J. Slonczewski, "Current-driven excitation of magnetic multilayers," *J. Magn. Magn. Mater.* **159**, pp. 1–7 (1996); doi: 10.1016/0304-8853(96)00062-5.
64. Y. B. Bazaliy, B. Jones, and S.-C. Zhang, "Modification of the Landau–Lifshitz equation in the presence of a spin-polarized current in colossal- and giant-magnetoresistive materials," *Phys. Rev. B* **57**, pp. R3213–R3216 (1998); doi: 10.1103/PhysRevB.57.R3213.
65. S. Petit, C. Baraduc, C. Thirion, U. Ebels, Y. Liu, M. Li, P. Wang, and B. Dieny, "Spin-torque influence on the high-frequency magnetization fluctuations in magnetic tunnel junctions," *Phys. Rev. Lett.* **98**, 077203 (2007); doi: 10.1103/PhysRevLett.98.077203.
66. S. Petit, N. de Mestier, C. Baraduc, C. Thirion, Y. Liu, M. Li, P. Wang, and B. Dieny, "Influence of spin-transfer torque on thermally activated ferromagnetic resonance excitations in magnetic tunnel junctions," *Phys. Rev. B* **78** 184420 (2008); doi: 10.1103/PhysRevB.78.184420.
67. M. Farle, "Ferromagnetic resonance of ultrathin metallic layers," *Rep. Prog. Phys.* **61**, pp. 755–826 (1998); doi: 10.1088/0034-4885/61/7/001.
68. S. E. Russek, R. D. McMichael, M. J. Donahue, and S. Kaka, "High speed switching and rotational dynamics in small magnetic thin film devices," in *Spin Dynamics in Confined Magnetic Structures II*, Topics in Applied Physics, eds. B. Hillebrands and K. Ounadjela, vol. 87, Springer, Berlin, 2003, pp. 93- 154.
69. D. Ralph and M. Stiles, "Spin transfer torques," *J. Magn. Magn. Mater.* **320**, pp. 1190–1216 (2008); doi: 10.1016/j.jmmm.2007.12.019.
70. G. Bertotti, I. D. Mayergoyz, and C. Serpico, *Nonlinear Magnetization Dynamics in Nanosystems*, Elsevier Science, Amsterdam, 2008.

CHAPTER 5

MAGNETIC RANDOM-ACCESS MEMORY

Bernard Dieny[1,2,3] *and I. Lucian Prejbeanu*[1,2,3]
[1]INAC-SPINTEC, Université Grenoble Alpes, Grenoble, France
[2]INAC-SPINTEC, CEA, Grenoble, France
[3]SPINTEC, CNRS, Grenoble, France

5.1 INTRODUCTION TO MAGNETIC RANDOM-ACCESS MEMORY (MRAM)

5.1.1 Historical Perspective

Magnetism has been used for a long time to store information in the form of a magnetization orientation in magnetic materials. This has been used in tape recording since 1928, in magnetic hard disk drives since 1956, and in a number of magnetic solid-state memory devices, starting with magnetic core memory (1955–1975), bubble memory in the 1970s and early 1980s, followed by several magnetoresistive memory technologies. The latter were first based on the anisotropic magnetoresistance phenomenon (variation of resistivity of a few percent at room temperature in magnetic transition metals as a function of the angle between magnetization and current) and subsequently on the giant magnetoresistance (GMR) effect, following its discovery in 1988 (magnetoresistance amplitude in the range 15–20% in GMR structures of practical interest).

The interest in magnetic random-access memory (MRAM) was renewed after the first successful attempts in 1995 to fabricate magnetic tunnel junctions (MTJs) that exhibit large tunneling magnetoresistance (TMR) amplitude of several tens of percentages at room temperature using amorphous AlO_x barriers (see Chapters 1 and 2). Besides their larger magnetoresistance amplitude, MTJs are more suitable than GMR metallic structures for memory applications due to their larger impedance (adjustable to several kilohms), which allows an easier integration with complementary metal–oxide–semiconductor (CMOS) components.

Two other breakthroughs further boosted research and development in MRAM. The first was the discovery that MTJs based on crystalline MgO barriers associated with crystalline magnetic electrodes exhibit much larger TMR amplitude, in the range

Introduction to Magnetic Random-Access Memory, First Edition. Edited by Bernard Dieny, Ronald B. Goldfarb, and Kyung-Jin Lee.

of 150–600% at room temperature (1,2), than their counterparts based on amorphous alumina tunnel barriers. This larger TMR provides a much improved read margin and faster read in memory devices. The second was the possibility to switch the magnetization of a magnetic nanostructure by a spin-polarized current, thanks to the spin-transfer torque (STT) effect. This effect was first measured in metallic structures (3) and later in magnetic tunnel junctions (4). It provided a new write scheme in MRAM with much better downsize scalability than in the first generations of MRAM, as will be explained later in this chapter. Currently, most of research and development in MRAM are focused on STT-MRAM with perpendicular anisotropy (i.e., the magnetization in the ferromagnetic electrodes is oriented out of the plane of the layers) since STT-MRAM seems to be the most promising in terms of scalability down to and beyond the 16 nm technological node. Thermal assistance during the write process is an additional factor also used in other magnetic storage technology, in particular in heat-assisted magnetic recording (HAMR) in hard disk drive technology. Thermal assistance allows magnetization to switch easily during the write process while achieving a very good thermal stability of the magnetization in standby mode. Besides extending the downsize scalability, this technology also offers new functionalities (see Section 5.6).

5.1.2 Various Categories of MRAM

Since 1996, various families of MRAM have been developed, benefiting from the progress in spintronics research, particularly the giant TMR of MgO tunnel junctions and the STT phenomenon (see Fig. 5.1). All of these MRAM use MTJs as elementary

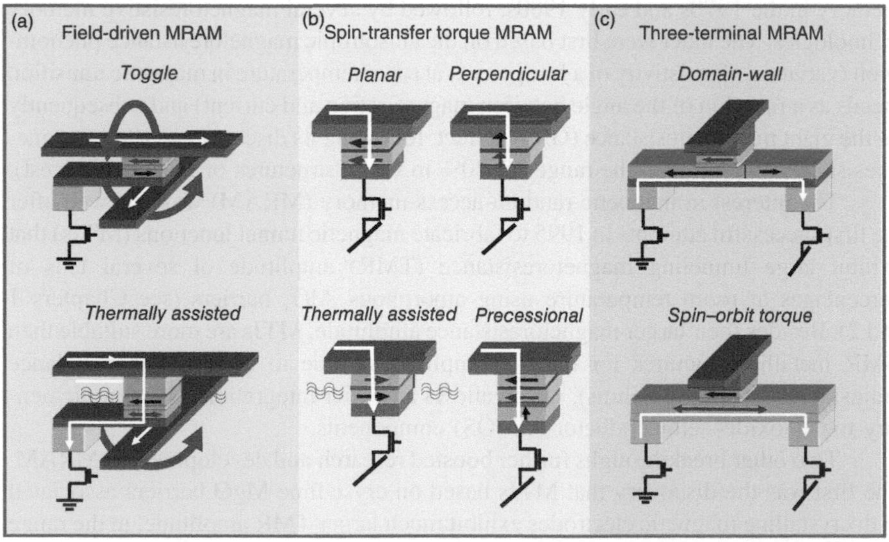

Figure 5.1 Various categories of MRAM developed since 1996. (a) This column corresponds to MRAM written by magnetic fields. (b) This column illustrates MRAM based on STT writing. (c) This column contains three-terminal MRAM based on domain wall propagation or spin–orbit torque switching.

storage cells. MTJs essentially consist of two ferromagnetic layers (typically 1–2.5 nm thick) separated by a thin insulating barrier (typically 1–1.5 nm thick). The resistance of the MTJ in the direction perpendicular to the plane of the layers depends on the relative orientation of the magnetization in these two ferromagnetic layers. The amplitude of the resistance change between the antiparallel magnetic configuration (high resistance state) and the parallel magnetic configuration (low resistance state) defines the TMR amplitude. In most MRAM embodiments, the magnetization of one of the ferromagnetic layers is fixed in a predetermined direction to provide a reference direction for the spin of the electrons. This layer is called the reference layer. The magnetization of the other layer, called the storage layer, can be switched between two stable states: Its magnetization can be set either parallel or antiparallel to that of the reference layer. As illustrated in Fig. 5.1, the magnetization in the ferromagnetic electrodes can lie in the plane of the layers or point perpendicular to the plane depending on the particular choice of materials. For small devices, typically below 50 nm, MTJs with perpendicular magnetization are preferred because their higher magnetic anisotropy provides better stability of the written information (see Section 5.2). However, these perpendicular-to-the-plane magnetized materials are generally more difficult to grow than the in-plane magnetized materials.

In most embodiments, each MTJ is connected in series with a selection transistor used as a switch, which controls the current through the MTJ. During read, a moderate current is sent through the MTJ (corresponding to a bias voltage across the MTJ on the order of 0.1–0.2 V). The change of resistance between parallel and antiparallel magnetic configurations allows the magnetic state of the storage layer to be sensed and, therefore, the stored information to be read (see Section 5.3). The various families of MRAM displayed in Fig. 5.1 differ by the way the information is written in the memory.

Between 1996 and 2004, most research and development focused on MRAM written by field (Fig. 5.1a). Until the discovery of STT switching and its gradual implementation in MTJ after 2004, the only known way to manipulate the magnetization of a magnetic nanostructure (here the MTJ storage layer) was indeed with use of a magnetic field. The magnetic field is created by pulses of current flowing in conducting lines located below and above the MTJ. These approaches, which will be described in more detail in Section 5.4, actually resulted in the commercialization of the first MRAM products (1, 4, 8, and 16 Mbit MRAM chips) by Freescale Semiconductor and its spin-off Everspin Technologies in 2006. An extension of field-written MRAM is represented by thermally-assisted MRAM (TA-MRAM) mainly developed by Crocus Technology. In the latter, the write selectivity is achieved by a combination of temporary heating of the selected cell produced by the tunneling current flowing through the cell and a single pulse of magnetic field. The power consumption to write these memory elements is significantly reduced compared to conventional field-written MRAM, thanks to the possibility of using lower magnetic fields and of sharing each field pulse among several cells so as to write several bits at once (see Section 5.6). Field-written technology is robust and is already used in a variety of applications where reliability, endurance, and resistance to radiation are important features, such as in automotive and space applications. However, the downsize scalability provided by field-writing in conventional technology is limited to MTJ dimensions on the order of 60 nm × 120 nm due to electromigration in the conducting lines used to generate the field. In addition, in field-writing, the write field extends all along the conducting line where it is produced and decreases relatively gradually in space, inversely proportional to the distance to this line. As a result, unselected bits adjacent to selected bits may sense a significant fraction of the write field, which may yield accidental switching of these unselected bits. In TA-MRAM, smaller dimensions can be reached, thanks to the field sharing approach, the lower required amplitude of the write field, and the different mechanism of write selectivity (see Section 5.6).

Since the first observation of STT-induced switching in GMR metallic spin-valve pillars (3), the interest in using STT as a new write approach in MRAM has increased, motivated by the fact that STT-writing (middle column in Fig. 5.1) offers a much better downsize scalability than field-writing. This is due to the fact that the critical current required to write by STT decreases in proportion to the cell area, whereas it tends to increase in the case of field-writing. Furthermore, STT provides very good write selectivity since the STT current flows through only the selected cells. As will be explained in Section 5.5, the greatest interest in STT-MRAM is now focused on out-of-plane magnetized STT-MRAM because it requires significantly less write current than its in-plane counterparts for a given criterion of memory retention.

Thermal assistance can also be combined with the STT write approach. In this case, one takes advantage of the Joule heating produced by the STT write current flowing through the MTJ tunnel barrier to facilitate the magnetization switching (see Section 5.6). This allows circumvention of a classical difficulty in data storage: the trade-off between the memory writability and retention.

A third category of MRAM under research and development is represented in Fig. 5.1c. These are three-terminal MRAM cells (see Section 5.7). The purpose of these embodiments is to separate the write and read current paths and thereby increase the reliability of the memory. These approaches open new possibilities in the design of the memory architecture or for logic applications. The writing is here based on current induced domain wall propagation or another phenomenon called spin–orbit -torque (see Section 5.7).

The two following sections discuss two common functions of the various categories of MTJ-based MRAM shown in Fig. 5.1: the storage function (Section 5.2) and the read function (Section 5.3).

5.2 STORAGE FUNCTION: MRAM RETENTION

5.2.1 Key Role of the Thermal Stability Factor

In memory applications, a key characteristic is the retention of the memory, that is, how long the memory chip is capable of keeping the information that has been written in it. The specification depends on the application but is, for instance, on the order of 10 years for mass storage application such as in hard disk drives (HDD). In MRAM, the information may get corrupted by unintended switching of the magnetization of the storage layer due to thermal fluctuations. The failure rate in an MRAM chip of N bits in standby mode can be estimated as follows. The magnetization of the storage layer of the memory cell can be described as a bistable system, the two stable states being separated by an energy barrier ΔE. ΔE is determined by the magnetic material's properties and the shape and dimensions of the magnetic element, that is, the MTJ storage layer, as is explained immediately below. At a temperature T, the characteristic thermally activated switching time is given by an Arrhenius law:

$$\tau = \tau_0 \exp\left(\frac{\Delta E}{k_B T}\right), \tag{5.1}$$

where k_B is the Boltzmann constant and τ_0 is an attempt time of the order of 1 ns (see Chapter 3). For a given bit, the probability of not having accidentally switched after a time t is

$$P_{noswitch}(t) = \exp(-t/\tau). \tag{5.2}$$

For N bits, the probability for the set of N bits of not having experienced any switching event after a time t is

$$P^N_{noswitch}(t) = [P_{noswitch}(t)]^N = \exp(-Nt/\tau). \tag{5.3}$$

Consequently, the probability of having experienced at least one switching event after a time t, that is, the failure rate in standby mode, is given by

$$F(t) = 1 - \exp(-Nt/\tau) = 1 - \exp\left[\frac{-Nt}{\tau_0}\exp\left(-\frac{\Delta E}{k_B T}\right)\right]. \tag{5.4}$$

This expression clearly shows that the factor $\Delta = \Delta E/k_B T$, often called the thermal stability factor, plays a key role in the failure rate of MRAM chips in standby mode, that is, memory retention failure. Figure 5.2 shows the variation of the failure rate

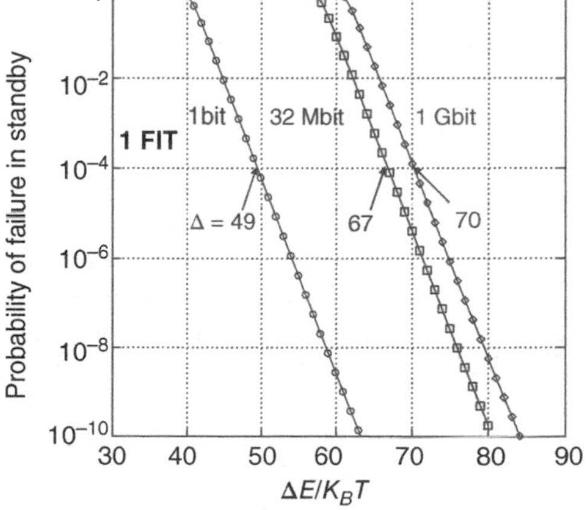

Figure 5.2 Failure rate during 10 years in standby mode for MRAM chips of 1 bit, 32 Mbit, or 1 Gbit as a function of thermal stability factor.

during 10 years in standby mode (not during a write or read operation) as a function of the thermal stability factor (**Δ**) for a 32 Mbit and a 1 Gbit MRAM chip. In order for the probability of experiencing one failure in time (FIT) during 10 years in standby mode to be below an acceptable level of 10^{-4} (this number depends on whether the application is memory or logic, and on the possible use of error correction codes), the thermal stability factor must be greater than 67 for the 32 Mbit chip and greater than 70 for the 1 Gbit chip. The higher the memory capacity, the larger the thermal stability factor has to be.

5.2.2 Thermal Stability Factor for In-Plane and Out-of-Plane Magnetized Storage Layer

In magnetic materials, the barrier height ΔE, which determines the thermal stability factor, is most often created by a magnetic anisotropy that can have different origins. This anisotropy can be of magnetocrystalline origin or can be due to stress or to some electronic hybridization effects taking place at the interfaces in magnetic multilayers. It can be also due to the shape of the patterned magnetic element: a magnetic nanostructure having an elongated shape has an easy axis of magnetization along the long dimension of the structure (see Chapter 3). Whatever the origin of this anisotropy, it can generally be described by an energy per unit volume $E_{anisotropy} = -K(\hat{n} \cdot \hat{M})^2$, where $\hat{n}$ is a unit vector along the easy axis of magnetization and $\hat{M}$ is a unit vector parallel to the magnetization direction. The barrier height separating the two opposite stable states along the easy axis direction is then given by $\Delta E = KV$, where V is the volume of the magnetic nanostructure, that is, the volume of the storage layer in a considered memory cell.

Commonly used in-plane magnetized materials in MRAM have in general a relatively weak magnetocrystalline anisotropy. Their magnetic anisotropy mostly originates from their shape. The MTJs, and in particular their storage layers, are patterned in the form of elliptical elements. With the notation given in Fig. 5.3a, the barrier height is given by

$$\Delta E = \frac{1}{2}\mu_0 (N_y - N_x) M_s^2 V, \tag{5.5}$$

where N_x and N_y are the storage layer demagnetizing coefficients, respectively, along the short and long axis of the ellipse, M_s is the storage layer magnetization, and μ_0 is the vacuum permeability equal to $4\pi\ 10^{-7}$ H/m (SI).

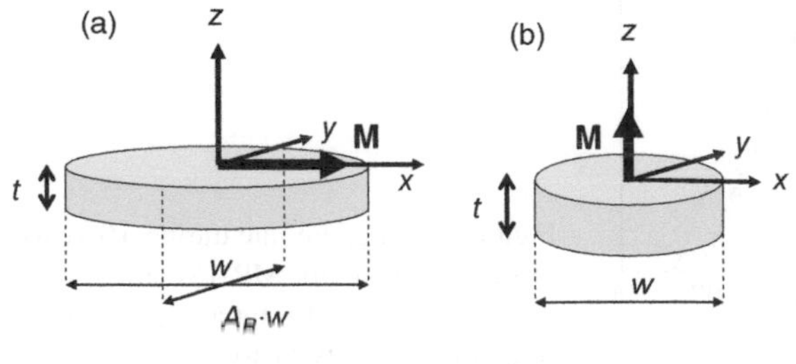

Figure 5.3 Schematic representation of a storage layer with its characteristic dimensions. (a) Case of an in-plane magnetized storage layer with an elliptical prism shape. (b) Case of an out-of-plane magnetized storage layer having the shape of a flat cylinder.

Unfortunately, there are no simple expressions for the demagnetizing coefficients N_x and N_y of an elliptical cylinder as a function of its characteristic dimensions (thickness t, width w, length $A_R \cdot w$, where A_R is the in-plane aspect ratio). These coefficients are functions of elliptic integrals and must be tabulated or numerically calculated (5,6). Let's mention that for elliptical cylinders with small t, the structure is often approximated as a uniformly magnetized ellipsoid, for which closed-form solutions exist. Qualitatively, for the purpose of understanding the general influence of the storage layer dimensions, the thermal stability factor Δ of in-plane magnetized storage layer can be approximated by

$$\Delta = \frac{\Delta E}{k_B T} = \frac{\mu_0}{2} \frac{M_s^2 t^2 (A_R - 1) w}{k_B T}. \tag{5.6}$$

The Δ values given by Eq. (5.6) only differs by a few percent from those provided by the more exact expressions from Ref. 6. Equation (5.6) shows that for in-plane magnetized system, Δ scales with the lateral dimension w (typically the technological node), aspect ratio as $(A_R - 1)$, square of saturation magnetization M_s,, and square of thickness t. Practically, aspect ratios between 1.5 and 2.5 are used. For larger aspect ratio, the magnetization switching no longer occurs via a coherent rotation of the magnetization (i.e., as a whole magnetic block) but proceeds via nucleation of a reversed domain and propagation of domain wall. Therefore, there is no benefit in terms of thermal stability of the magnetization in increasing the aspect ratio above ~2.5. From Eq. (5.6), one can estimate that the smallest dimensions that an in-plane CoFeB storage layer ($M_s \approx 10^6$ A/m) 2.5 nm thick can have while keeping a thermal stability factor above 70 is of the order of 60 nm × 150 nm.

Perpendicular-to-plane magnetized storage layer can in general have better thermal stability than their in-plane counterparts at smaller dimensions because of their much larger intrinsic anisotropy. This is actually one of the reasons why the hard disk drive industry switched from in-plane magnetized magnetic media to perpendicular media in 2004. The perpendicular magnetic anisotropy (PMA) in thin film magnetic materials can have different origins. It can originate from the bulk of the material as in FePt or FePd $L1_0$ ordered alloys due to a tetragonal distortion of the lattice cell along the growth direction. This bulk anisotropy is usually described by a characteristic energy per unit volume (in J/m^3)

$$E_{anisotropy}^{Bulk} = -K_V (\hat{n} \cdot M)^2, \tag{5.7}$$

where $\hat{n}$ is a unit vector normal to the plane of the layer. The PMA may also be of interfacial origin due to interfacial electron hybridization or interfacial stress (for instance at Co–Pd, Co–Pt, or CoFe–MgO interfaces, see Section 5.5.5.3). This interfacial PMA is usually expressed in terms of a volume energy (in J/m^3) as

$$E_{anisotropy}^{Interface} = -\frac{K_S}{t} \left(\hat{n} \cdot \hat{M}\right)^2, \tag{5.8}$$

where K_S is the surface energy (J/m^2) and t the thickness of the storage layer. However, this PMA is partially counterbalanced by the demagnetizing energy, that is, the cost of magnetostatic energy required to pull the magnetization of a thin film out of the plane. Indeed, from a pure magnetostatic viewpoint, the thickness of a magnetic thin film is most often its smallest dimension. The direction along the thickness is therefore usually the hard axis of magnetization (in terms of magnetostatics). If the magnetization is nevertheless pulled out of plane by other sources of anisotropy, these other sources must overcome the demagnetizing energy. For an extended thin film, the demagnetizing energy per unit volume is written as

$$E_{demagnetizing} = +\frac{1}{2}\mu_0 M_s^2 (\hat{n}\cdot\hat{M})^2. \tag{5.9}$$

In perpendicular STT-MRAM, the MTJ and therefore its storage layer is usually patterned in the form of a flat cylinder, since there is no longer a need for in-plane shape anisotropy to insure thermal stability of the magnetization. A cylindrical shape allows an increase in areal density. If the width of the cell is larger than 50 nm, the demagnetizing energy expression Eq. (5.9) is a sufficiently good approximation. However, for smaller dimensions, it is necessary to take into account the exact values of N_x, N_y, N_z demagnetizing coefficients. Since $N_x + N_y + N_z \approx 1$ (see Chapter 3) and $N_x = N_y$ due to the cylindrical shape, the demagnetizing energy is expressed as

$$E_{demagnetizing} = +\frac{1}{4}\mu_0 (3N_z - 1) M_s^2 (\hat{n}\cdot\hat{M})^2. \tag{5.10}$$

The expression of N_z can be found for instance in Ref. 6 for flat disks. It can be approximated to 1 for storage layer of diameter larger than ~50 nm.

The total magnetic anisotropy energy per unit volume of the patterned storage layer is then the sum of these various anisotropy contributions:

$$\begin{aligned} E_{anisotropy} &= E_{anisotropy}^{Bulk} + E_{anisotropy}^{Interface} + E_{demagnetizing} \\ &= -\left[K_V - \frac{1}{4}\mu_0 (3N_z - 1) M_s^2 + \frac{K_s}{t}\right](\hat{n}\cdot\hat{M})^2. \end{aligned} \tag{5.11}$$

For the storage layer magnetization to remain stable in the out-of-plane direction, the term

$$K_{eff} = K_V - \frac{1}{4}\mu_0 (3N_z - 1) M_s^2 + \frac{K_s}{t} \tag{5.12}$$

called the effective anisotropy, must be positive, meaning that the bulk and interfacial perpendicular anisotropy must exceed the demagnetizing energy. To estimate the thermal stability factor of the storage layer, a cylindrical storage layer of diameter w in a perpendicular MTJ, one must consider which magnetization switching process prevails. As explained in Chapter 3, in out-of-plane magnetized nanostructures of dimensions typically >40 nm, magnetization reversal tends to proceed by nucleation of a reversed domain, most often at the edge of the nanostructures and propagation of the domain wall throughout the nanostructure. In this case, the barrier height for reversal is determined by the energy required for the nucleation of the reversed domain. In contrast, for smaller structures (typically <30 nm in diameter), the magnetization does not have enough space to nucleate a reversed domain and consequently switches as a rigid magnetic block (often called a "macrospin"). In this macrospin approximation, which is valid for technological node below ~30 nm, the thermal stability factor of a cylindrical storage layer of diameter w in a perpendicular MTJ is expressed as

$$\Delta = \frac{\Delta E}{k_B T} = \frac{\left[(K_V - (1/4)\mu_0(3N_z - 1)M_s^2)t + K_s\right]\frac{\pi}{4}w^2}{k_B T}. \tag{5.13}$$

As an example, for a CoFeB storage layer, 1.2 nm thick, with typical CoFeB/MgO interfacial anisotropy of 1.2×10^{-3} J/m^2 (see Section 5.5.3), and a magnetization on the order of 1000 kA/m: $\Delta = 107$ for $w = 30$ nm, $\Delta = 77$ for $w = 25$ nm, $\Delta = 52$ for $w = 20$ nm, and $\Delta = 32$ for $w = 15$ nm. A 10-year retention for a MTJ with $w \leq 20$ nm therefore requires an increase in the interfacial anisotropy, a decrease in the storage layer thickness while maintaining its TMR amplitude, or a decrease in its saturation magnetization.

For larger structures, as mentioned above, the magnetization switching proceeds by nucleation of a reversed domain and propagation of a domain wall. In this case, the thermal stability factor weakly depends on the cell area as illustrated in the right part of Fig. 5.4. Figure 5.4 clearly shows the crossover from the macrospin regime (coherent switching of the storage layer magnetization) at cell width below ~30 nm to the regime of nucleation of reversed domain and propagation of domain wall above ~40 nm. High density p-STT-MRAM (perpendicular, that is, out-of-plane magnetized STT-MRAM) cells of sub-25 nm diameter will likely switch in the macrospin switching mode, also called the coherent rotation regime.

Finally, another factor that can influence the thermal stability factor of a MRAM cell is the presence of a static field, which, depending on its direction relative to the storage layer magnetization, can further stabilize or on the contrary destabilize the magnetization. Of particular relevance is the stray field from the reference layer. To minimize it, synthetic antiferromagnetic reference layers are often used. These consist of two ferromagnetic layers antiferromagnetically coupled through a Ru spacer layer (see Chapter 2). Thanks to the partial compensation of the magnetic moment of these

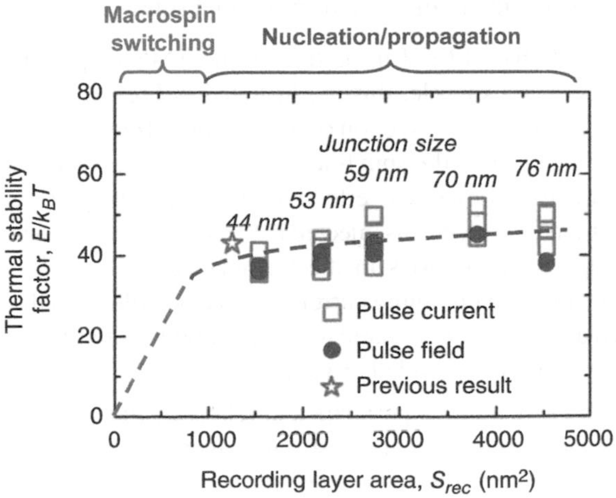

Figure 5.4 Experimentally measured thermal stability factor of the storage layer in perpendicular MTJ of composition Ta 5 nm/$Co_{20}Fe_{60}B_{20}$ 0.9 nm (reference layer)/MgO 0.9 nm/$Co_{20}Fe_{60}B_{20}$ 1.5 nm (storage layer)/Ta 5 nm. (Adapted from Ref. 7 with permission from AIP Publishing LLC.)

two antiparallel coupled layers (i.e., a path of flux closure), their overall stray field on the storage layer is reduced. Quantitatively, if H_s is the static field acting on the storage layer magnetization, then the thermal stability factor is enhanced or reduced as

$$\Delta_{H_s} = \Delta \left(1 \pm \frac{H_s}{H_k^{eff}}\right)^2, \tag{5.14}$$

where H_k^{eff} is the effective anisotropy field related to the effective anisotropy by (8) $H_k^{eff} = 2(K_{eff}/(\mu_0 M_s))$. K_{eff} is mainly determined by the shape anisotropy for in-plane magnetized systems and is given by Eq. (5.12) for out-of-plane magnetized systems. The static field H_s acting on the storage layer can be determined by performing a minor hysteresis loop measurement on the storage layer. In the presence of a nonzero H_s, the storage layer hysteresis loop is shifted along the field axis by the quantity H_s (see Figure 2.13).

5.3 READ FUNCTION

5.3.1 Principle of Read Operation

The general principle of the read operation in MTJ-based MRAM consists in exploiting the change of resistance between P and AP magnetic configurations to determine the magnetic state of the junction and therefore the written information. This principle is illustrated in Fig. 5.5. During read, a read current flows through the MTJ to sense its magnetic state from the value of the cell resistance. Usually, the output resistance value is compared to a reference cell whose resistance is the average value of the low- and high-resistance state (Fig. 5.5c). To discriminate between the two possible MTJ states in a fast and reliable way, the cell-to-cell distribution of low resistance and high-resistance states must be separated by at least $6(\sigma_{low} + \sigma_{high})$, where σ_{low} and σ_{high} are, respectively, the half-width of the distributions of low- and high-resistance states over an MRAM chip (Fig. 5.5b).

During read, the read current is chosen so that the voltage across the MTJ during read is in the range 0.1–0.2 V. This choice is motivated by two reasons:

1. In MTJs in general and in MgO-based MTJ in particular, the TMR amplitude decreases with bias voltage. As an example, this is illustrated in Fig. 5.6. This TMR decrease versus V is explained by a reduction of the spin polarization as the bias voltage increases due to the fact that the spin filtering mechanism associated with the symmetry of the wave becomes less effective at higher voltage (see Chapters 1 and 2). Also, hot tunneling electrons injected in the receiving electrode after tunneling generate magnetic excitations in the ferromagnetic electrodes, which can contribute to depolarize the tunneling current. With read voltage in the range of 0.1–0.2 V, the decrease of TMR compared to the maximum amplitude at very low voltage is no more than 10% in relative value.
2. The second reason for which the reading is performed at relatively low voltage is to avoid spin-transfer torque disturbance of the storage layer magnetic state by the read current as explained in the following section.

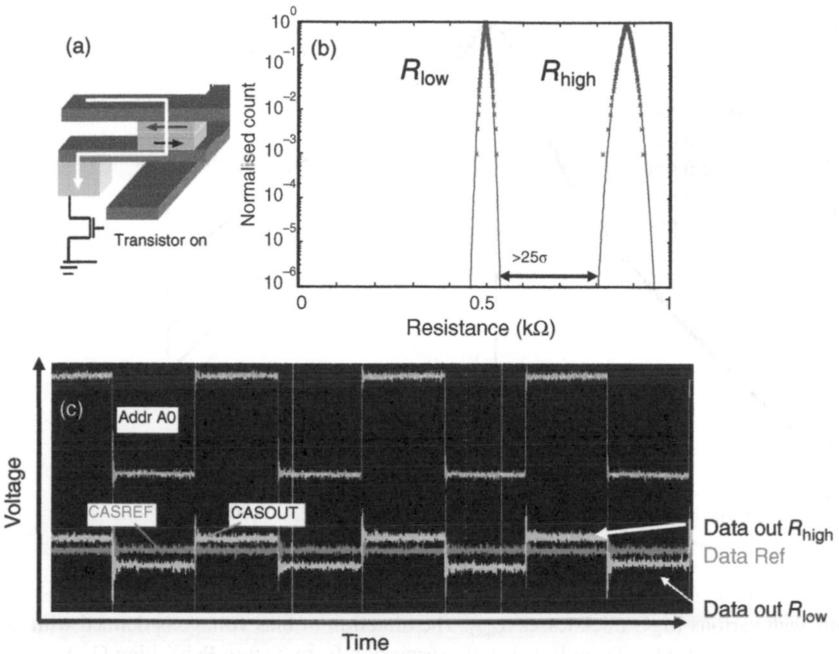

Figure 5.5 Principle of read operation. (a) During read, a current flows through the MTJ to determine the resistance state. (b) The distributions of R_{low} corresponding to P state and R_{high} corresponding to AP state must be sufficiently separated to allow fast and reliable discrimination between the two memory states. Here, data from Crocus Technology on a 1 Mbit TA-MRAM chip in 130 nm technology. (c) Example of readout signal from the same 1 Mbit chip as in part (b): top curve, address signal switching between two cells in opposite magnetic states; low curves, comparison of the readout signal from the read cells with a reference cell having an intermediate resistance $(R_{low} + R_{high})/2$.

5.3.2 STT-Induced Disturbance of the Storage Layer Magnetic State During Read

Since during read a current flows throughout the MTJ, it exerts a STT on the storage layer magnetization very similar to the write current in STT-MRAM. In order to avoid writing during read, the read current must be low enough compared to the STT critical current for switching. The failure rate during read associated with accidental STT switching of the storage layer magnetization during read can be estimated as follows. Let I_0^{STT} indicate the critical current for switching by STT. The barrier height for magnetization switching under a read current $I_{read} < I_0^{STT}$ is decreased due to the STT influence according to

$$\Delta E = \Delta E_0 \left(1 - \frac{I_{read}}{I_0^{STT}}\right)^{\delta}, \quad (5.15)$$

where ΔE_0 is the barrier height at zero current, as given by Eqs. (5.5) or (5.13). The exponent δ is usually taken equal to 1. However, recent experimental studies and

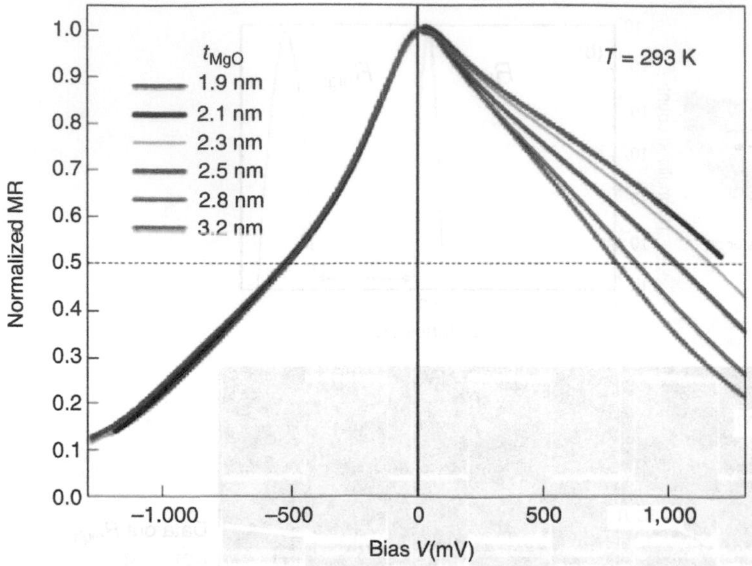

Figure 5.6 Dependence of the TMR amplitude in Fe(0 0 1)/MgO(0 0 1)/Fe(0 0 1) tunnel junctions with various MgO thicknesses t_{MgO}. The direction of bias voltage is defined with respect to the top electrode. (From Ref. 1 with permission from Nature Publishing Group.)

theories based on the Fokker–Planck equation (9) have concluded that δ should be taken equal to 2. With the same notation as for expression (5.4), this yields the following failure rate during read:

$$F(t) = 1 - \exp(-Nt/\tau) = 1 - \exp\left[\frac{-Nt}{\tau_0}\exp\left(-\frac{\Delta E}{k_B T}\left(1 - \frac{I_{read}}{I_0^{STT}}\right)^{\delta}\right)\right]. \qquad (5.16)$$

Figure 5.7 shows how the read current has to be adjusted with respect to the STT critical current for switching depending on the storage layer thermal stability factor and on the required failure rate. In STT-MRAM, the write voltage on the order of 0.5 V and the read voltage on the order of 0.15 V are typically used, corresponding to a I_{read}/I_0^{STT} ratio between 1/3 and 1/4.

5.4 FIELD-WRITTEN MRAM (FIMS-MRAM)

Two categories of field-induced magnetic switching MRAM (FIMS-MRAM) are described in this section. These are MRAM based on the Stoner–Wohlfarth model and an improved version: the “toggle” MRAM, which reached the market in 2006.

5.4.1 Stoner–Wohlfarth MRAM

The Stoner–Wohlfarth MRAM (SW-MRAM) was the first developed category of MTJ-based MRAM. The research and development in this area has been very useful

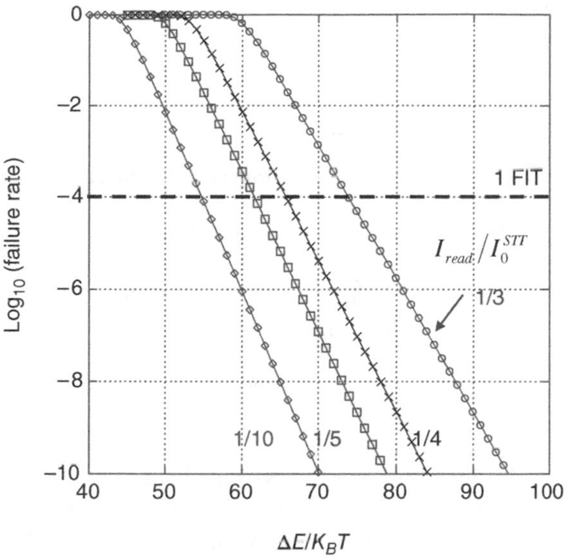

Figure 5.7 Calculated failure rate during read as a function of the thermal stability factor in standby for several values of the I_{read}/I_0^{STT} ratio (assumption $\delta = 1$). The plot assumes that 8 bits are read simultaneously during 10% of the time for 10 years. The dashed line corresponds to an example situation where one FIT would happen in 10 years with a probability of 10^{-4}.

in starting the development of hybrid CMOS/MTJ technology but it did not yield a product because of write selectivity problems and poor downsize scalability, as explained below.

SW-MRAM consists of an array of MTJs in which each MTJ is connected in series with a selection transistor (Fig. 5.8). The MTJs are sandwiched between two

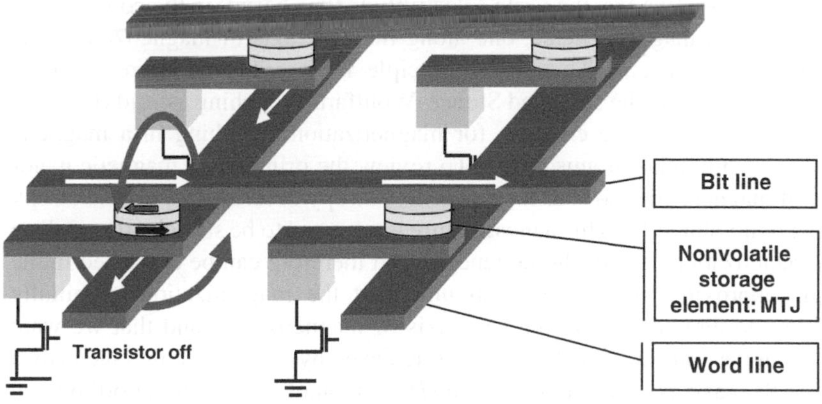

Figure 5.8 Schematic representation of a SW-MRAM array. The MTJs are patterned as elliptical cylinders. They are in-plane magnetized with one layer of fixed magnetization pinned along the ellipse long axis ("reference layer," black arrow) and one layer of switchable magnetization having two stable states along the ellipse long axis ("storage layer," red arrow). To address the memory element located at the front left of the array, two pulses of current (represented by white arrows) are simultaneously sent in the bit line and word line that cross each other at the addressed memory point. These pulses generate two perpendicular magnetic fields (represented by orange arrows), which add as two vectors at the addressed memory point.

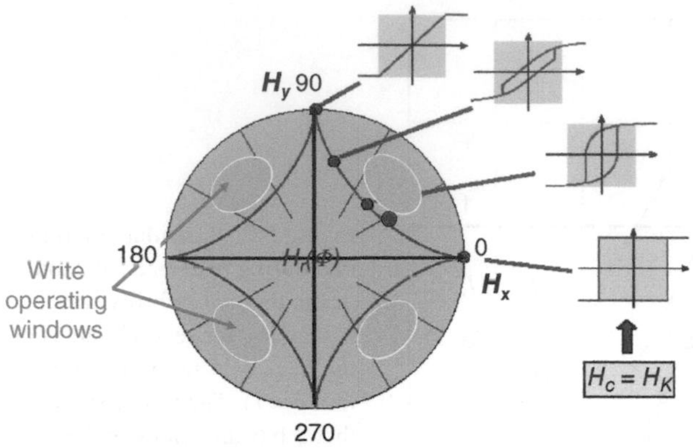

Figure 5.9 Stoner–Wohlfarth astroid (in red). Examples of hysteresis loops measured for various directions of field in the (x, y) plane. The operating windows for SW-MRAM are indicated in green.

sets of orthogonal conducting lines (bit lines and word lines) aimed at creating local magnetic fields on the MTJ's storage layer when current flows are sent along them. To write at a particular addressed cell, two simultaneous pulses of current are sent in the bit line and world line, which cross each other at the addressed MTJ cell. These currents must be adjusted so that the resulting field at the addressed cell is locally large enough to switch its storage layer magnetization in the desired direction while not switching the storage layer magnetization in the other memory points located further along the same bit line or the same world line. Indeed these other memory cells also feel the field created by the current pulse but not the two perpendicular fields simultaneously. They are called half-selected bits.

In these SW-MRAMs, the write selectivity is thus based on the combination of two orthogonal magnetic fields, one along the easy axis of magnetization, the other along the hard axis. This write principle is described in more detail in Chapter 4. It is based on the so-called Stoner–Wohlfarth switching astroid (Fig. 5.9) that provides a quantitative criterion for magnetization switching in a magnetic nanostructure with uniaxial anisotropy. To review the principle, a magnetic nanostructure of magnetization M_s has a uniaxial anisotropy described by an anisotropy energy per unit volume K_u. This nanostructure is assumed to be sufficiently small so that its magnetization remains homogeneous and therefore can be described in the macrospin approximation. We now assume that the magnetization is initially oriented in one direction along the easy axis of magnetization and that we want to switch it in the opposite direction. This is achieved by applying simultaneously a field along the easy axis of magnetization (H_x) and another one in the orthogonal direction, that is, along the hard axis of magnetization (H_y). It was demonstrated that in this situation, the condition for switching is that the total field vector of components (H_x, H_y) must fall out of the Stoner–Wohlfarth astroid, which is defined by the relationship

$$H_x^{2/3} + H_Y^{2/3} = \left(\frac{2K_u}{M_s}\right)^{2/3}. \tag{5.17}$$

This sets a lower limit to the amplitude of the write field in a SW-MRAM chip.

In order to avoid half-selected bits to switch, the easy axis field must be lower than the anisotropy field H_k given by $H_K = (2K_u/M_s)$, otherwise this field alone would switch all bits located along the corresponding word line. Thus, taking into account the lower limit set by the SW astroid and upper limit set by the anisotropy field, this defines the ideal operating window for SW-MRAM. However, in practical devices, several factors actually restrain the size of this operating window:

1. The SW astroid in elliptic MTJ of typical dimensions 100 nm × 200 nm is often distorted due to micromagnetic distortions of the magnetization.
2. The switching criterion given by the SW astroid is actually valid only at 0 K. For devices operating at ambient temperature, thermal activation can significantly assist the magnetic switching so that the operating window must be pushed further away from the theroretical astroid.
3. Due to variability in the patterning process, cell-to-cell distributions in effective anisotropy field lead to cell-to-cell distribution in the shape and size of the SW astroid. As a result of these phenomena, it was very difficult to find any write operating window in SW-MRAM chips, even those of moderate capacity (e.g., 1 Mbit). That is, no write conditions exist in which all bits could be properly written without writing any half-selected bits.

Fortunately, a solution to this problem was found by Savtchenko et al. from Freescale Technologies called "toggle writing (10)."

5.4.2 Toggle MRAM

5.4.2.1 Toggle Write Principle Toggle MRAM are also written with magnetic fields but the structure of the MTJ storage layer and the synchronization of the two orthogonal pulses of magnetic field differ from SW-MRAM. The magnetization in the MTJ ferromagnetic layers is still in-plane and the cells are patterned in elliptical shape, providing a uniaxial shape anisotropy with easy axis along the long axis of the ellipse. In toggle MRAM, the ellipses are oriented at 45° with respect to the bit lines and word lines to optimize the write process, as explained below. Concerning the storage layer composition, the latter consists of a compensated synthetic antiferromagnet, that is, two ferromagnetic layers of same magnetic moment antiferromagnetically coupled through a thin Ru spacer layer. At rest, the magnetic moments of these two layers lie antiparallel along their easy axis of magnetization. In toggle-MRAM, the resistance state is determined by the magnetic orientation of the magnetic layer, which is in contact with the tunnel barrier.

When a moderate field is applied to such a structure, a magnetic transition takes place at a field called "spin–flop field" between a configuration at low fields, wherein the magnetic moments of the two layers lie antiparallel along their common anisotropy axis, and a configuration above the spin–flop field, where they lie symmetrically with respect to the field direction in a scissor configuration (see Fig. 5.10). This scissor configuration is called the spin–flop configuration. As a result, at low fields, the system has no net magnetic moment, whereas above the spin–flop field, it acquires a net magnetic moment.

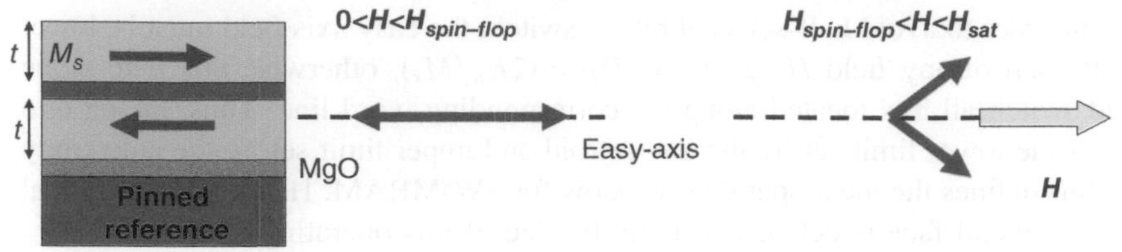

Figure 5.10 Schematic explanation of the spin–flop transition that takes place in the synthetic antiferromagnetic storage layer of a toggle MRAM cell.

The spin–flop field $H_{spin-flop}$ is given by the following expression (11):

$$\mu_0 M_s H_{spin-flop} = 2\sqrt{K_{eff}\left(\frac{A}{t} + K_{eff}\right)}, \tag{5.18}$$

where M_s (A/m) is the saturation magnetization of the two ferromagnetic layers (assumed here to be identical), t (m) is their thickness, K_{eff} (J/m^3) is their effective anisotropy per unit volume mainly of shape origin, and A (J/m^2) is the amplitude of the interfacial antiferromagnetic coupling through the Ru spacer. Therefore, the spin–flop field can be adjusted by varying the cell aspect ratio, which determines K_{eff} or the ferromagnetic layers thickness or the Ru thickness, which determines the amplitude of the antiferromagnetic coupling. Typical values of this spin–flop field in toggle MRAM are on the order of $\mu_0 H_{spin-flop} = 5$ mT. The write toggle operation then proceeds as represented in Fig. 5.11.

In the initial state, the two ferromagnetic layers are in antiparallel magnetic configuration along the long axis of the elliptical cell. At t_1, a current is sent in the x-line generating a magnetic field on the storage layer in the y-direction. This field is typically on the order of 7–10 mT, above the spin–flop transition. The two magnetizations then scissor in the direction of this field and get in spin–flop configuration. The applied field is then gradually rotated by two steps of 45°. This is achieved by applying simultaneously a current along the x-line and y-line between t_2 and t_3, then only along the y-line between t_3 and t_4. During these steps, the spin–flop magnetic configuration rotates with the field. The y-current is then stopped. Since no more magnetic field is applied, the magnetization of the ferromagnetic magnetic layers then

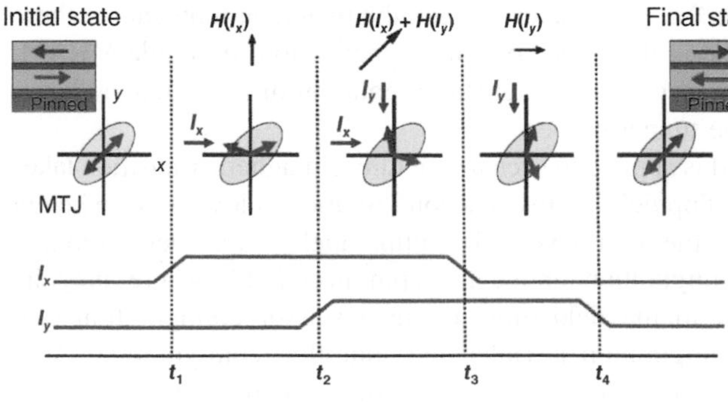

Figure 5.11 Toggle write operation sequence. (After Ref. 12 with permission from IEEE.)

relax back to the antiparallel configuration along their easy axis of magnetization. In the final state, both layers have rotated by 180°. This writing process takes between 10 and 20 ns (12).

In order to maximize the field produced by a given current in the word and bit lines and better confine the field at the memory element, these lines are coated on their faces opposite to the memory element with a soft magnetic material (cladding) (13). This material gets polarized by the Ampere field created by the current flowing in the line and helps focus the field on the storage layer. Typically, a factor 2 gain in field amplitude is obtained, thanks to the cladding. The application of this sequence of current pulses systematically rotates the storage layer magnetization by 180° whatever the initial state of the system (i.e., independent of whether a "0" or a "1" is written in the memory element). Therefore, it is necessary to read before write to determine whether the magnetic state has to be changed or not.

5.4.2.2 Improved Write Margin Toggle writing provides a much wider operation window than SW writing as illustrated in Fig. 5.12 on a 4 Mbit toggle chip from Everspin Technologies. In toggle writing, the energy barrier to switch half-selected bits is many times larger than for switching fully selected bits (14). Thanks to these improved write performances, Everspin Technologies succeeded in launching the first MRAM products in 2006 (15).

5.4.2.3 Applications of Toggle MRAM Toggle MRAM have many interesting features that combine the advantages of static random-access memory (SRAM) and Flash memory. Toggle MRAM is nonvolatile as in the case of Flash, with a projected retention of more than 10 years. Their read–write cycle is on the order of 35 ns. They can work in applications at various ranges of temperature: commercial (0–70 °C), industrial (−40–85 °C), extended (−40–105 °C), and automotive AEC-Q100 Grade 1 (−40–125 °C). They occupy a relatively small surface area, for a capacity of up to 16 Mbit in one chip. They combine the functions of several types of memory: Flash, SRAM, EEPROM, nvRAM, and battery-backed SRAM. Their radiation hardness makes them interesting for space and avionic applications. Toggle MRAM is already

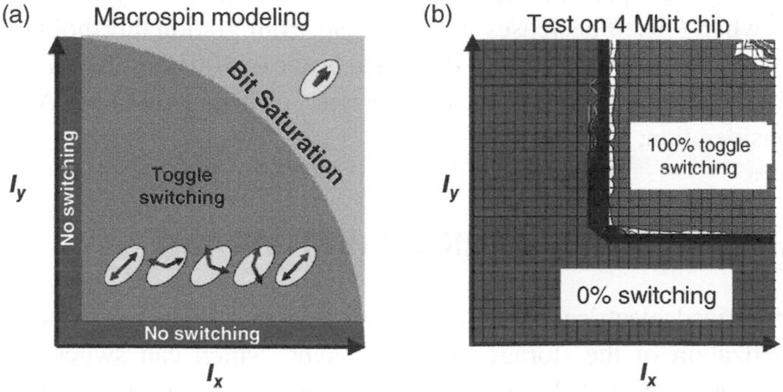

Figure 5.12 Toggle operation windows determined from macrospin modeling (a) and from testing a 4 Mbit chip (b). (After Ref. 12 with permission from IEEE.)

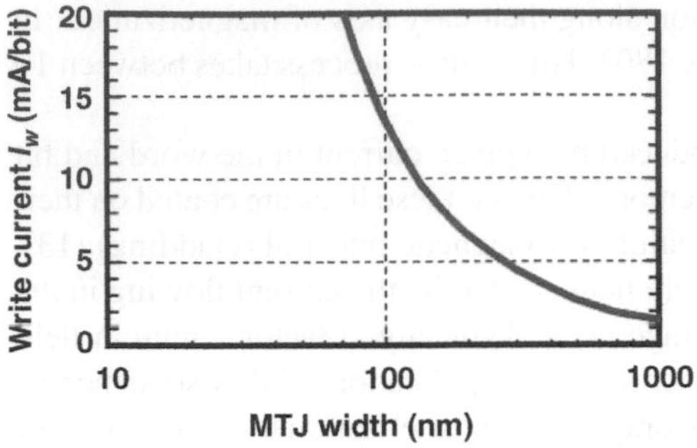

Figure 5.13 Write current versus MTJ short axis dimension (aspect ratio is typically between 2 and 2.5). The write current is here defined as the sum: $I_w = I_{word\ line} + I_{bit\ line}$.

used in a variety of applications: microcontrollers, automotive industry, motorcycles, slot machines, satellites, airplanes, and so on. However, their power consumption remains relatively large due to the large current required to generate the pulses of magnetic field (~10 mA).

5.4.3 Limitation in Downsize Scalability

In field-written MRAM, downsize scaling is mainly limited by electromigration taking place in the field-generation word and bit lines. This can be understood as follows: In order to ensure sufficient memory retention, the volume of the storage layer must fulfill the relationship $K_{eff}V > \sim 70\, k_BT$, in which V is the volume of the storage layer and K_{eff} is its effective anisotropy (see Section 5.2.2). As the feature size F decreases, the volume of the storage layer decreases as F^2 so that the effective anisotropy must be increased as F^{-2}. In toggle writing, the spin–flop field given by Eq. (5.18) roughly scales as $(K_{eff})^{1/2}$, that is, as F^{-1}. This spin–flop field is the lower limit of the write field in toggle MRAM. Because the distance between the center of the field-generating line and the storage layer does not vary significantly as the technology shrinks, the current required to generate the write field is proportional to the write field amplitude and therefore roughly scales as F^{-1}, as illustrated in Fig. 5.13. In terms of current density, if only the width of the field generating lines decreases while their height is approximately constant, the current density in these lines increases as F^{-2} when F itself decreases. Since the electromigration threshold in Cu is $\sim 10^7$ A/cm^2, this limits the MTJ width to dimensions above 100 nm. Therefore, the downsize scalability of toggle MRAM and of other types of field-written MRAM is limited by this consideration.

5.5 SPIN-TRANSFER TORQUE MRAM (STT-MRAM)

In STT writing, a polarized electrical current exerts a torque called spin-transfer torque on the magnetization of the storage magnetic layer, which can switch its direction for a large enough current density. For high density and fast read–write operation, STT-MRAM provides a good alternative to FIMS-MRAM. The STT-MRAM cell is simplified in comparison to FIMS-MRAM cell since no field lines are required. Each cell consists of only a MTJ connected in series with a selection transistor. The writing is performed with bipolar pulses of current. The reading is performed at lower current to avoid write errors during read.

Writing a "0" (i.e., a parallel configuration of the magnetization in the storage and pinned layers) can be achieved by sending a current pulse through the stack, the electrons flowing from the pinned layer to the storage layer. Writing a "1" can be achieved by sending a pulse of current of opposite polarity. This approach clearly solves the write selectivity problem of the SW-MRAM since the write current flows only through the addressed cell so there is no risk of writing an unselected cell. Another major advantage of the STT write approach is the downsize scalability. Indeed, the critical current for which the magnetization switches due to the STT influence is determined by a critical current density (3) so that the total current required to write a memory cell scales as the area of the cell, assuming a constant free layer thickness. As a result, the STT writing approach offers much better downsize scalability than FIMS-MRAM. However, at very small dimensions, there are still problems concerning the thermal stability of the information written in the cell, as explained in Section 5.4.1. These limit the downsize scalability of in-plane magnetized STT-MRAM down to about 60 nm × 150 nm and for out-of-plane magnetized STT-MRAM to dimension on the order of 20 nm as explained further.

5.5.1 Principle of STT Writing

When a spin-polarized current flows through a magnetic nanostructure, the STT results from the interaction between the spin of the conduction electrons and those responsible for the nanostructure magnetization. This torque is exerted on the local magnetization and tends to switch it toward a direction parallel or antiparallel to that of the spin polarizing layer depending on the current direction. The physical origin of the STT was explained in detail in Chapter 1. It can be summarized as follows: Each conduction electron has angular momentum $\hbar/2$. Considering that the current has a finite spin polarization (typically between 40 and 80%), and taking into account the details of the interactions between the conduction electrons and the local magnetization, yields a STT efficiency coefficient η $(0<\eta<1)$ so that the average angular momentum carried by each electron becomes $\eta(\hbar/2)$. In a multilayered pillar geometry, the number of electrons per unit time and unit area flowing perpendicular to the plane of the layers is given by J/e (J being the current density and e the absolute value of the electron charge) so that the corresponding flow of angular momentum per unit area is $\eta(J/e)(\hbar/2)$. When such a spin-polarized incoming flow of electrons penetrates in a magnetic layer, the spin of these electrons gets reoriented in the direction parallel to the local magnetization within ~1 nm. This corresponds to an interfacial absorption of the incoming flow of transverse angular momentum. By conservation of overall angular momentum, this absorption results in a torque exerted on the local magnetization, the spin-transfer torque given by

$$\Gamma_{STT} = -\gamma\eta\frac{J}{e}\frac{\hbar}{2}\frac{1}{\mu_0 M_s d}\hat{m}\times(\hat{m}\times\hat{p}), \tag{5.19}$$

where $\hat{m}$ and $\hat{p}$ are unit vectors, respectively, along the magnetic nanostructure magnetization and incoming spin polarization (often set by the direction of magnetization of the reference layer in a MTJ), d is the nanostructure thickness along the current direction, M_s its magnetization, γ is the gyromagnetic ratio, and μ_0 the vacuum permeability. This expression was first established by Slonczewski in 1996 (16,17). This term is often called Slonczewski torque or in-plane torque (since it lies in the plane defined by $\hat{m}$ and $\hat{p}$) or damping-like torque (since it has the same form as the damping term in the LLG equation, see below). In magnetic tunnel junctions, it was shown that the STT also introduces a second term called field-like term, but which has an amplitude of only 10–25% of the Slonczewski STT torque. This second term most often is considered as having a negligible role in the design of STT-MRAM (18).

As explained in Chapter 2, taking into account the STT, the general equation governing the dynamics of magnetization of the magnetic nanostructure is written as

$$\frac{d\hat{m}}{dt} = -\gamma\left(\hat{m} \times \vec{H}_{eff}\right) + \alpha\left(\hat{m} \times \frac{d\hat{m}}{dt}\right) - \frac{\gamma\hbar}{2\mu_0 M_s}\frac{1}{d}\frac{\eta J}{e}\hat{m} \times (\hat{m} \times \hat{p}). \tag{5.20}$$

In this equation, an extension of the famous Landau–Lifshitz–Gilbert (LLG) equation (see Chapter 3), the first term describes a precessional motion of the magnetization around the local effective field $\vec{H}_{eff}$, which contains contributions from the applied field, the demagnetizing field, the anisotropy field, and the STT field-like term (if it is taken into account). This first term is conservative, that is, conserves energy. The second term is the Gilbert damping term that describes the magnetic dissipation in the system, that is, the fact that in the absence of STT influence, the magnetization tends to gradually relax toward the local effective field. This term always dissipates energy. The third term is the STT previously discussed. Interestingly, this term is also nonconservative. Furthermore, depending on the current direction through the magnetic nanostructure (e.g., the storage layer in a STT-MRAM cell), this term can absorb or dissipate energy, which means that it either behaves as a damping or antidamping term. If the current has the proper direction and the current density is large enough, the STT antidamping effect can exceed the natural Gilbert damping. Very peculiar magnetization dynamics effects then arise, such as STT-induced magnetization switching (3,4) or magnetization steady-state oscillations (19). STT offers a new way to manipulate the magnetization of magnetic nanostructures. STT-induced magnetization switching provides a new way to write the information in MRAM or logic devices. STT-induced steady-state magnetic oscillations allow the generation of RF voltage and thus new types of frequency-tunable RF oscillators.

The possibility to use the STT to switch the magnetization of a free layer in a magnetoresistive stack was first demonstrated in metallic pillars called spin-valves[3] traversed by a current flowing perpendicular to the plane of the layers. The latter comprise a thick magnetic layer of fixed magnetization acting as a current polarizer and a second magnetic layer of switchable magnetization that reacts to the STT influence from the fixed layer magnetization. These two magnetic layers are magnetically decoupled, thanks to a thin, nonmagnetic, metallic spacer, typically Cu 3–5 nm thick. The current density required to switch the magnetization in these metallic pillars was on the order of 2×10^7 A/cm^2. A few years later, thanks to the progress made in

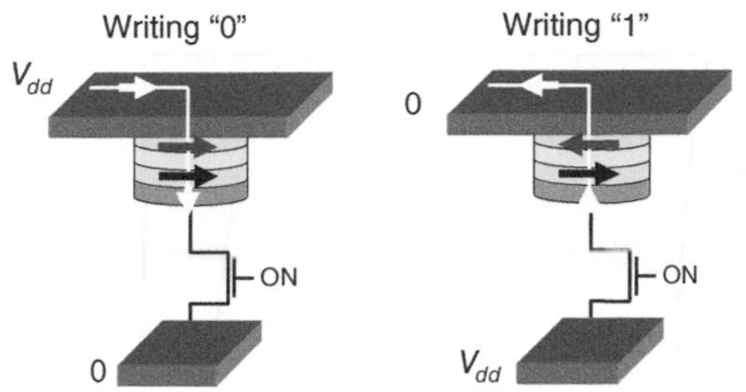

Figure 5.14 Principle of writing in STT-MRAM. Each STT-MRAM cell consists of an MTJ connected in series with a transistor. To write the parallel magnetic configuration, a current flow of density larger than $j_{cAP \to P}$ is sent through the MTJ from the storage layer (red arrow) to the pinned reference layer (black arrow) (i.e., electrons tunnel from the pinned reference layer to the free layer). To write the antiparallel magnetic configuration, a current flow of density larger than $j_{cP \to AP}$ is sent through the MTJ from the reference layer (red arrow) to the storage layer.

the development of low *RA* (resistance × area product) magnetic tunnel junctions, magnetization switching induced by STT was also observed (4). Interestingly, the current density required to write in MTJ was lower than in metallic pillars (in the range 2–6 × 10^6 A/cm^2 in MTJ) mainly due to a higher effective spin polarization in MTJ compared to spin-valves. Since then, the interest in STT in MTJs for STT-MRAM applications has kept on increasing. STT indeed provides a powerful write scheme in MRAM for several reasons:

- In STT-MRAM there is no need to create pulses of magnetic field. Each cell is directly written by the current flowing through the stack. As a result, the cells are much more compact than in field-written MRAM, as illustrated in Fig. 5.14.
- During write of a selected cell, the corresponding selection transistor in series with the MTJ is closed so that the write current flows only through the selected cell. This provides excellent write selectivity, much better than in field-written MRAM.
- As explained in Chapter 1, in STT writing, the condition for magnetization switching is set by a critical current density j_c (actually two critical current densities, $j_c^{P \to AP}$ and $j_c^{AP \to P}$, as will be explained later). The magnetization of the storage layer switches if the current density of proper direction exceeds j_c. This provides very good downsize scalability since the total current required to write scales like the cell area down to very small dimensions where it becomes limited by the thermal stability factor.

One can define switching voltages associated with the critical switching current densities. They are given by $V_c^{P \to AP} = (RA)_P(V) j_c^{P \to AP}$ and $V_c^{AP \to P} = (RA)_{AP}(V) j_c^{AP \to P}$, where *RA* is voltage-dependent resistance × area product of the MTJ in the corresponding magnetic configuration. Interestingly, in MTJ, whereas the critical current $j_c^{P \to AP}$ can be typically twice the critical current $j_c^{AP \to P}$ due to a weaker STT efficiency in P than in AP configuration, the critical voltage for switching from P to AP and from AP to P: $V_c^{P \to AP}$ and $V_c^{AP \to P}$ are very close in absolute values due to the larger *RA* value in the AP than in the P configuration.

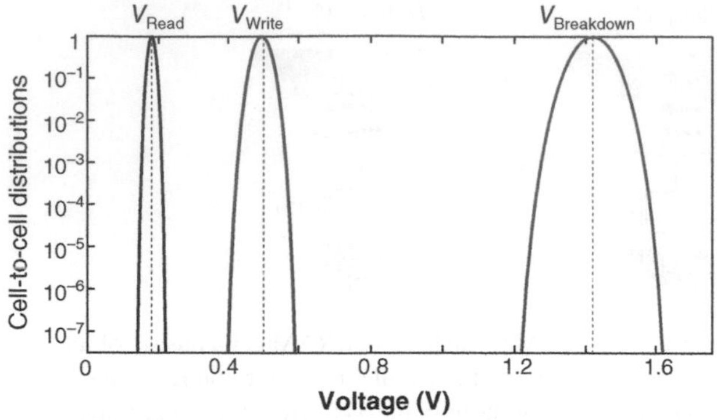

Figure 5.15 Schematic representation of the three voltage distributions (read, write, and breakdown) specific for STT-MRAM functioning.

5.5.2 Considerations of Breakdown, Write, Read Voltage Distributions

In conventional STT-MRAM, the write and read current paths are the same. In order to avoid write disturbance during read, the read voltage must be chosen low enough compared to the critical write voltage. There are therefore three voltage cell-to-cell distributions in an MRAM chip that need to be well separated for proper functioning and reliability of the chip. This is schematically illustrated in Fig. 5.15. These three distributions are as follows:

1. *The breakdown voltage distribution.* The MTJ tunnel barrier is a thin dielectric oxide layer (MgO ~1 nm thick). When exposed to an excessively large voltage, this barrier may experience dielectric breakdown. The breakdown voltage in the MTJ depends on the voltage pulse duration, the number of pulses, and even on the delay between pulses (20). Compared to CMOS dielectrics, these ultrathin MgO barriers are relatively resistant to breakdown. This mainly comes from the fact that the tunneling through the barrier is direct in normal working conditions, in contrast to Fowler-Nordheim tunneling as in Flash memories. As a result, upon cycling, less defects are generated in the MgO barrier than in CMOS oxides for Flash memory. For voltage pulse width in the range of 10 ns, the voltage breakdown is usually larger than 1.2 V even for $RA \sim 5\,\Omega\cdot\mu m^2$.
2. *The write voltage distribution.* At each write event (and to lesser extend read event), the tunnel barrier is exposed to an electrical stress that may cause electrical breakdown. To avoid breakdown failure, the highest write voltage in the distribution must be sufficiently low compared to the weakest MTJ in terms of breakdown. By adjusting the MTJ's stack composition and their *RA*, one tries to get this write voltage distribution centered around 0.5 V and be as narrow as possible (typical width as illustrated in Fig. 5.15). The distribution width mainly originates from fluctuations in the shape and particularly in edge defects associated with the patterning process.

3. *The read voltage distribution.* The read voltage distribution originates from the variation in the resistance of the selection transistor that is connected in series with the MTJ. The read voltage across the MTJ is typically in the range 0.1–0.15 V. As explained in Section 5.3, it must be low enough compared to the write voltage in order to avoid any write disturbance during read caused by the STT from the read current. However, the lower the read voltage, the slower the read-out process. Therefore, a trade-off must be found.

5.5.3 Influence of STT Write Pulse Duration

The global picture of STT switching versus the current pulse width is shown in Fig. 5.16. As commonly observed in microelectronics, the larger the voltage (i.e., the current density), the faster the operation. Two distinct switching modes have been found, namely, thermal activation and precessional switching (21,22) (Fig. 16) (23). At finite temperature, thermal activation plays an important role in reducing the switching current at long current pulses (>10 ns). In this slow thermally activated switching regime, the switching current depends on the current pulse width τ and thermal stability factor $\Delta = K_u V / k_B T$ of the free layer (24,25):

$$J_c(\tau) = J_{c0}\left[1 - \frac{k_B T}{K_u V}\ln\left(\frac{\tau}{\tau_0}\right)\right], \qquad (5.21)$$

where $\tau_0 \sim 1$ ns is the inverse of the attempt frequency and $K_u V$ is the anisotropy energy.

J_{c0} is the so-called critical current that corresponds to the linearly extrapolated value of the switching current density for a pulse duration of 1 ns.

For fast precessional switching in the nanosecond regime (less than a few nanoseconds), the required switching current is several times greater than the critical

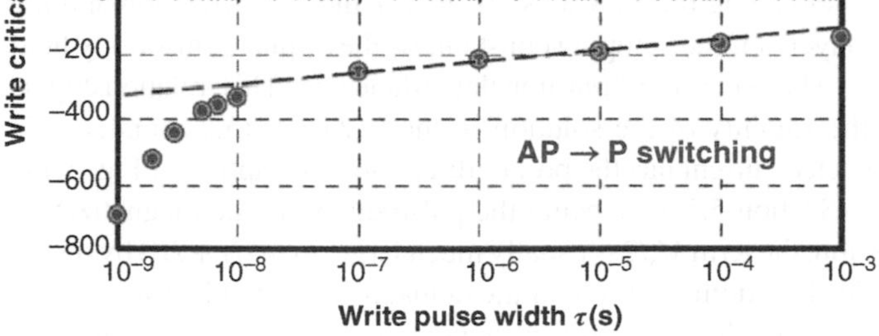

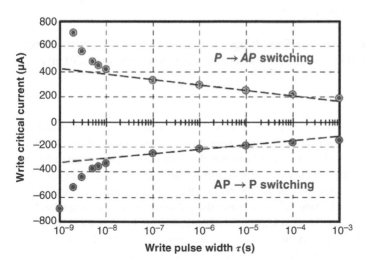

Figure 5.16 Typical values of STT switching current density (normalized by J_{c0}) as a function of write pulse width (After Ref. 23 with permission from IEEE).

current. In this regime, the switching time varies as (24)

$$\tau_{switching} = \frac{(1+\alpha^2)}{\gamma H_K} J_{c0} \frac{Ln(\pi/\theta_0)}{((J/J_{c0})-1)}, \quad (5.22)$$

where α is the Gilbert damping, H_k the anisotropy field, γ the gyromagnetic ratio, and θ_0 is the angle between the storage layer magnetization and the current spin polarization at the onset of the current pulse. This initial angle is often due to random thermal fluctuations that may cause stochasticity in the magnetization switching and slow down the write process. Some particular embodiments such as one with orthogonal polarizers, as described in Section "Perpendicular Polarizer," may help to improve the reproducibility (deterministic character) of the switching dynamics. Equation (5.22) illustrates that the larger the current density, the faster the switching, that is, the shorter the write current pulse. In this regime, the tunnel barrier may however be exposed to a too large electrical stress that may result in accidental electrical breakdown.

Typical write speeds in operation in STT-MRAM are around 5–10 ns, which corresponds to the intermediate regime between thermally activated and precessional switching.

5.5.4 In-Plane STT-MRAM

The first investigated MTJs for STT-MRAM were in-plane magnetized MTJ, meaning that both magnetic electrodes have in-plane magnetization. The reason is mainly historical since, besides the tunnel barrier itself, their configuration is fairly close to that of spin-valves that had been developed for hard disk drive magnetoresistive heads since 1991. However, as will be explained further, their performance and downsize scalability is not as good as for STT-MRAM based on out-of-plane magnetized MTJ. But the latter are more difficult to prepare so that their optimization is still on-going.

5.5.4.1 Critical Current for Switching The expression of the critical current for switching obtained in a macrospin model and at zero temperature is (26) given as

$$J_{c0} = \frac{2e\alpha\mu_0 M_s t_F (H + H_k + (M_s/2))}{\hbar\eta}, \quad (5.23)$$

where H is the applied field along the easy axis, M_s and t_F the magnetization and the thickness of the storage layer, α is the damping constant, H_k the in-plane anisotropy field (magnetocrystalline and shape), η the spin-transfer efficiency. This relationship is derived by analyzing the stability of the solution of the LLG Eq. (5.20) under STT influence. When the injected current has the proper direction (see caption of Fig. 5.14) and is larger than $J_c(\tau)$ (Section 5.5.3), τ being the pulse duration, the magnetization reverses. In this expression, the term $M_s/2$ is usually much larger than $(H+H_k)$ by one or two orders of magnitude. This dominant role of the demagnetizing field term ($\mu_0 M_s/2$) comes from the fact that during the STT-induced switching of the magnetization of the in-plane magnetized layer, the magnetization has to precess out-of-plane, which increases the demagnetizing energy. As a result, a good approximation of J_{c0} is given by $J_{c0} = (e\alpha\mu_0 M_s^2 t_F)/\hbar\eta$. It is also important to note that because of the dominance of the demagnetizing field term in (5.23), the critical current for STT writing weakly depends on H_k, the in-plane anisotropy field that determines the thermal stability of the magnetization at rest. In other words, in in-plane magnetized MTJs, the barrier for STT switching is mainly related to the demagnetizing energy, whereas the barrier for thermal stability of the magnetization, that is, the memory retention, is determined by the in-plane shape anisotropy. The first barrier is usually much larger than the second one, which means that the in-plane magnetized configuration is not efficient in terms of compromise between thermal stability and writability. This is the reason why the out-of plane magnetized MTJ is more favorable, as will be explained in Section 5.5.

5.5.4.2 Minimization of Critical Current for Writing The reduction of the intrinsic critical switching current J_{c0} can be realized by materials and magnetic anisotropy engineering (lower M_s, higher spin polarization factor η, and perpendicular anisotropy) and MTJ and free layer structures improvement (dual stack, perpendicular polarizer). The main challenge is to reduce the critical current while maintaining a large enough thermal stability.

Dual Magnetic Tunnel Junction Dual MTJs consist of two MgO tunnel barriers of different resistances, two pinned reference layers aligned antiparallel, and a free storage layer sandwiched between the two tunnel barriers. In such dual structure, STT effect is enlarged by the double spin-torque current acting on both sides of the free layer (21). Moreover, this type of structure reduces the asymmetry of the writing current, which is helpful for a wider read–write margin.

Perpendicular Polarizer The efforts to reduce the energy to write one cell have mostly concentrated on reducing the threshold of critical current density. This reduction also reduces the current value that the selection transistor has to sink/drive, and therefore smaller transistor sizes can be achieved. Another possibility to reduce the energy used to write one bit is to shorten the time that current/power needs to be applied to switch the magnetic configuration of the storage element. To achieve sub-nanosecond switching times, it is possible to take advantage of the precessional motion of the magnetization around the effective field direction before it aligns itself with this direction. The precession frequency can be calculated from Kittel's formula, $(2\pi f/\gamma\mu_0)^2 = H_{eff}(M_s + H_{eff})$, where f is the frequency, γ the gyromagnetic ratio, and M_s the saturation magnetization assumed to be parallel to the effective field H_{eff}. Typically, f is in the gigahertz frequency range for common ferromagnetic materials.

The precessional switching approach has already been explored in magnetic field driven switching concepts (27,28). In this case, a magnetic field pulse is applied in the hard axis direction. The magnetization starts precessing around the hard axis direction and, by doing so, acquires an out-of-plane component. Even at small angles, this out-of-plane magnetization generates a large demagnetizing field in the out-of-plane direction, becoming the new dominant field direction. The magnetization then precesses around this out-of-plane field direction. At each half-precession period the magnetization rotates by 180° and therefore changes from the initial magnetization direction to the opposite direction, all this at gigahertz frequencies. By properly timing the field pulse duration to stop at a half precession, it is possible to achieve sub-nanosecond switching of the magnetization. In magnetic field-driven precessional switching, this half precession corresponded to 140 ps. Field precessional switching was not pursued in actual MRAM demonstrations because the currents required to generate the hard axis field are large and cannot be scaled down by cell size reduction. Another problem is the need to accurately control the pulse duration with an accuracy of typically ±50 ps in a large capacity chip.

A more promising way to achieve precessional switching in the sub-nanosecond range with real scaling possibilities was proposed (29,30) in parallel to studies on spin-transfer oscillators. These spin-transfer oscillators use a perpendicular spin polarizer to induce the precession of the magnetization of an in-plane ferromagnetic layer, without an external applied field (Fig. 5.17) (31). In this case, the precession frequency can be controlled by the current density flowing through the device. The first demonstrations of precessional switching were realized in full metallic current-perpendicular-to-plane GMR systems (32,33). The structure in this case corresponded to a bottom perpendicular spin polarizer based on a Co/Pt multilayer separated from a NiFe 3/Co 0.5 nm storage layer by a Cu spacer. The reference layer was Co 3 nm pinned by an IrMn layer. The storage layer magnetization precesses around the normal to the layers for current densities higher than $j_c = (2e/\hbar)(M_s t/g)(H_k/2)$, H_k being the in-plane anisotropy field (32). The precession frequency increases linearly with the current density up to a maximum equal to $2\gamma M_s$. Figure 5.17 illustrates this concept of the spin-transfer precessional switching.

The storage layer changes its magnetic orientation from the initial direction to the opposite 180° direction within 250 ps and back to the initial direction after 500 ps. The switching probability of the initial precession is 100%, while for subsequent precessions there is a gradual lack of precession coherence due to thermal

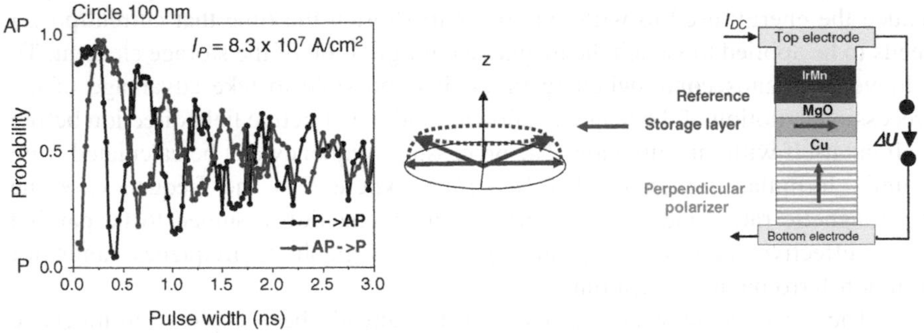

Figure 5.17 Illustration of the concept of spin-transfer precessional switching. The stack (right) includes a perpendicular spin polarizer, the precessing storage layer, and the reference layer. This latter layer is aimed to convert the storage layer direction change into a resistance change. The graph shows how the probability of magnetic switching of the storage layer oscillates between parallel and antiparallel alignment as a function of current pulse duration with a frequency of ~1 GHz for a current density of 8×10^7 A/cm^2. (After Ref. 32 with permission from AIP Publishing LCC.)

fluctuations, which is not detrimental to the application, since only the initial half precession is actually required to realize the switching. The precession can be induced by a current density of ~8×10^7 A/cm^2 and it is actually independent of the current direction. This is shown by the two different traces, one starting at the parallel configuration, while the other starts in the antiparallel configuration. This is important, since the transistor to be connected to these cells can be optimized for one current polarity. This means that higher currents per transistor lateral size has also been achieved, ultimately allowing for smaller cells. The same switching process has also been used with magnetic tunnel junction cells (34–36). In these studies, swithing in less than 200ps were demonstrated with associated energy down to about 200fJ. The remaining difficulties with this technology are associated with i) the write endurance since pulses of voltage around 0.7 - 1V are used for ultrafast switching, ii) the assymetry between the P to AP and AP to P transition due to the STT from the in-plane analyzer, iii) improve the write error rate by an accurate control of the precessional motion of the storage layer magnetization.

Reduced Demagnetizing Field In the case of thin films with in-plane magnetization, the shape anisotropy energy that determines the value of the critical STT current can be reduced by adding an interface anisotropy term to the volume anisotropy term (37). This interface term can even be the leading term, thus spontaneously orienting the magnetization out-of-plane below a critical magnetic thickness. This phenomenon can be used to develop planar MTJ's with reduced demagnetizing field using magnetic metal-oxide combinations. In conventional MTJ, the demagnetizing field of the magnetic layer is $-\mu_0 M_s$. As explained earlier, it is usually the dominant contribution to the critical STT switching current linked to the in-plane anisotropy. However, the relative role of demagnetizing and anisotropy energy can be strongly modified in the presence of either volume or interfacial PMA. Indeed, in that case, the demagnetizing and anisotropy terms have opposite signs, leading to an expected significant reduction in critical current density (38,39). Since there are different ways to obtain interfacial out-of-plane anisotropy, several research teams have proposed structures based on this concept (40–42).

The first low demagnetizing structure was a spin-valve (40) that used interfacial anisotropy from Co/Ni multilayers, a very versatile system in which the PMA can be tuned over a wide range by changing the thickness of each layer and/or the number of repeats. The zero-thermal-fluctuation critical current Jc_0 is reduced by a factor of 5–6 ($Jc_0 = 2 \times 10^6$ A/cm^2) compared to control samples with high demagnetizing energy and the same total magnetic moment, while the thermal stability was almost the same. Further reduction in critical current is expected by optimizing the spin polarization using a magnetic tunnel junction rather than a spin-valve. The same team (43) fabricated MgO magnetic tunnel junctions using similar Co/Ni switching layers combined with a FeCoB insertion layer to reduce the effective demagnetizing field (thanks to interface anisotropy) to 0.2 T, still keeping the tunnel magnetoresistance ratio as high as 106%. However, the use of Co/Ni or Pt/Co materials in the switching layer is not recommended because of increased damping and difficulties in maintaining high TMR in MTJs because such materials with fcc (1 1 1) texture are incompatible with the (1 0 0) texture of the MgO barrier.

Large interfacial out-of-plane anisotropy has also been evidenced at Co(Fe)–oxide interfaces (44–46). This interfacial anisotropy is able to orient the magnetization of a 1 nm thick CoFeB layer out-of-plane (47). This makes it possible to obtain a free layer with a low effective demagnetizing field in planar MTJ just by increasing the layer thickness in order to decrease the relative anisotropy contribution of the CoFeB–MgO interface. The effect of the CoFeB composition on device characteristics (anisotropy and critical thickness) was investigated in full MTJ stacks with bottom pinned $Co_{80-x1}Fe_{x1}B_{20}$ and top free $Co_{80-x2}Fe_{x2}B_{20}$ electrodes with $x_1 = 20$, 40, 60, and $x_2 = 20$, 60 (41). MTJs with a free layer thickness of 1.8 nm were compared with standard devices, resulting in a 40% reduction in the average quasi-static switching current density from 2.8 to 1.6 MA/cm^2 when the free layer was changed from a Co-rich ($x_1 = 40$ and $x_2 = 20$) to a Fe-rich ($x_1 = 40$ and $x_2 = 60$) composition. It was believed that the reduction of current density by a factor 2.4 was related to the interfacial anisotropy increase since the layer thickness was reduced only by a factor of 1.2 (from 2.03 to 1.69 nm). For a 1.8 nm free layer thickness of Fe-rich CoFeB, the TMR ratio obtained was still high, around 120%.

Work on STT spin-valves (40) or classical STT-MTJ (41) showed that PMA contribution greatly reduces the required switching current by decreasing the demagnetizing field effect. Switching current densities and perpendicular anisotropy are strongly dependent on the free layer thickness for in-plane magnetized structures with PMA contribution (41). Critical current densities as low as 2×10^6 A/cm^2 were obtained in planar MTJ with low demagnetizing field. Fast switching, of the order of hundreds of picoseconds, was observed in in-plane systems comprising an additional perpendicular polarizer (32–35) or with reduced demagnetizing field (40). At the end of 2012, based on these results, the company EVERSPIN Technologies released a 64Mbit STT-MRAM product implementing such in-plane magnetized MTJs with reduced demagnetizing field in the storage layer (48).

5.5.5 Out-of-Plane STT-MRAM

Devices that show a magnetic anisotropy normal to the film surface hold great promise for faster and smaller magnetic bits in data-storage applications. They have many advantages compared to in-plane magnetized MTJs:

1. The switching current density is expected to be significantly reduced because the two terms present in the expression of the critical current density partially cancel each other (49).
2. The thermal energy barrier is provided by this large effective perpendicular anisotropy instead of in-plane shape anisotropy (47,50). For comparison, in MRAM systems based on field induced magnetic switching, high coercive fields and large anisotropy were problematic. However, in STT-MRAM, these same properties become an advantage, since they enable thermally stable elements beyond the 45 nm technology node. As a consequence, elongated cell shapes are no longer needed and the perpendicular MTJs can be patterned in circular shape. This facilitates manufacturability at smaller technology nodes and leads to lower switching current for a given memory retention.

3. Finally, dipole field interaction between neighboring cells can also be reduced in high bit density layouts.

The perpendicular magnetic anisotropy can have different origins, either bulk (in hcp CoCrPt, heavy rare-earth–transition metal, or $L1_0$ FePt ordered alloys) or interfacial (in Pt/Co, Pd/Co, or Co/Ni multilayers). Besides, a quite large PMA can be induced at the interface between a ferromagnetic electrode and an oxide (44–46).

The growth of good quality perpendicular MTJs is a more difficult task than for their in-plane counterparts for reasons that will be explained below.

The first report on perpendicular magnetic tunnel junctions dates from 2002 (51). In this publication, the authors presented results on electrodes made of rare-earth–transition metal (RE-TM) alloys exchange-coupled to a 1.0 nm thick CoFe layer, separated by an Al_2O_3 barrier. In pillars 300 nm in size, they obtained a TMR ratio of up to 55%. At that time, RE-TM alloys were probably the best choice, since alumina barriers require only moderate annealing to optimize their transport properties. The first results on TMR properties in MgO-based structures (64% TMR) were published in 2008 (52), with electrodes made of RE-TM alloys exchange-coupled to a Fe layer. The same year, Toshiba presented the first STT switching results in MTJs with perpendicular anisotropy having a typical stack composed of TbCoFe/CoFeB/MgO/CoFeB/TbCoFe (as shown in Fig. 5.18) and patterned into a circular shape (50). Electrical measurements have shown a low 15% TMR, while the critical current densities extracted from switching curves at current pulse width from 30–100 ns were in the range of 5×10^6 A/cm^2. Coercive fields of 0.12 T yield a thermal stability factor $\Delta = 107$, large compared to less than $40k_BT$ in in-plane anisotropy cells at similar critical current densities.

Since 2009, different materials have been tested in order to induce perpendicular anisotropy in the magnetic electrodes on both sides of the MgO barrier (RE-TM alloys, CoCrPt alloys, ordered $L1_0$ alloys, and (Co/Pt) and (Co/Pd) multilayers) (53). All materials have both their advantages and their disadvantages. RE-TM alloys can be relatively easily fabricated, but their anisotropy properties greatly depend on alloy

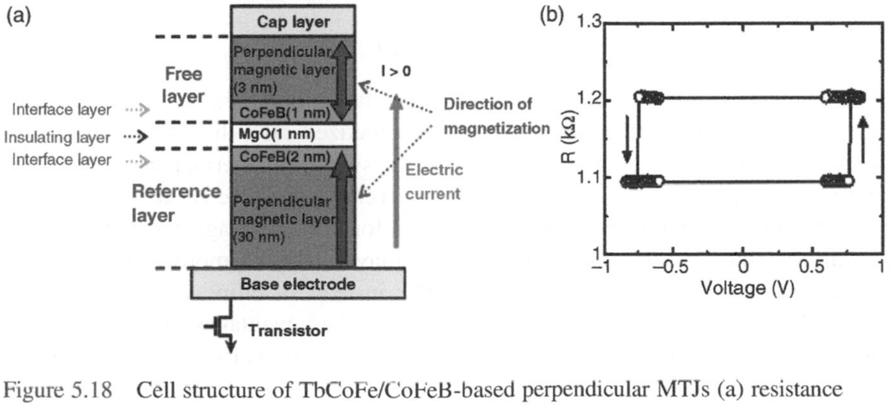

Figure 5.18 Cell structure of TbCoFe/CoFeB-based perpendicular MTJs (a) resistance versus voltage pulses for the perpendicular MTJ structure schematically shown (b). (From Ref. 50 with permission from AIP Publishing LLC.)

composition, annealing temperature and operation temperature, which makes their optimization rather difficult. Furthermore, they are rather susceptible to corrosion.

Nevertheless (54), critical switching current densities as low as 5×10^5 A/cm^2 have been reported in TbCoFe/CoFeB/MgO/CoFeB/L1_0-FePd structures. These results show that critical current densities smaller than those in in-plane magnetized MTJ can be achieved in such out-of-plane magnetized junctions.

The great breakthrough came in 2010 when structures based on interfacial magnetic metal/oxide perpendicular anisotropy (44–46) with Ta/CoFeB electrodes were proposed (47), very similar to their in-plane counterparts but with thinner storage layer. The problem of the bcc (1 0 0) texturation of the CoFeB electrode upon crystallization was thus solved. In these structures, the perpendicular anisotropy mainly arises from the CoFeB–MgO interface. After patterning into sub-micrometer pillars, two well-separated parallel and antiparallel states were obtained, leading to simultaneously high TMR ratio (120%), low *RA* product (18 Ω µm^2) and low STT switching current density (4 MA/cm^2). It is important to emphasize that this approach based on the interfacial PMA at magnetic metal/ oxide interface (see section 5.5.5.3) has now become the mainstream one in the development of perpendicular MTJ for STT-MRAM.

5.5.5.1 Benefit of Out-of-Plane Configuration in Terms of Write Current The critical current for spin-transfer reversal of the storage layer obtained from the LLG equation (Eq. 5.20) in the out-of-plane configuration is given by

$$I_{c0} = \frac{2 \cdot e}{\hbar} \cdot \frac{\alpha \cdot A \cdot t \cdot \mu_0 M_s}{\eta} \cdot H_{eff}, \tag{5.24}$$

where A is the area of the magnetic element, e is the electron charge, $\hbar$ is the reduced Planck constant, μ_0 the vacuum permeability, α is the Gilbert damping coefficient, M_s and t are the saturation magnetization and thickness of the storage layer, η the spin-transfer torque efficiency that depends on the relative orientation of the magnetizations ($\theta = 0$ or π) and on the polarization P, and H_{eff} the effective switching field.

In magnetic junctions with out-of-plane magnetization, the effective field is given by

$$H_{eff} = H_{K\perp} - M_s, \tag{5.25}$$

where $H_{K\perp}$ is the perpendicular anisotropy field that pulls the magnetization out-of-plane ($H_{K\perp} > M_s$). In contrast to the in-plane magnetized case, the energy barriers to overcome to switch by STT during write or to switch by thermal fluctuations in standby are here identical. Therefore, for a given memory retention, the critical switching current can thus be much smaller than for in-plane magnetized electrodes. In addition, since shape anisotropy does not play a role anymore in the thermal stability, magnetic cells can be made circular instead of elliptical, which makes downsize scaling easier. Several papers addressed the advantages of perpendicular magnetic tunnel junctions (48,55–57). By introducing the thermal stability factor

$$\Delta = \frac{KV}{k_B T} = \frac{\mu_0 M_s H_{eff} A t}{2 k_B T}, \tag{5.26}$$

where K is the anisotropy, $V = A \cdot t$ is the volume of the storage layer, k_B is the Boltzmann's constant, and T is the absolute temperature, Eq. (5.23) can be rewritten as

$$I_{c0} = \frac{4 \cdot e}{\hbar} \cdot \frac{\alpha \cdot k_B T}{\eta} \cdot \Delta. \tag{5.27}$$

This relation expresses that in the macrospin approximation (valid at dimensions below typically 40 nm, as discussed regarding Fig. 5.4), a direct proportionality exists between the thermal stability factor (i.e., the memory retention) and the write critical current. This is the classical dilemma between retention and writability often encountered in memory technology: the more stable the information, the more difficult it is to change it. To account for Eq. (5.27), a figure of merit has been proposed (58), which is the ratio of the thermal stability factor to the critical switching current, Δ/I_{C0}. Following the discussion of Section 5.4.1 and of this section, the figures of merit obtained with out-of-plane magnetized MTJ are expected to be much better (larger Δ/I_{C0}) than with in-plane magnetized MTJs. This is however true only if the Gilbert damping constant α, to which I_{c0} is proportional can be maintained as low as in in-plane magnetized material, which is not easy, as explained below.

5.5.5.2 Trade-off Between Strong Perpendicular Anisotropy and Low Gilbert Damping With perpendicular MTJ, current density on the order of 5×10^5 A/cm^2 to 3×10^6 A/cm^2 were obtained, values of the same order of magnitude as the values obtained for in-plane switching (~2×10^6 A/cm^2). These rather large values are associated to the large perpendicular anisotropy of the investigated materials (e.g. Rare-Earth/transition metal alloys, FePt or FePd ordered alloys) but also to the relatively large values of the Gilbert damping constant commonly obtained in these materials (59). Indeed, both Gilbert damping and magnetic anisotropy derive from the spin–orbit interactions that basically couple the electron spin to the lattice. As a result, materials that exhibit large anisotropy (for instance FePt ordered alloys, CoPt multilayers and alloys) also exhibit large Gilbert damping constant α (in the range 0.05–0.2). In contrast, in-plane magnetized materials such as CoFeB have weak anisotropy and low damping (in the range 0.007–0.01).

A promising route to circumvent the problem of large damping in out-of-plane magnetized materials consists in using a storage layer where the perpendicular anisotropy does not arise from a bulk contribution as in Co/Pd or Co/Pt multilayers or as in FePt ordered alloys, but from a large interfacial perpendicular anisotropy that exists at the interface between the magnetic electrode and the tunnel barrier, as explained below.

5.5.5.3 Benefit from Magnetic Metal/Oxide Perpendicular Anisotropy In magnetic thin films, the interfaces have in general a high impact on their magnetic properties. This is particularly the case in some systems wherein a very strong perpendicular magnetic anisotropy can originate from the interface between a magnetic and a nonmagnetic layer, as predicted by Néel (60) and experimentally observed several years later (61). In a very simple picture, the measured effective anisotropy energy (K_{eff})

has two major contributions: a volume anisotropy (K_V), which itself can have different origins (magnetocrystalline, magnetoelastic, demagnetizing), and a surface–interface anisotropy (K_S), whose relative contribution increases as the thickness t of the magnetic film decreases. Taking into account these different contributions, the effective anisotropy per unit volume is given as $K_{eff} = K_V + \frac{K_s}{t}$.

Remarkably, a strong PMA interfacial contribution exists at magnetic metal–oxide interfaces (Co–AlO_x, CoFe–MgO, and Co–CrO_x) (45–47) despite the weak spin–orbit coupling of the involved materials. Experiments showed that this phenomenon is quite general (44–46,62,63) since it was obtained at the interface between various magnetic transition metals and oxide layers and is independent of the crystalline structure of the oxide layers (amorphous as alumina tunnel barriers or crystalline as MgO barriers). In addition, X-ray photoemission spectroscopy (63) showed that oxygen plays an essential role in the PMA at the magnetic metal–oxide interface by the formation of Fe(Co)–oxygen bonds. For a better understanding of the PMA origin at the magnetic metal–oxide interfaces, an *ab initio* study was realized taking into account the weak spin–orbit coupling (64). The origin of the large PMA is ascribed to the combination of several factors:

1. Lift of degeneracy of out-of-plane 3d orbitals due to the spin–orbit coupling,
2. Hybridizations between dz^2 and dxz and dyz 3d orbitals induced by spin–orbit interactions
3. Hybridizations between Fe(Co)-3d and O-2p orbitals at the interface between the transition metal and the insulator.

This interfacial anisotropy makes possible to pull out-of-plane the magnetization of materials having low Gilbert damping such as CoFeB alloys ($\alpha_{CoFeB} \sim 0.01$, (65) compared with $\alpha_{(Co/Pd)} \sim 0.1$ and $\alpha_{(Co/Pt)} \sim 0.2$). Using the interfacial anisotropy therefore makes it possible to achieve simultaneously large perpendicular anisotropy required for high thermal stability of the magnetization and the weak Gilbert damping required for low STT switching current density. The interdiffusion between layers, caused by the nature of the materials (immiscibility) or by annealing, is always detrimental to the interface anisotropy. For example, it was experimentally observed (45,66) and calculated (64) that over/under-oxidation of the magnetic material–oxide interface strongly reduces the interface anisotropy. The remaining presence of B along the CoFe–MgO interface after annealing is also detrimental for the interfacial PMA. A lot of care must therefore be taken in the growth, oxidation and annealing conditions of these out-of-plane magnetized MTJ. Since the first demonstrations of STT switching in MTJ with large TMR using CoFeB/MgO interface anisotropy (48), tremendous progress have been made in the development of these MTJs to reach the specifications required in practical products (67), in particular sufficiently low bit-error rates (68,69). To further take advantage of the magnetic metal/oxide interface anisotropy, it was proposed to sandwich the storage layer between two oxide layers thus benefiting from two metal/oxide interfaces (70). The storage layer typically consists of a CoFeB 1.5 nm/Ta0.4 nm/CoFeB 1.2 nm composite structure sandwiched between two MgO layers wherein the thin Ta layer is intended to absorb the B away from the MgO interfaces upon annealing thus allowing the crystallization of the

CoFeB layers into bcc structure and increasing the interface PMA at MgO/CoFe and CoFe/MgO interfaces (70,71). In these double barrier MTJs, the top electrode is usually non-magnetic so that there is no TMR associated with the top MgO barrier. The bottom MgO barrier provides the TMR signal required to determine the magnetic state of the storage layer. It is made thicker (i.e. has a higher RA product) than the top MgO layer to minimize the TMR decrease associated with the serial resistance of the top MgO layer. Very successful demonstrations of fully functional STT-MRAM chips were made following the optimization of these stacks (72–74). In 2016, the company EVERSPIN Technologies announced the launching of 256Mbit STT-MRAM products based on this technology. In 2016, other companies made also announcements on their move towards embedded Flash or SRAM replacement by this type of perpendicular STT-MRAM.

5.5.5.4 Downsize Scalability of Perpendicular STT-MRAM For high-density applications, below the 45 nm technological node and *a fortiori* at even more advanced nodes, it is mandatory to use out-of-plane magnetized MTJ to get long enough memory retention in STT-MRAM. But how small can we hope to go? With presently known materials, the optimized structures should look like the one represented in Fig. 5.19.

The general structure of this stack is a double tunnel junction with out-of-plane magnetization wherein the storage layer is sandwiched between two tunnel barriers of different *RA* product, themselves sandwiched between two polarizers of opposite magnetization. This dual configuration has three main advantages:

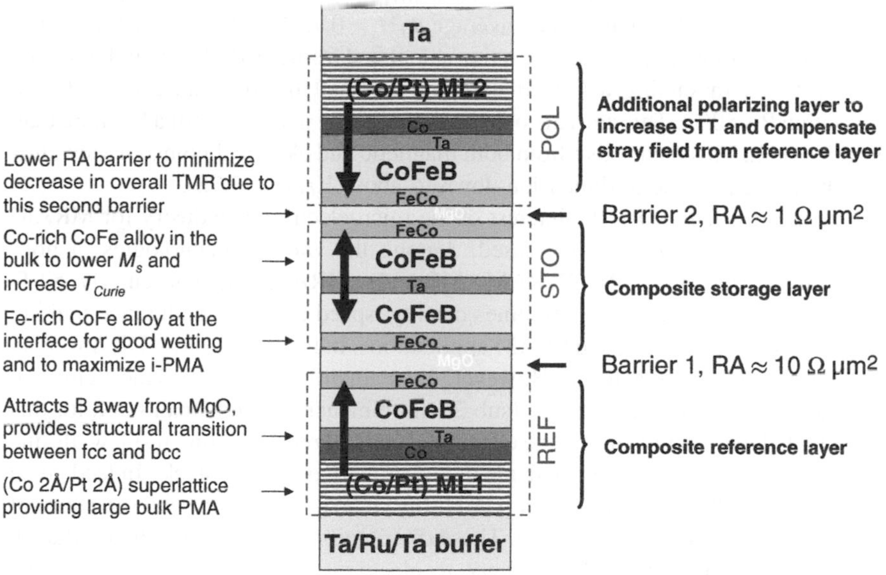

Figure 5.19 Expected optimized stack for downsize scalability of STT-MRAM to sub-20 nm node (75,76).

1. The STT efficiency on the storage layer magnetization is more than doubled, thanks to the additive contributions from the two antiparallel polarizing layers.
2. The interfacial PMA on the storage layer magnetization is doubled by the presence of two CoFeB–MgO interfaces instead of only one in single MTJ stacks.
3. The stray field from the two polarizing layers on the storage layer magnetization is cancelled due to the antiparallel alignment of the two polarizing layers so that both states of the storage layer have the same thermal stability.

Since the antiparallel orientation of the two polarizing layers is rather unstable, the stability of this configuration can be improved by pinning the magnetization of these layers by antiferromagnetic layers (on both layers or one or the other only).

To further optimize the stack, each layer has a composite structure. The thin Ta insertions are intended to attract the B away from the MgO–CoFeB interfaces since B next to the MgO interface reduces the TMR as well as the PMA. The Co/Pt superlattices in the polarizing layers are increasing the overall PMA of these layers and thereby their magnetic stability. The storage layer is also composed of several layers. The bulk of the storage layer is made of a Co-rich CoFeB alloy to lower its magnetization (compared to Fe-rich alloys) and increase its exchange stiffness, thanks to the higher Curie temperature of Co than of Fe. This layer can be doped with some V to further lower its magnetization, thereby reducing its demagnetizing energy and thus increasing the effective perpendicular anisotropy of this layer. The layers in contact with the MgO barrier is Fe-rich to maximize the interfacial PMA and to allow good wetting of this layer during growth.

From the known properties of the involved materials, we can estimate that the magnetization of the storage layer should be stable with a thermal stability factor above 70 at RT down to diameter on the order of 10 nm (assuming interfacial anisotropy of 1.45 mJ/m^2 at FeCo–MgO interfaces, average $\mu_0 M_s = 0.75$ T in the storage layer). The current required to switch the magnetization calculated from Eq. (5.27) should be on the order of 13 μA (assuming a Gilbert damping constant $\alpha \sim 0.01$and a current spin polarization P ~80%). This current could be delivered by a transistor of a bit more than 10 nm in width. This means that from both magnetic and electrical points of view, this stack should be scalable to dimension down to about 10 nm (77,78).

In the past few years, the interest of the microelectronics industry for MRAM technology has tremendously increased. Among the various technologies of non-volatile memories (PCRAM, ReRAM, MRAM), MRAM and particularly STT-MRAM is the only one which combines density, speed and long endurance making it so promissing for e-Flash and SRAM applications. In the longer term, DRAM may also be replaced by STT-MRAM. However, the main difficulty remains the etching of the MTJ stacks at small feature size (sub-20 nm dimensions) and small pitch for this type of high density application. The edge damages and redeposition on sidewalls due to etching affect the transport and magnetic properties of the stack. Indeed, etch damages yield cell-to-cell dispersion (variability) in *RA*, TMR, and switching current as well as in retention and endurance. These distributions need to be further reduced. The reliability associated with resistance to electrical breakdown needs also to be further improved. Cyclability in STT-MRAM of 10^{12} cycles has already been demonstrated (79) and studies in accelerated conditions have shown that 10^{16} cycles can be achieved provided the density of electron trapping sites in the barrier is reduced (20,80). These trapping sites can be oxygen or Mg vacancies in the MgO barrier, dislocations associated with the 4% crystallographic mismatch between CoFe and MgO, BO formation along the MgO interface if the diffusion of B away from the interface during annealing is not properly controlled, or water molecules absorbed in the MgO barrier if the water partial pressure during MgO growth is not low enough. These various challenges can be met by an optimization of the growth and oxidation conditions of the stack. The objective is to bring the write endurance above 10^{16} cycles so that MRAM could become a universal memory, combining the advantages of DRAM (density), SRAM (speed), and Flash (nonvolatility).

In order to obtain minimal power consumption and optimized reliability, many tricks have been proposed (21). For example, subnanosecond switching with optimized power consumption can be achieved by tuning the switching current pulse width (81). Resonant STT switching was proposed to increase the reliability of the writing: (82) a weak microwave signal is applied to the junction in combination with the switching DC pulse. Such a microwave signal triggers the magnetization reversal by exciting its main precession modes, decreasing thus the writing consumption. Other assistance methods have been proposed, such as electric field (83–85), or thermal assistance (86,87). The idea is to decrease the anisotropy of the storage layer during write, either by radio-frequency fields or heating, so that it becomes easier to switch the junction either by magnetic field or spin-transfer torque. By disconnecting the anisotropy requirements under standby and writing conditions, it becomes possible to scale down to even smaller dimensions. Thermally-assisted MRAM are described in detail in the next section. In addition to allowing further downsize scalability, thermal assistance during write opens new functionalities particularly interesting for security or router applications.

5.6 THERMALLY-ASSISTED MRAM (TA-MRAM)

5.6.1 Trade-off Between Retention and Writability; General Idea of Thermally-Assisted Writing

It was shown in the previous sections that if writing is performed by field, then the field required to switch the magnetization of an element of anisotropy K and saturation magnetization M_s is $\mu_0 H_{write} = 2K/M_s$. Therefore, the larger the anisotropy, the larger the write field and therefore the larger the current required to create this magnetic field, meaning larger write power consumption. Now, if the writing is performed by spin-transfer torque, a similar problem occurs. Indeed, for spin-transfer writing, it is preferable to use magnetic tunnel junctions that are magnetized in the out-of-plane direction since these junctions require less current density to switch the magnetization by STT than their in-plane magnetized counterparts. In these perpendicular MTJs, the switching current is directly proportional to the thermal stability factor Δ, which controls the memory retention (Eq. 5.27). Therefore, both for field and STT writing, increasing the memory retention of the cell systematically yields an increase in the energy required to write.

The same problem is actually also encountered in magnetic HDD technology, wherein the increase in areal density of information stored on HDD requires a reduction in the grain size in the storage media in order to maintain a sufficiently large signal-to-noise ratio. However, to maintain sufficient stability of the magnetization of the grains, their anisotropy must be increased. But then the required field to write becomes too large to be produced by the magnetic pole of the write head. This is known as the magnetic storage "trilemma."

This general difficulty validates the interest in using a thermally-assisted write approach. Indeed, in magnetic materials, it is known that, in the vast majority of cases, it is easier to switch the magnetization of a magnetic element at elevated temperature than at low temperature. This originates from the fact that the magnetic anisotropy decreases with temperature so that the barrier height for switching (ΔE) decreases with temperature. Furthermore, the higher thermal activation itself may help the magnetization to switch above the barrier. The concept of thermally-assisted writing thus consists in storing the information at a standby temperature at which the anisotropy and therefore the thermal stability factor are very large, thereby temporarily increasing the temperature of the magnetic element during each write event to reduce the barrier height and ease the switching of the magnetization.

5.6.2 Self-Heating in MTJ Due to High-Density Tunneling Current

In MRAM, the heating of the storage layer can be produced in a simple way by taking advantage of the Joule dissipation around the tunnel barrier. Actually the heating in MTJ is not a simple Joule heating as in metallic systems because we are here dealing with tunneling instead of ohmic transport. Rather, the heating in MTJ is due to the inelastic relaxation of tunneling hot electrons, which takes place when the tunneling electrons penetrate in the receiving electrode after their ballistic tunneling across the tunnel barrier. However, the heating power per unit area has the same expression as usual Joule heating: $P = jV$, where j is the current density flowing through the tunnel barrier and V the bias voltage across the barrier. In MTJs having RA product on the order of $30\,\Omega\,\mu m^2$, with heating current density on the order of $10^6\,A/cm^2$, a rise in temperature ΔT on the order of 200 °C within 5 ns is typically achieved (88).

5.6.3 In-Plane TA-MRAM

5.6.3.1 Write Selectivity Due to a Combination of Heating and Field In 2001, a thermally-assisted switching (TAS) MRAM concept (TA-MRAM) was proposed to improve the thermal stability, write selectivity, and power consumption of MRAM cells (89). A convenient approach to TA-MRAM is to heat directly with a current flow through the MRAM cell. In this scheme, a conventional MRAM junction is modified by replacing the simple ferromagnetic storage layer by an exchange-biased storage layer, that is, a ferromagnetic storage layer exchange coupled to an antiferromagnetic layer (89). The write procedure requires heating above the storage layer blocking temperature and cooling in the presence of a magnetic field. We remind here that the blocking temperature in a ferromagnetic/antiferromagnetic bilayer is the temperature at which the loop shift induced by the exchange coupling across the interface between these two layers vanishes. This temperature is usually slightly below the Néel temperature of the antiferromagnet (see Chapter 2). In TA-MRAM, the reference and the storage layer must be exchange-biased with antiferromagnets having sufficiently different blocking temperatures. Typically, PtMn with blocking temperature of 350 °C is used in the reference layer, whereas the storage layer antiferromagnet is chosen with a blocking temperature in the range 180–250 °C depending on the requirements on the device operating temperature range. The main advantages of this scheme are (i) the use of a single field selection line instead of two for toggle writing, (ii) an important reduction in write power consumption as will be explained later, and (iii) realization of cell with high thermal stability, that is, much improved retention due to the exchange pinning. For a given heating pulse duration, the temperature increase in the TA-MRAM cell is proportional to the dissipated power density P_d. The temperature increase can actually be directly derived experimentally from the measurements of the storage layer exchange bias field versus heating current. The total power density is effectively determined by the RA product (resistance × area product) of the junction and the current density j as $P_d = RA\, j^2$. Typical RA used in TA-MRAM are in the range 20–30 Ω μm^2 and heating current density around 10^6 A/cm^2 comparable to the lowest used in STT writing. One way to increase the heating efficiency and reduce the power density is to insert low-thermal conductivity materials at both ends of the magnetic tunnel junction stack, serving as thermal barriers between the junction and the electrical leads. This confines the heat to the junction volume, preventing lead heating and possible thermal crosstalk (90). Typical heating time in TA-MRAM are in the range 3–10 ns. This heating time is primarily influenced by the heating power dissipated at the tunnel barrier. The cooling rate depends on the heat diffusion constant toward the top and bottom of the MTJ stack and on the specific heat of the MTJ. It is typically in the range 10–20 ns (91).

A bit write sequence, illustrated in the bottom drawing in Fig. 5.20, starts from a given initial orientation of the magnetization of the exchange-biased storage layer, for instance representing a low resistance state "0". The corresponding storage layer loop is shifted around a negative field, as seen in Fig. 5.20 in the hysteresis cycle before the heating pulse is applied (red curve). The reversal of the storage layer bias is achieved by heating the AF layer above its blocking temperature with a current pulse and applying simultaneously an external magnetic field H_{sw} larger than the coercive field of the storage layer. The field is applied in the direction parallel or antiparallel to the reference layer magnetization depending on whether a "0" or a "1" is to be written. In the example of Fig. 5.20, a "0" is written. The heating current pulse is then stopped so that the storage layer cools in a magnetic field. This maintains the storage layer magnetization in the field-cooling direction during its freezing. The result is a reversal of the pinning orientation of the storage layer magnetization and a bit state change to a low resistance "0". As a result, the storage layer loop is now shifted toward positive values as shown in Fig. 5.20 (blue curve).

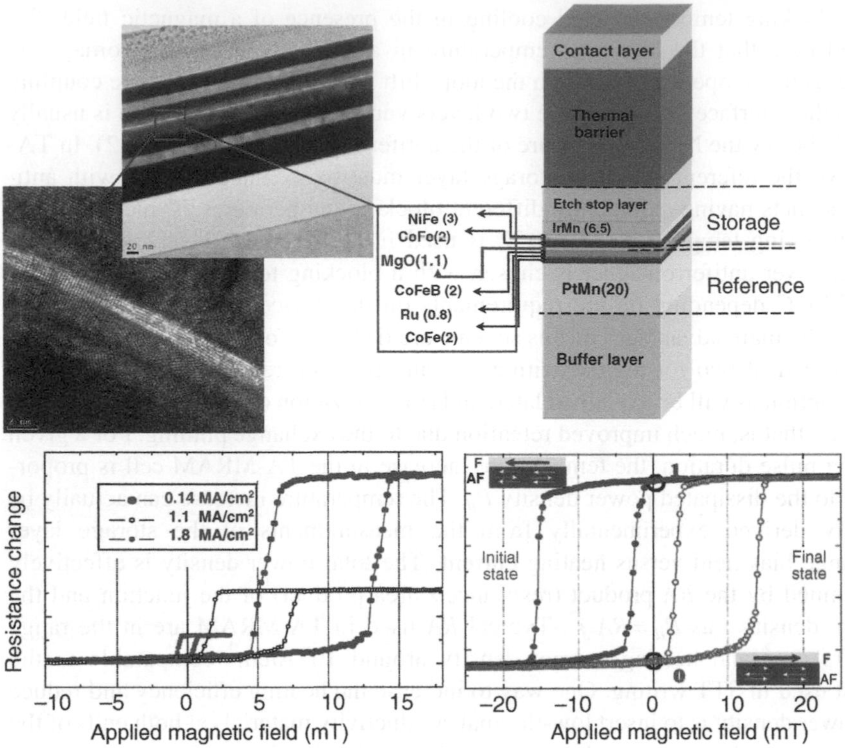

Figure 5.20 *Top:* Composition of the stack and cross-sectional transmission electron microscopy images of the MTJs. *Bottom left:* Hysteresis loops of the storage layer for various heating current density showing the decrease in the pinning as the heating current increases. At the highest heating current density, the loop is centered, meaning that the exchange bias has vanished and low fields of a few millitesla are then sufficient to switch the storage layer magnetization. The drop in amplitude of the hysteresis loop is due to the dependence of the TMR on heating bias voltage. *Bottom right:* Resistance hysteresis loops of the exchange-biased storage layer before (red) and after (blue) writing, showing the inversion of the hysteresis loop shift. Note that in both cases, only one remanent state is stable at zero field: either the low or the high resistance state, meaning that the memory is very stable against perturbing external fields (92).

Several demonstrations of TA-MRAM were realized first in micrometer-size junctions (2 µm × 2 µm) under DC currents (93), then on submicrometer cells using heating pulses down to 10 ns (94). The company Crocus Technology in collaboration with TowerJazz has produced fully functional 1 Mbit demonstrators using this technology.

5.6.3.2 Reduced Power Consumption, Thanks to Low Write Field and Field Sharing In conventional TA-MRAM, the write selectivity is achieved by a combination of temporary heating of the storage layer that lowers its magnetic pinning energy together with the application of a pulse of magnetic field. The efficiency of this write selectivity scheme is illustrated in Fig. 5.21. Repeated successful writing can be achieved only when heating and field pulse are combined. This therefore solves the write selectivity problems that were present in Stoner–Wohlfarth first MRAM.

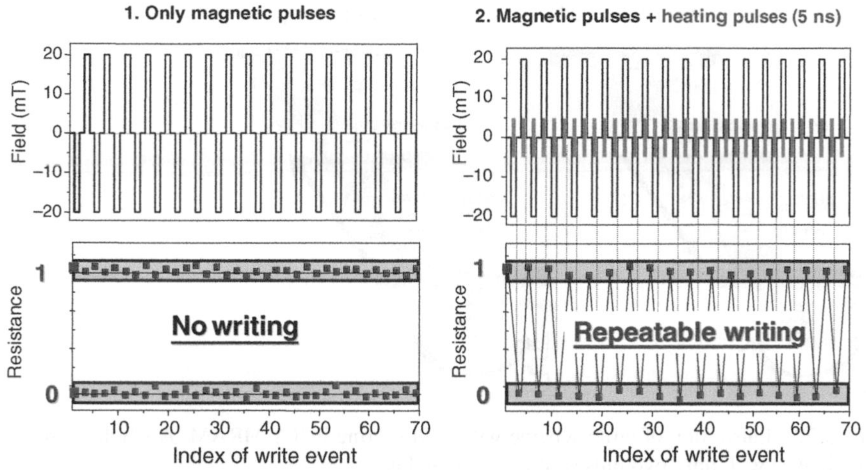

Figure 5.21 Illustration of the write selectivity in TA-MRAM resulting from the combination of heating pulse and application of a magnetic field. *Left:* Application of pulses of magnetic field alone (no writing). *Right:* Combination of heating pulses and alternating pulses of magnetic field (repeatable writing of *P* and *AP* (0 and 1) magnetic states).

Another advantage of this write scheme is the protection against field erasure in standby. As seen in Fig. 5.20 (right), in standby, only one state of the storage layer is stable at zero field. This means that even if the TA-MRAM chip is exposed to a perturbation field, this field may temporarily switch the magnetic configuration of the memory but the latter will spontaneously return to its original state before perturbation once the perturbation disappears (92). This, however, would not be true during write. If a perturbation field that is a significant fraction of the write field is applied during write, a write error may occur. A read-after-write scheme may be used to verify that no such error has occurred during write.

Field writing combined with thermal assistance is also quite advantageous in terms of power consumption in MRAM chips by offering the possibility to share the pulse of magnetic field among numerous bits. Indeed, in a RAM, the bits are usually not written randomly one by one but word by word. Each word may contain 32 or 64 bits. In TA-MRAM, to write a full word, only two pulses of magnetic field are required. The scheme for writing a word is illustrated on Fig. 5.22. In a first step, all bits in the word that have to be written to "0" are heated simultaneously by sending a heating current through the corresponding MTJ. The "0" field is then applied and the heating currents are switched off so that the heated cells cool down in the "0" magnetic field. As a result, these cells freeze in the "0" configuration. The cells that are not heated do not switch so that they remain in their original state. In a second step, all bits that have to be written to "1" are heated simultaneously and the "1" magnetic field is applied by sending a current in the corresponding word line (Fig. 5.22). The heating currents are then switched off, while the "1" field is still on so that the heated cells now freeze in the "1" configuration. At the end, the whole word has been written with 32 pulses of heating current (assuming a 32-bit word). The corresponding energy is ~1 pJ per dot plus two pulses of magnetic field.

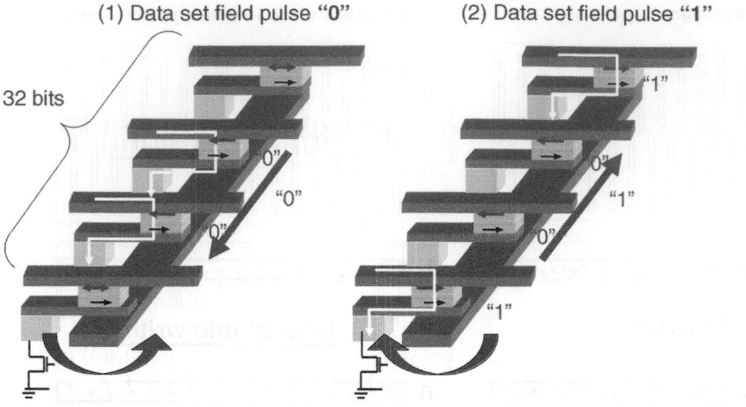

Figure 5.22 Illustration of write scheme with field sharing in TA-MRAM. A whole word can be written with only two pulses of magnetic fields.

Furthermore, in TA-MRAM, since the thermal stability of the storage layer magnetization in standby is provided by exchange coupling to an adjacent antiferromagnetic layer, the cell can have a circular shape. As a result, once the writing is enabled by heating the cell above the blocking temperature of the antiferromagnetic layer, only a low field of a few millitesla (2–5 mT) is sufficient to switch the storage layer magnetization since the cell has no shape anisotropy.

For comparison, consider the case of toggle MRAM. Writing a bit in a toggle MRAM requires two orthogonal pulses of magnetic fields. To write a 32 bits word, one of these pulses can be shared by the 32 bits of the same word by sending a pulse of current along the corresponding word line. However, 32 pulses of magnetic field must be independently generated along the corresponding 32 bit lines. This means that 33 pulses of magnetic field of about 10 mT each must be generated in toggle MRAM instead of two pulses of about 5 mT for TA-MRAM.

5.6.4 TA-MRAM with Soft Reference: Magnetic Logic Unit (MLU)

In a second possible implementation of TA-MRAM, a soft reference (SR) layer is made of a material with easily switchable magnetization. The storage layer is still exchange-biased by an adjacent antiferromagnetic layer as in standard TA-MRAM (see Fig. 5.23). The writing of the storage layer is achieved similar to the thermally-assisted MRAM devices by the application of a combination of an external field and a heating pulse that allows the storage layer to be switched and then pinned in the opposite direction as the system cools back below its antiferromagnetic blocking temperature. The reading is performed in two steps: the SR magnetization is set in a first direction, and the MTJ resistance is measured. The SR magnetization is then switched to the opposite direction and the new resistance is measured. The resistance variation between the two measurements yields the magnetic orientation of the storage layer. In this approach, the read cycle is longer (~50 ns) but the tolerance to process variation is greatly enhanced since each bit is self-referenced. Indeed, for the standard reading scheme, the two resistance state distributions have to be well separated in order to avoid read errors. This implies good dot size and shape control.

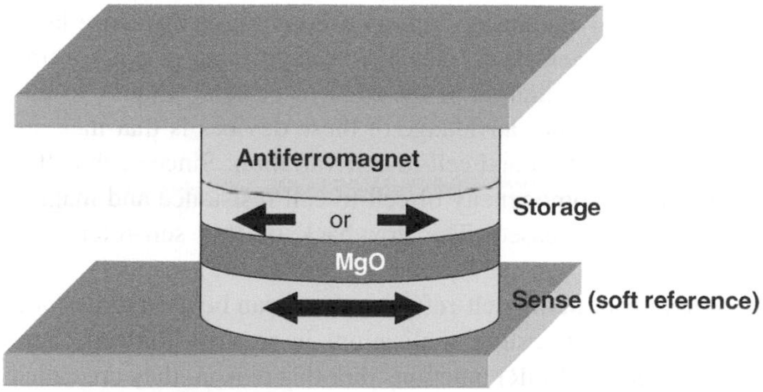

Figure 5.23 Schematic representation of a self-referenced MRAM with thermally-assisted write.

In contrast, for the SR reading scheme, the difference of resistance between the two states is used to read the junction, and is thus not sensitive to dot size variation. This is particularly useful for small technological nodes where the dot to dot resistance and magnetoresistance variability tend to significantly increase.

5.6.4.1 Principle of Reading with Soft Reference The reading scheme consists in switching the free layer in a first predetermined direction (along the stable directions of storage layer magnetization) and then in the opposite direction. The change of resistance associated with these two states due to the TMR effect, either increasing or decreasing, is used to infer the storage layer magnetization direction. A typical normalized R(H) cycle is shown in Fig. 5.24. This reading can be performed

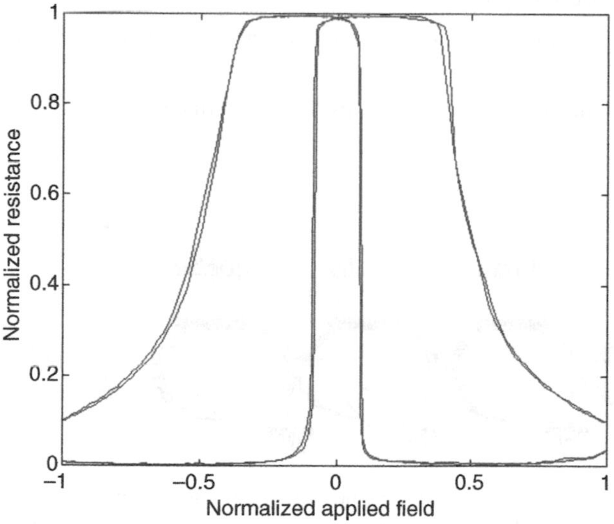

Figure 5.24 Normalized resistance response as a function of the external field for a self-referenced MRAM at standby temperature in state “0” (blue) and “1” (red). The resistance transition at low field is associated with the switching of the sense layer magnetization.

in a two-step quasi-static way or dynamically with an oscillation of the sense layer magnetization away from a 90° orientation, with the storage layer magnetization yielding an upward or downward oscillation of the MTJ resistance. Thanks to this self-referenced readout scheme, a major advantage of these devices is that they are much more tolerant of process variation and cell-to-cell variation. Since each cell is self-referenced, the constrains on homogeneity of cell-to-cell resistance and magnetoresistance amplitude are much released. The drawback of this self-referenced scheme is a slower readout process.

Besides its MRAM application, this self-referenced cell can be used to perform logic operations with particularly interesting applications in security. Indeed, these devices combine memory and logic (XOR) functions. For this reason, they are called Magnetic Logic Units (MLU™) (95). In addition to being storage devices, self-referenced MRAM intrinsically allows performing logic comparison functions. Considering the storage and the sense layer magnetizations as two inputs, this magnetic stack outputs the exclusive OR function of these logic inputs through the magnetoresistance level. Both inputs are set using pulses of magnetic field, with first the storage layer being written with a heating pulse and then the sense layer switched at room temperature. By using a set of self-referenced MRAM cells connected in series to form a NAND chain, a Match-In-Place™ engine can be created, as illustrated in Fig. 5.25, wherein a pattern applied to the sense layers is compared to the written pattern stored in the storage layers (96). Any mismatch between the two patterns systematically results in a higher resistance than for a perfect match. The expected advantages of this architecture are as follows:

- The stored patterns are never read, and never exposed to potential hackers.
- The matching cycles are very quick and could be orders of magnitude faster than existing methods and use less power.
- Match-In-Place engines could act as a hardware accelerator, and simplify the overall chip.
- The cost structures of the secure chips are considerably reduced.

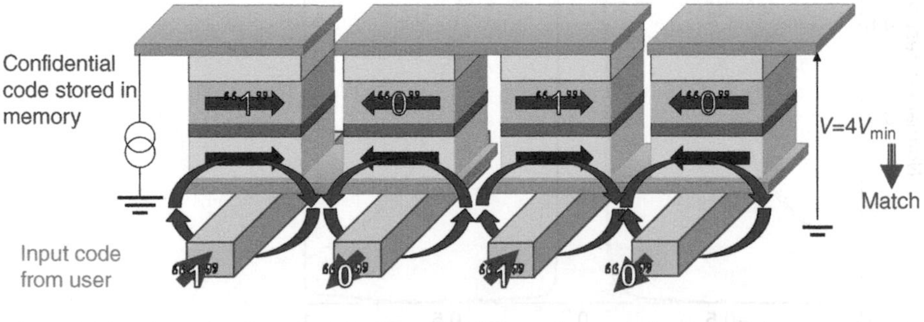

Figure 5.25 Example of a Match-In-Place™ (MIP) engine made by the serial connection of four MLUs in a NAND chain. The input binary pattern "1 0 1 0" is compared with the stored binary pattern "1 0 1 0". When the input pattern matches the stored pattern, the serial resistance of the chain is at its minimum level. If the input pattern does not perfectly match the stored pattern, the resistance of the MIP is larger.

The basic Match-In-Place architecture can be built at three levels: The single cell for direct matching at the bit level, NAND chains that combine multiple individual cells in series for linear Match-In-Place engines, and a matrix that combines multiple NAND chains in parallel to match one input pattern to a stack of stored patterns.

When the input bit matches a stored bit (a "0" with a "0" or "1" with a "1"), the cell measures a match. Conversely, if the input data do not match the cell's contents (a "0" with a "1" or "1" with a "0"), the cell measures a mismatch. Unlike a RAM-based content-addressable memory (CAM), an MLU-based Match-In-Place cell retains the stored bit when power is removed. This avoids any need to reload the stored information, a step that exposes the information to hacking. An MLU-based Match-In-Place engine can handle unlimited write cycles, and a read or match cycle does not disturb the stored information. The read cycle of an individual cell is extremely fast; a compare can occur in as little as 5 ns. Program cycles can be implemented through a separate path with a cycle time of about 20 ns.

5.6.4.2 Content-Addressable Memory These Match-In-Place engines can be assembled to design fast associative memories and notably CAM, which have applications in search engines and routers for instance, where incoming data packets are directly routed to the recipient address. An example of such a CAM is illustrated in Fig. 5.26. Fields of use of these new architectures are quite wide and include secure microcontrollers, subscriber identity module (SIM) cards, banking cards, biometric authentication chips, near-field communication, and hardware acceleration. Expected benefits are enhanced security, lower cost, and faster response time.

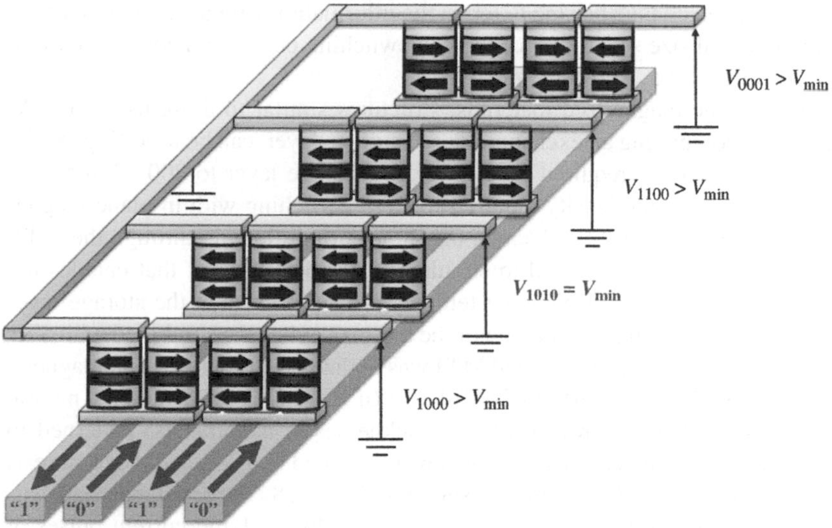

Figure 5.26 Example of a content-addressable memory (CAM) with four Match-In-Place engines. The input binary pattern "1 0 1 0" is compared with the memory content and matched with the lowest voltage for a set reading current value.

5.6.5 Thermally-Assisted STT-MRAM

So far in Section 5.6, MRAM using write scheme based on thermal assistance combined with pulses of magnetic field have been described. Alternatively, thermal assistance can also be used in combination with spin-transfer torque to extend the scalability of STT-MRAM both with in-plane and out-of-plane magnetized materials. This is explained in the following sections.

5.6.5.1 In-Plane STT Plus TA-MRAM As explained earlier, there are several advantages in using a thermally-assisted MRAM concept, the main one being the decoupling of the thermal stability (memory retention) and the write power consumption (writeability), since the bit can have simultaneously a low write field at the write temperature and be stable in standby in the whole operating temperature range. The problem with the field-driven writing of TA-MRAM cells is still that the magnetic field needs to be generated by a current line with current pulses of a few milliamperes. TA-MRAM requires a single magnetic field and lower field values compared to the toggle MRAM approach, thus lowering the total power consumption. However, the write field does not scale with cell size and can be at best kept constant, unlike STT-MRAM where the write current scales with cell size. In TA-MRAM, the heating current is not the bottleneck, since the use of thermal barriers has already demonstrated a heating current density in the $1–2 \times 10^6$ A/cm^2 range, similar to low values of spin-transfer torque MRAM cells (STT-MRAM). Alternatively, it is possible to get rid of the field line by still using the thermally-assisted concept but combining it with spin-transfer torque to switch an exchange-biased storage layer (89). In this case, the same current flowing through the cell is used both to heat up the cell and switch the storage layer magnetization by STT. It is thus possible to combine the added stability obtained from the exchange biasing to retain the information with the reduction of the current through cell size scaling, since the cell switching occurs at a constant current density, typically in the 10^6 A/cm^2 range.

With in-plane magnetized materials, structures similar to those used for TA-FIMS-MRAM, comprising an exchange-biased storage layer, can be used (Fig. 5.27). Since the current density required for heating the storage layer to 200 °C in 5 ns is comparable to the current density required for STT switching with in-plane magnetized materials (both ~$2–4 \times 10^6$ A/cm^2), the same current flowing through the MTJ, downward or upward, can be used to simultaneously heat the cell that enables the writing by reaching the AFM blocking temperature and to switch the storage layer magnetization. The first implementation of the thermal assistance combined with STT (TAS + STT) with in-plane magnetized MTJ was realized in 2009 (97). The magnetic stack was Ta 3/CuN 30/Ta 5/PtMn 20/CoFe 2/Ru 0.74/CoFeB 2/MgO 1.1 natural oxidation/CoFeB 2/NiFe 3/IrMn 6.5/Ta 5 (thicknesses in nanometers) patterned to 140 nm diameter circular cells. The storage layer is pinned by IrMn and the hysteresis cycles show a pinning field of ~5 mT, as shown in Fig. 5.28. The resistance state was switched between the low- and high-resistance states by applying current pulses of alternating polarity across the junction. Each pulse first creates a temperature increase above the antiferromagnet blocking temperature. With the ferromagnetic layer no longer pinned, the spin-polarized current simultaneously exerts a torque on the

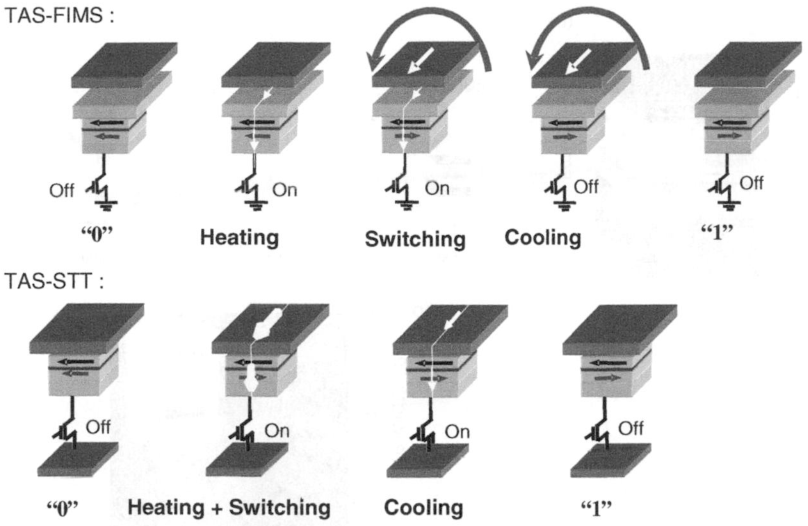

Figure 5.27 TA write principle combined with FIMS or STT. In TA-FIMS, the temporary heating of the MTJ is combined with a pulse of magnetic field. In TA-STT, the current flowing through the MTJ both heats the MTJ and exerts the magnetic torque that switches the magnetization.

ferromagnetic storage layer, reversing its magnetization direction depending on the current direction. The write voltage is then gradually decreased to zero so that the junction cools down while STT is still on. The antiferromagnet then freezes the new storage layer magnetization direction. Figure 5.28 shows this writing process with 30 ns pulses applied in a sequence of positive and negative polarity resulting in a zero-field resistance that is alternating between the low- and high-resistance states. Additionally, there is also a change in the storage layer pinning direction as can be seen from the alternating positive and negative exchange bias field values. The critical current density required for switching the exchange pinned layer was 8×10^6 A/cm^2 and 13×10^6 A/cm^2, respectively, for the antiparallel to parallel and the parallel to antiparallel transitions. Similar structures without exchange biasing resulted in critical current densities of 8×10^6 A/cm^2 and 13×10^6 A/cm^2, virtually identical to thermally-assisted cells, meaning that the heating is not the limiting factor and that the exchange pinning does not increase the critical current density for STT switching. This shows how the thermally-assisted approach can benefit the STT-MRAM concept.

5.6.5.2 Out-of-Plane STT Plus TA-MRAM Ultimately, thermally induced anisotropy reorientation (TIAR) (86) can be used to assist the spin-transfer torque switching in perpendicular magnetic tunnel junctions. The TIAR provides a way to reorient the direction of the free layer magnetization with heating. Indeed, the magnetization of an out-of-plane magnetized layer at room temperature can reorient in the thin film plane when heated due to the different temperature dependence of the out-of-plane anisotropy and demagnetizing energy (98). One should notice that

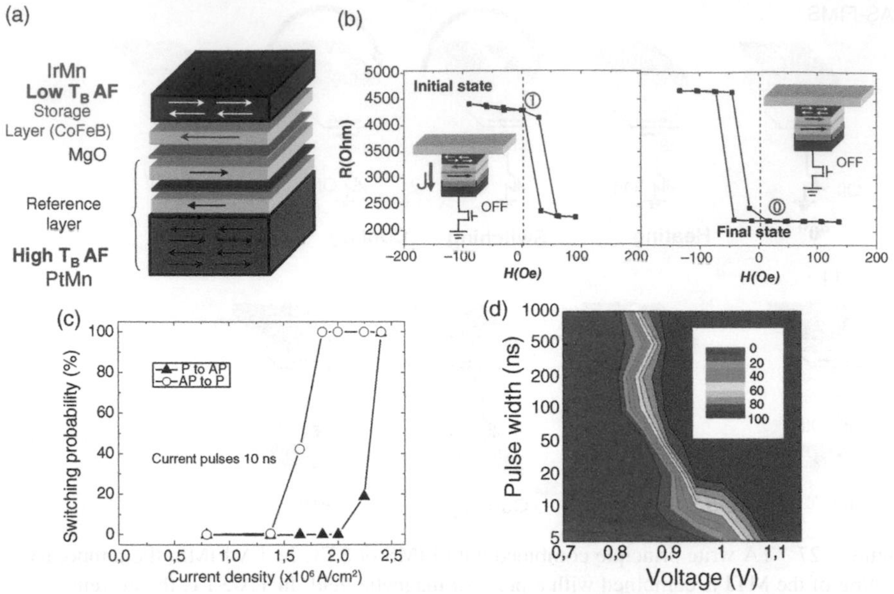

Figure 5.28 Illustration of STT + TA switching using exchange-biased in-plane magnetized storage layer. (a) Stack used. (b) Magnetoresistance loops showing the reversal of the exchange pinning direction after applying 30 ns pulses. The inset represents the exchange bias field after pulses of alternating current direction. (c): Switching probability from *P* to *AP* and *AP* to *P* versus current density for current pulses 10 ns long. (d) Color map showing the switching probability versus voltage pulse amplitude and voltage pulse duration.

heating is always occurring when a current is injected in a MTJ during writing. This heating assists STT switching, but materials used in MTJ are usually not optimized to take advantage of this effect. Oppositely, TIAR-assisted switching consists in optimizing the temperature dependence of the MTJ magnetic properties to decrease the STT critical current while keeping a satisfying thermal stability under standby conditions. In that latter case, the current is responsible for three phenomena: STT, heating, and TIAR. Such a switching method is promising to achieve highly scalable MRAM cells with low power consumption and reliable writing.

Figure 5.29 presents the proposed writing scheme: the MTJ exhibits a large thermal stability factor in standby on the operating temperature window. During write, a current pulse is sent through the junction. Due to the inelastic relaxation of tunneling electrons, heating occurs so that the PMA decreases. As a result, the free layer magnetization falls into the thin film plane, while the reference layer is engineered so as to keep its out-of-plane magnetization. This latter electrode polarizes the current in the out-of-plane direction. The STT, due to this spin-polarized current, pulls the free layer magnetization in the upper or lower hemispheres depending on the current direction and induces large angle out-of-plane precessions (31). In such a configuration, the STT effect is highest since the spin polarization is almost perpendicular to the magnetization. It results in a more reproducible switching since no thermal fluctuations are required to initiate the reversal. The injected current is then

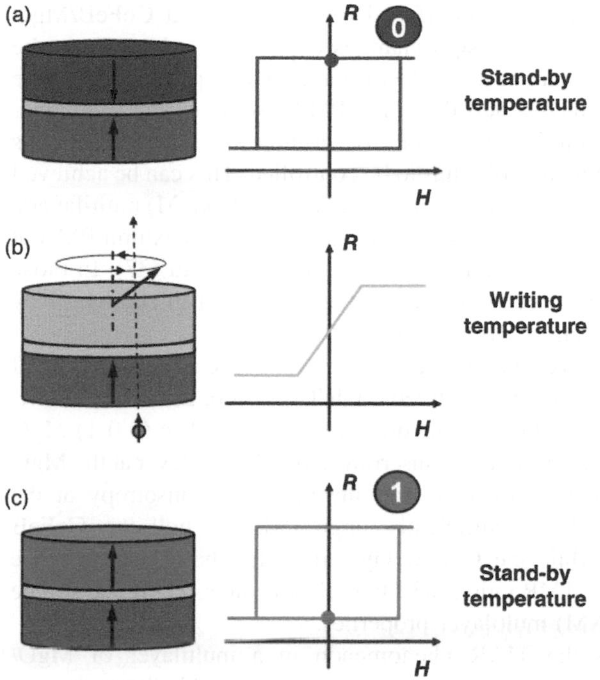

Figure 5.29 Principle of TIAR-assisted switching. (a) The junction exhibits a large PMA at standby temperature. (b) A current is sent through the junction. The FL magnetization undergoes a TIAR due to the heating. The STT induces a large-angle precessional motion of the FL magnetization and pulls it upward or downward depending on the current direction. (c) The FL recovers its PMA during the cooling when the current is gradually decreased to zero. Its magnetization rotates out-of-plane in the direction corresponding to the hemisphere in which it was pulled in by STT.

gradually decreased so that the junction cools while STT is still effective. The free layer recovers its PMA, and its magnetization gets reoriented out-of-plane, in the direction defined by the hemisphere into which it was previously pulled by STT. This switching approach reduces the write consumption as soon as the required current to heat the junction to the anisotropy reorientation temperature T_K is lower than the critical switching current in a standard MTJ (typically a few 10^6 A/cm^2). This condition is fulfilled in most common cases. In addition, the heating current density can be further reduced by adding in-stack thermal barriers on both sides of the MTJ.

In this particular writing scheme, the voltage required to switch the junction is mainly determined by the anisotropy reorientation temperature T_K, which sets the lower limit of the writing window. T_K can be estimated theoretically by calculating K^{eff} as a function of temperature. For a cubic or uniaxial anisotropy, it can be written to first order as

$$K(T) = K_{\perp}(0)\left(\frac{M_s(T)}{M_s(0)}\right)^m - (N_{perp} - N_{plan})\frac{\mu_0}{2}M_s(T)^2, \tag{5.28}$$

where $K_{\perp}(0)$ is the perpendicular magnetic anisotropy constant at 0 K, $M_S(T)$ is the free layer magnetization, and $-(N_{perp} - N_{plan})(\mu_0/2)M_s(T)^2$ is the demagnetizing energy, N_{perp} and N_{plan} being the demagnetizing factors of the pillar. The exponent m depends on the origin of the PMA. For a single ion bulk anisotropy, $m = i(i+1)/2$, where i denotes the anisotropy constant order, equal to 2 to first order in the most common case of a uniaxial anisotropy so that $m = 3$. However, this exponent can differ in the case of an anisotropy of interfacial origin such as at CoFeB/MgO interface. From this equation, the K^{eff} sign changes beyond T_K, where T_K is the temperature where $K^{eff} = 0$. In that situation, the in-plane anisotropy induced by the demagnetizing energy becomes larger than the PMA, so that the free layer magnetization falls into the thin film plane. In order to tune T_K, Eq. (5.25) shows that the evolution of M_S as a function of T has to be controlled. This can be achieved by tuning the Curie temperature T_C of the magnetic material. (Co/NM) multilayers, where NM is a nonmagnetic metal, provide a wide range of T_C, and exhibit PMA at room temperature when NM is a noble metal such as Pt, Pd, or Au (37). In those systems, the T_C can be easily tuned by changing the Co or NM thickness, or by doping Co with Ni or nonmagnetic elements (99–101).

However, such multilayers cannot be used alone as magnetic electrodes in MTJ for MRAM applications, which require relatively high TMR ratio. Indeed, (Co/NM) multilayers present an fcc (1 1 1) texture that does not match the bcc (0 0 1) MgO texture (102). This problem can be solved by inserting a thin CoFeB layer at the MgO interface: the (Co/NM) multilayer together with the interfacial anisotropy at the CoFeB–MgO interface provides a sufficiently large PMA to pull the CoFeB magnetization out-of-plane, while the CoFeB layer provides the appropriate bcc texture required to obtain large TMR ratio (103). In such structures, TIAR can still be obtained by tuning the (Co/NM) multilayer properties.

Figure 5.30 illustrates the TIAR phenomenon in a multilayer of MgO/CoFeB1.5/Ta0.2/(Pd1.2/Co0.3) × 3 (thickness in nanometers). In this figure, hysteresis loops have been measured on a full sheet sample by polar magneto-optical Kerr effect as a function of temperature, with an out-of-plane applied field. At room temperature, the loops have square-loop and coercive behavior, indicating out-of-plane magnetization. When the temperature increases, the coercivity decreases while the signal amplitude decreases due to the decrease of magnetization. At 150 °C, perpendicular domains appear, leading to a loop with no remanence. The anisotropy reorientation occurs at 175 (±20) °C. Above this temperature, the loops exhibit the reversible linear behavior expected from a measurement along the hard axis direction. When the sample is cooled down, the loop recovers its squareness and coercivity, indicating that the TIAR phenomenon is reversible.

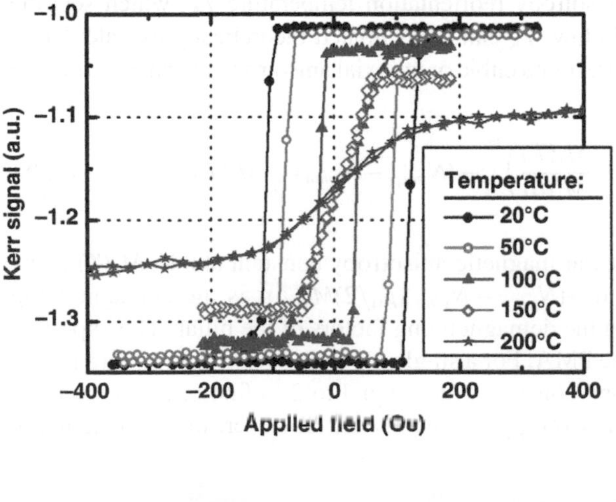

Figure 5.30 Hysteresis loops of CoFeB 1.5/Ta0.2/(Pd 1.2/Co 0.3) × 3/Pd 2 for different measurement temperatures with out-of-plane applied field.

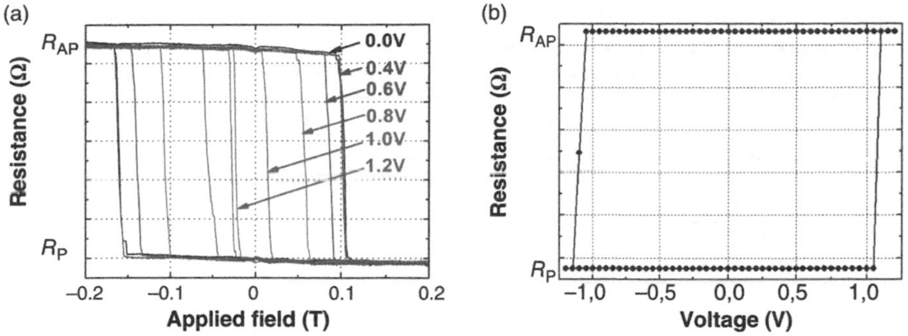

Figure 5.31 (a) Hysteresis loop of a 110 nm diameter pillar with 30 ns-long applied pulses of various amplitudes. (b) STT + TAS resistance switching.

The magnetic tunnel junction also requires a reference electrode. For that electrode, a (Co/Pt)-based synthetic antiferromagnet (SAF) can be used. Such a SAF is important for MRAM technologies, since they create a reduced stray field on the storage layer magnetization, thus avoiding different thermal stabilities Δ in the P and AP magnetic configurations as well as asymmetric P to AP and AP to P write currents (104,105). In order to study the effect of heating, the hysteresis loops were then measured while applying 30 ns long pulses of square shape of increasing voltage at each field step (0.2 mT). The resistance of the junction was measured after each pulse (see Fig. 5.31a). Figure 5.31 illustrates the very strong decrease in switching field associated with the heating of the storage layer due to the voltage pulse. The possibility to switch the magnetization without field just by combining spin-transfer torque with heating is then shown in Fig. 5.31b (86). With 30 ns long pulses, the switching voltage is 1.15 ($\pm$0.05) V, corresponding to a current density of $j_w = 4.6 \times 10^6$ A/cm^2. At the time of this study (2011), this current was of the same order as the values reported in the literature using a standard Ta/CoFeB/MgO free layer and no thermal writing assistance. However, in order to calculate the switching efficiency, the figure of merit Δ/I_{c0} gives a better idea of the writing efficiency as discussed for Eq. (5.27).

Table 5.1 gives a comparison of this figure of merit Δ/I_{c0} between the standard STT switching in perpendicular MTJ and the TIAR-assisted switching at the time of this study. This table shows reported write current densities corresponding to 30 ns long pulses. The thermal stability at 300 K given here are extrapolated for $d = 40$ nm diameter junctions and 30 ns pulse width (47,86,106).

Under these assumptions, it is clear that TIAR-assisted STT switching has improved efficiency compared to the standard STT switching. This is even more striking regarding the damping constant α of the different kinds of free layer: whereas it is equal to 0.013–0.03 in a Ta/CoFeB/MgO stack, it is of about 0.15 in a CoFeB/(Pd/Co) stack (107). The efficiency of the TIAR-assisted STT writing could be further optimized by implementing thermal barriers in the stack. It has been shown that with this kind of barriers, a temperature of about 200 °C can be obtained with an RA product of about 30 Ω·μm^2 and a current density of about 10^6 A.cm^{-2} (97,108). The thermal stability Δ at operating temperature could be enhanced,

TABLE 5.1 Comparison of Writing Current Densities at 30 ns, j_{write}; Corresponding Current I_c for 40 nm Diameter Cell, Thermal Stability Factor Δ at 300 K for 40 nm Large Devices and Figure of Merit Δ/I_c of Standard STT Switching (47,106) and TIAR-Assisted STT Switching (86).

Reference	Material	Writing principle	j_{write} (MA/cm^2)	I_c(30 ns) (μA)	Δ(300 K)	Δ/I_c (μA^{-1})
106	Ta/CoFeB/MgO	STT	3.8	48	43	0.89
47	Ta/CoFeB/MgO	STT	7.4	93	50	0.54
86	CoFeB/(Pd/Co)	TIAR + STT	4.6	58	73	1.26

keeping a constant writing consumption by controlling T_K, leading to even more scalable MTJ.

In conclusion, TIAR plus STT writing could help to extend the downsize scalability of perpendicular STT-MRAM to even smaller dimensions than discussed in Section 5.5.5.4.

5.7 THREE-TERMINAL MRAM DEVICES

Among recent emerging nonvolatile memory, spin-transfer torque magnetic random-access memory (STT-MRAM) has been identified by the ITRS as one of the two most credible candidates. However, STT-MRAM still suffers from a lack of speed for the fastest cache applications and from potential endurance limitations due to the large current injected through the tunnel barrier for switching the magnetization. Indeed, at each write event, the tunnel barrier is submitted to a pulse of voltage on the order of 0.5 V, 10 ns long in conventional STT-MRAM, 0.7–1 V in thermally-assisted schemes that use larger *RA* than STT-MRAM. Since the barrier is typically 1 nm thick, this represents an electrical field across the oxide barrier on the order of 10^9 V/m, which is approaching the values for which electrical breakdown of the barrier may occur. In order to circumvent this problem, several groups have proposed concepts of three-terminal STT-MRAM cells (represented in Fig. 5.1c) in which the write current does not flow across the tunnel barrier, thereby limiting the electrical stress during write. These approaches also open new possibilities in the design of the memory architecture or for logic applications since the write and read current paths are different.

5.7.1 Field versus Current-Induced Domain Wall Propagation

Figure 5.32 gives an experimental comparison of the domain wall (DW) dynamics that can be induced in a magnetic stripe by application of a magnetic field or by a current flowing in the stripe. The blue and red colors correspond to magnetic domains magnetized in opposite direction along the wire. In Fig. 5.32b, a field is applied parallel to the magnetization of the blue colored domains and correspondingly antiparallel to the red colored domain. The field therefore favors the blue domain magnetic orientation. As a result, the blue domains expand as a function of

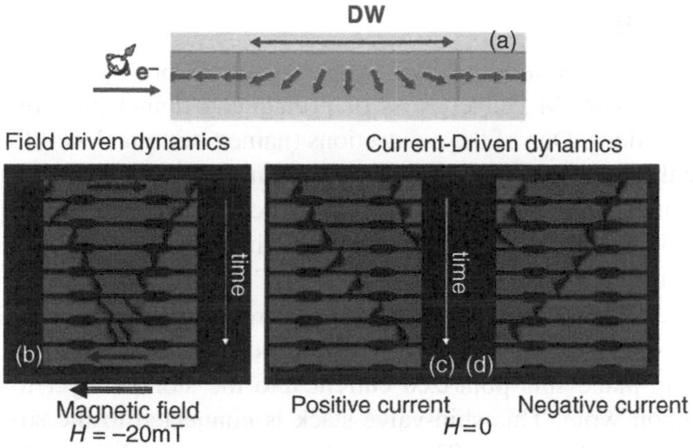

Figure 5.32 (a) A domain wall is a transition region separating two magnetic domains magnetized in different directions. The wall has a certain width to minimize the cost in exchange energy associated with the change in the direction of the magnetization. The domain wall width is determined by a balance between exchange energy and anisotropy energy. (b–d) The blue and red regions represent domains of opposite magnetization along a magnetic stripe. Part (b) illustrates the domain wall dynamics induced by a pulse of magnetic field. Parts (c and d) illustrate the dynamics induced by a current flowing through the magnetic stripe. (Courtesy of Stuart Parkin, IBM, unpublished.)

time, while the red domain shrinks to the point where the red domain disappears. The domain size is therefore not conserved under field induced dynamics. In contrast, Fig. 32c and d illustrate the domain wall dynamics induced by a positive or a negative current flowing through the magnetic stripe. The current gets spin polarized when traversing the domains and exert a spin-torque on the walls separating the domains. The spin-transfer torque on the domain wall contains two terms called adiabatic STT and nonadiabatic STT. The adiabatic STT corresponds to the situation where the spin of the electrons continuously follow the local magnetization direction (dominant for large walls and strong exchange interaction between the spin of the conduction electrons and the spin of the electrons responsible for the local magnetization), whereas the nonadiabatic STT corresponds to situation where the spin of the conduction electrons gets re-oriented on a length scale much longer than the domain wall width. The dominant effect of the adiabatic STT term is to distort the shape of the domain wall, whereas the effect of the nonadiabatic term is effectively to exert a pressure on the wall and induce its motion. Quite interestingly, all the walls located along the wires are submitted to the same pressure so that they move coherently in the same direction determined by the current direction. This is illustrated in Fig. 32c and d for two opposite current directions. Remarkably, because the walls move at the same speed under the current flow, the size of the domains does not vary and the whole domain pattern shifts coherently along the magnetic stripe. Domain wall speeds on the order of a hundred to several hundreds of meters per second (i.e., nanometers per nanosecond) are achieved with current density on the order of 1 to a few 10^7 A/cm^2.

5.7.2 Principle of Writing

One of the proposed concepts of three-terminal MRAM based on DW propagation is illustrated in Fig. 5.33: Each MRAM cell consists of a magnetic tunnel junction having three electrical connections. One of the connections (named bottom electrode in Fig. 5.33) is, as usual, at the bottom of the cell, and most often connects the bottom part of the MTJ (i.e., the underlayer of the antiferromagnetic pinning layer) to a selection transistor. In addition, two other contacts are connecting the top part of the MTJ, that is, its storage layer. One of the contacts is made of a Cu spacer plus another pinned ferromagnetic layer. This contact together with the free layer of the MTJ constitutes a so-called spin-valve stack. The pinned ferromagnetic layer of the spin-valve is used to inject an in-plane spin-polarized current into the storage layer to switch its magnetization upon write. This spin-valve stack is connected to the so-called top electrode (see Fig. 5.33a and b). The second top contact to the middle electrode is made of a simple nonmagnetic metal such as Cu or Al. Writing an information in the memory means, as for conventional MRAM cell, orientating the

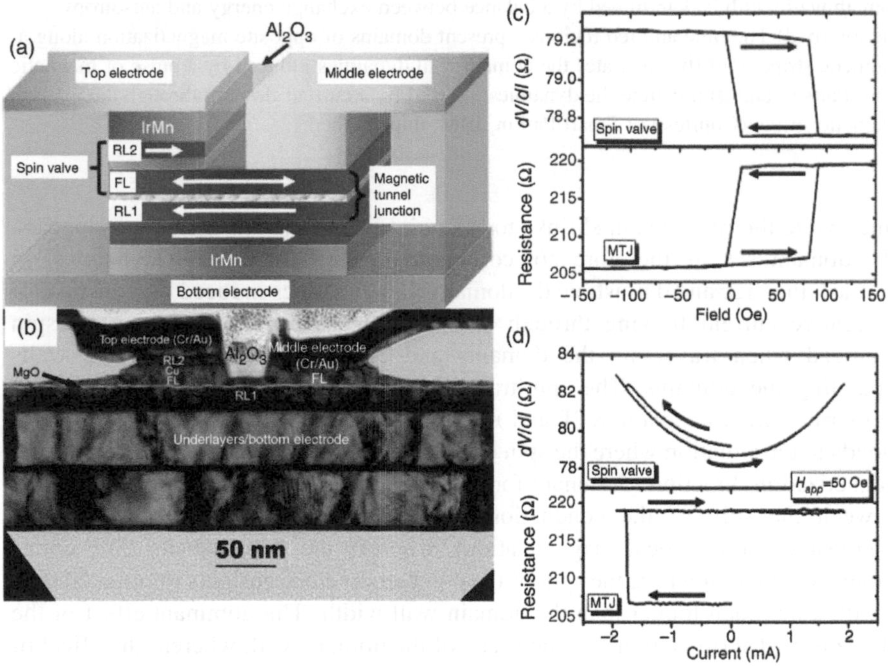

Figure 5.33 Example of three-terminal MRAM cell proposed by HGST (109). (a) Schematic representation of the MRAM cell. (b) Cross-sectional transmission electron microscopy view of the device. (c) Simultaneous measurements of the spin-valve differential resistance response versus field (measured between top and middle electrodes) and of the tunnel junction response (measured between the top and bottom electrodes). (d) Simultaneous measurements of the spin-valve differential resistance response versus current flowing from top and middle electrodes (measured between top and middle electrodes) and magnetoresistance of the tunnel junction response (measured between the top and bottom electrodes).

magnetization of the storage layer (FL in Fig. 5.33) parallel or antiparallel to that of the reference layer (RL1 in Fig. 5.33) of the MTJ. This is achieved by sending a pulse of current between the top electrode and the middle electrode. This current flows through the top contact, then in the plane of the storage layer, then in the middle contact to the middle electrode. By traversing the pinned layer of top contact, the conduction electrons become spin polarized. They then exert a spin-transfer torque on the left extremity of the storage layer which nucleates a reversed domain underneath the contact. At this point, the storage layer is divided into two magnetic domains, one nucleated underneath the top contact, and one in the original magnetization direction. These two domains are separated by a domain wall. Still, due to the spin-transfer effect, the current flowing in the free layer exerts a pressure on this domain wall, which then propagates across the entire free layer leading to a complete switching of its magnetization. The magnetization can then be switched back by using a write current of opposite polarity. By this mechanism, the free layer magnetization can be switched without current flowing through the barrier.

More recently, novel three-terminal memory concepts, called "spin–orbit torque" MRAM (SOT-MRAM) (109–113), were proposed and demonstrated. A current flowing in the plane of a magnetic multilayer with structural inversion asymmetry, such as Pt/Co (0.6 nm)/AlO_x, exerts a torque on the magnetization, due to the spin–orbit coupling. This torque can lead to magnetization reversal, with a switching time shorter than 500 ps (114) and a writing energy one tenth that in current STT-MRAM. The key advantage of the SOT-MRAM is that write and read are decoupled due to independent current paths. The SOT-MRAM thus naturally solves the reliability problems in current STT-MRAM with a potentially infinite endurance. This new concept is particularly interesting for cache-compatible high-speed applications (SRAM-type) for which density is not the primary concern. As future objectives, writing current needs to be further lowered to match advanced transistor outputs and remain much below electromigration threshold.

5.7.3 Advantages and Drawbacks of Three-Terminal Devices

The readout is performed as usual by sending a current from the top contact to the bottom contact and thereby measuring the resistance of the tunnel junction. This approach therefore allows for an SOT writing scheme without tunnel barrier wear-out issues while retaining the benefits for read-out of the large TMR signals from MTJs. In this approach, write and read current paths are separated which is not the case in STT-MRAM. The drawback is the increased cell size due to the requirement of making two isolated contacts with associated interconnects on top or bottom of each memory cell.

5.8 COMPARISON OF MRAM WITH OTHER NONVOLATILE MEMORY TECHNOLOGIES

5.8.1 MRAM in the International Technology Roadmap for Semiconductors (ITRS)

From an application perspective, the demand for on-chip memories has been increasing both on an absolute basis and relative to overall chip transistor counts.

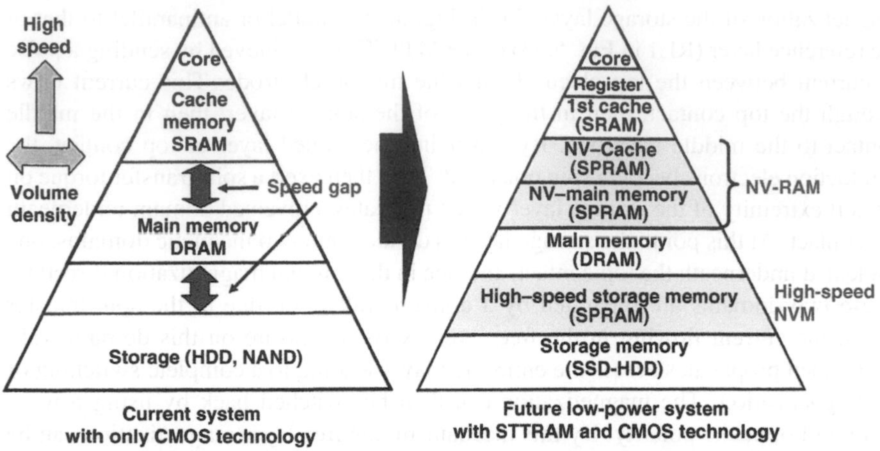

Figure 5.34 Need for introducing nonvolatility in the memory hierarchy (SPRAM stands for STT-MRAM) (115).

This is due to a confluence of factors such as the growth in demand for data storage and the increasing gap between processor and off-chip memory speeds. The move to multicore processors few years ago has only accelerated the demand for on-chip memory. As pointed out in the ITRS, one of the best solutions to limit power consumption and to fill the memory gap is the modification of the memory hierarchy by the integration of nonvolatility (NV) at different levels (storage-class memories, DRAM main working memory, SRAM cache memory), which would immediately minimize static power as well as pave the way toward normally off/instant-on computing (see Fig. 5.34).

The International Technology Roadmap for Semiconductors (ITRS) Emerging Research Devices and Emerging Research Materials Working Groups met in 2010 to evaluate eight emerging research memory technologies proposed to scale nonvolatile memory technology to and beyond the 16 nm node (ferroelectric gate field-effect transistor, nano-electromechanical switch, STT-MRAM, various types of resistive RAM, in particular redox RAM, nanothermal phase change RAM, electronic effects RAM, macromolecular memories, and molecular RAM). In conclusion, the ERD/ERM working groups "identified spin-transfer torque MRAM and redox RRAM as emerging memory technologies recommended for accelerated research and development leading to scaling and commercialization of nonvolatile RAM to and beyond the 16 nm generation (116)." Currently, there is an intense research and development effort in microelectronics on these two technologies, one based on spintronic phenomena, the other based on migration of vacancies or ions in an insulating matrix driven by oxidoreduction potentials. Both technologies could be used for standalone or embedded applications.

The potential benefits of spin-based memories are especially appealing when viewed in light of the exploding demand for on-chip memories. However, spin-based chips are still in an early stage of volume production. In order to realize their potential, there is a need to strengthen the technology maturity and for advances in circuit designs and innovative architectures.

5.8.2 Comparison of MRAM and Redox-RAM

The redox RAM combines good potential for scaling below 10 nm generation, fast read and write times, (<10 ns), and relatively low write current (in the microampere range). They should be stackable in three-dimensional architectures called cross-bar architectures and offer multilevel capabilities, thanks to the possibility to control the growth/dissolution of the conducting filaments, which determines the cell resistance level. However, the underlying physical mechanisms are based on statistical phenomena rather difficult to control at the scale of each dot leading to relatively large dot to dot variability. Predictive model of reliability are more difficult to establish for this technology. Finally, their endurance ($\sim 10^8$ cycles) is sufficient for Flash type applications but not enough for the working memory in microprocessors (which require $>10^{15}$ cycles). Therefore, these memory elements seem to be most suited for storage class memory applications (intermediate level between hard disk drives, solid-state drives (SSD), and DRAM) and memristor applications, thanks to the possibility that they offer to continuously vary their resistance between a minimum and a maximum value in a hysteretic way. This may open the path to neuromorphic architectures that mimic the working principle of the human brain.

The STT-MRAM appears today as the most credible candidate for embedded FLASH and SRAM replacement as it combines CMOS compatibility, high retention time (10 years), large endurance ($>10^{15}$ cycles), and fast write/read time (1–30 ns depending on the embodiments). A further goal is also DRAM replacement. However, this technology is still not yet mature enough for this high density application. The main difficulty remains associated with the difficulty to etch the MTJ stacks at small pitch (sub-20 nm) and with the related cell-to-cell variability. This variability is mainly caused by the edge defects generated during patterning of the cells. Whenever the MgO barrier is damaged by the patterning, this yields local changes in the barrier resistance, tunnel magnetoresistance, magnetic anisotropy, and correlatively, cell retention. Redeposition on the MTJ sidewalls during etching also contributes to the variability. Now that the number of laboratories (including major equipment suppliers) working on this technology has substantially increased, technological progress are much faster.

5.8.3 Main Applications of MRAM

According to recent market research reports, solid-state memory constitutes a market of over $50 billion, while the nonvolatile memory (NVM) segment is much smaller. Most laptops nowadays use NVMs, which retain data when the power is off, in the form of a solid-state drive in place of an HDD. The now ubiquitous smart cell (mobile) phones and other handheld devices also use NVM, but there is a trade-off between cost and performance. The cheapest NVM is Flash memory, which, among other uses, is the basis of small, portable Flash memory sticks. Flash memory, however, is slow and has a limited cyclability of 10^5 cycles, sufficient for a large number of storage applications ranging from memory sticks to digital camera memory to SSDs, but much lower than that of redox RAM or STT-MRAM.

Besides computers, today's portable electronics have become intensively computational devices as the user interface has migrated to a full multimedia experience. To provide the performance required for these applications, the portable electronics designer uses multiple types of memory: a medium-speed random-access memory for continuously changing data, a high-speed memory for caching instructions to the CPU, and a slower NVM for long-term information storage when the power is removed. Combining all of these memory types into a single memory has been a long-standing goal of the semiconductor industry. With such a memory, computing devices would become much simpler and smaller, more reliable, faster, and less energy-consuming.

As a result, advanced NVM chips are expected to see phenomenal growth in the next few years, with the global market increasing from \$209 million in 2012 to \$2,028 million by 2018 (117). This will occur in a number of applications: embedded system-on-chip (SOC) cards; radio-frequency identification (RFID) tags used in goods transported by high-speed detection conveyors; smart airbags used in automobiles; radiation-hardened memory in aerospace and nuclear installations; printed memory platforms (such as smart cards, games, sensors, display, storage-class memory network); and high-end smart mobile phones.

MRAM has the potential to become a "universal" memory device applicable to a wide variety of functions. As a matter of fact, MRAM can have similar performance to SRAM Cache 3 (switching time ~5 ns) but is nonvolatile. It has also similar density to DRAM but at lower power consumption since there is no need to refresh and with reduced leakage since MRAM can be powered off on standby. Besides, it is nonvolatile as Flash memory but much faster and suffers no degradation over time. It is this combination of features that makes it so attractive. Some suggest that it could replace SRAM, e-FLASH, DRAM, in embedded and standalone forms, some storage class memories, resulting in instant-on nonvolatile computers and tiny, super-fast and reliable portable devices.

In cell phones, handheld tablet computers, notebooks, personal digital assistants (PDAs), and other forms of mobile computing, MRAM is an attractive alternative to deploying both Flash and DRAM since it can save money and space. With software applications residing in memory, mobile devices could be instantaneously repowered up to exactly the same state in which they were when they were turned off. In general computing and networking, MRAM can be used to avoid boot-up delays and to provide faster access to hard drives and nonvolatile backup capabilities. At present, BIOS tend to use high-cost, low-density EEPROM or battery-backed-up SRAM—and volatile memory is used to alleviate I/O bottlenecks. In such applications, MRAM could prove much more economical.

In factory automation systems, microcontrollers and robots typically employ both RAM and PROMs/Flash. Lower costs will be achieved by replacing these two chips with one MRAM device. RFID tags need low-cost NVM, and a price point that makes MRAM economically viable for RFID applications will almost certainly push MRAM into other cost-sensitive areas. For aerospace use, MRAMs are radiation-hard, meaning that they can withstand ionizing radiation in contrast to most of semiconductor memories based on capacitor charge. This makes them suitable for use in airplanes (already in Airbus flight controllers), in satellites, and in spacecraft. They could also be used in nuclear environment such as nuclear power plants where conventional CMOS electronics devices fail when exposed to radiation.

5.9 CONCLUSION

MRAMs are expected to combine nonvolatility, high speed, moderate power consumption, infinite endurance, and radiation hardness, all at moderate cost and be easy to embed in devices. However, since its inception in the late 1990s, and the launching of Toggle MRAM products by EVERSPIN Technologies in 2006, MRAM has not yet reached large volume applications. The more recent advent of giant tunnel magnetoresistance in MgO-based MTJs and of spin-transfer torque, however, has shed a new light on MRAM with the promise of much improved performance and greater scalability to very advanced technology nodes. As a consequence, MRAM is now viewed as a credible replacement for existing technologies for applications where the combination of nonvolatility, speed, and endurance is key. Several start-up companies (e.g., Everspin Technologies, Crocus Technology, Avalanche, and Spin-Transfer Technology), large integrated device manufacturers (e.g., IBM, NEC, Toshiba, Hynix, TDK, TSMC, Renesas, Micron, Intel, and Samsung), and equipment suppliers (e.g., Canon Anelva, Applied Materials, Singulus, LAM, TEL and Keysight Technologies) are now actively developing STT-MRAM technology. The launch of 64 Mbit in-plane magnetized STT-MRAM products was announced by EVERSPIN Technologies at the end of 2012 as well as the forthcoming launch of 256 Mbit out-of-plane magnetized STT-MRAM products in 2016.

STT-MRAM has the potential of delivering the high density (stacking the magnetic cell with the selection transistor) and a scalable technology down to ~10 nm by using out-of-plane magnetized MTJs, taking advantage of the perpendicular magnetic anisotropy that exists at the CoFeB–MgO interface. The main difficulty for high density MRAM products currently remains the eching of the MTJ stack at small feature size and small pitch. Progress is also steadily being made in the composition of the stack to maximize the TMR amplitude, particularly in the perpendicular MTJ configuration (now above 240%), and in improvements in the temperature operating range (to minimize the decrease of the PMA with operating temperature).

Further downsize scalability in MRAM could be provided by combining thermally-assisted switching and spin-transfer torque writing to achieve the benefits of both: the stability of TAS and the full scalability of STT. This approach is promising for high-density, low-energy consumption, such as DRAM replacement.

Longer term approaches to reduce the power consumption in MRAM are envisioned based on a different write approach which relies on the voltage control of interface anisotropy but this is still at the research level (118,119).

ACKNOWLEDGMENTS

We thank all our colleagues from Crocus Technology and SPINTEC for fruitful discussions and for their contributions to some of the results presented in this chapter:

Stéphane Auffret, Sebastien Bandiera, Bertrand Cambou, Yann Conraux, Marie-Thérèse Delaye, Clarisse Ducruet, Erwan Gapihan, Ken Mackay, Lucien Lombard, Jean-Pierre Nozieres, Céline Portemont, Liliana Buda-Prejbeanu, Bernard Rodmacq, Quentin Stainer, and Maria Souza. Some of the results presented in this chapter were obtained in projects supported by the French *Agence Nationale de la Recherche* (ANR SPIN, RAMAC, and PATHOS) and the European Commission (ERC Adv Grant HYMAGINE n° 246942, ERC Adv Grant MAGICAL n° 669204).

REFERENCES

1. S. Yuasa, T. Nagahama, A. Fukushima, Y. Suzuki, and K. Ando, "Giant room temperature magnetoresistance in single-crystal Fe/MgO/Fe magnetic tunnel junctions," *Nat. Mater.* **3**, pp. 868–871 (2004); doi: 10.1038/nmat1257.
2. S. S. P. Parkin, C. Kaiser, A. Panchula, P. M. Rice, B. Hughes, M. Samant, and S.-H. Yang, "Giant tunnel magnetoresistance at room temperature with MgO (1 0 0) tunnel barriers," *Nat. Mater.* **3**, pp. 862–867 (2004); doi: 10.1038/nmat1256.
3. J. A. Katine, F. J. Albert, R. A. Buhrman, E. B. Myers, and D. C. Ralph, "Current-driven magnetization reversal and spin-wave excitations in Co/Cu/Co pillars," *Phys. Rev. Lett.* **84**, pp. 3149–3152 (2000); doi: 10.1103/PhysRevLett.84.3149.
4. Y. Huai, F. Albert, P. Nguyen, M. Pakala, and T. Valet, "Observation of spin-transfer switching in deep submicron-sized and low-resistance magnetic tunnel junctions," *Appl. Phys. Lett.* **84**, pp. 3118–3120 (2004); doi: 10.1063/1.1707228.
5. J. A. Osborn, "Demagnetizing factors of the general ellipsoid," *Phys. Rev.* **67**, pp. 351–357 (1945); doi: 10.1103/PhysRev.67.351.
6. M. Beleggia, M. De Graef, Y. T. Millev, D. A. Goode, and G. Rowlands, "Demagnetization factors for elliptic cylinders," *J. Phys. D Appl. Phys.* **38**, pp. 3333–3342 (2005); doi: 10.1088/0022-3727/38/18/001.
7. H. Sato, M. Yamanouchi, K. Miura, S. Ikeda, H. D. Gan, K. Mizunuma, R. Koizumi, F. Matsukura, and H. Ohno, "Junction size effect on switching current and thermal stability in CoFeB/MgO perpendicular magnetic tunnel junctions," *Appl. Phys. Lett.* **99**, 042501 (2011); doi: 10.1063/1.3617429.
8. H. Sato, M. Yamagouchi, S. Ikeda, S. Fukami, F. Matsukura, and H. Ohno, "Perpendicular anisotropy CoFeB-MgO magnetic tunnel junctions with a MgO/CoFeB/Ta/CoFeB/MgO recording structure," *Appl. Phys. Lett.* **101**, 022414 (2012); doi: 10.1063/1.4736727.
9. H. Tomita, S. Miwa, T. Nozaki, S. Yamashita, T. Nagase, K. Nishiyama, E. Kitagawa, M. Yoshikawa, T. Daibou, M. Nagamine, T. Kishi, S. Ikegawa, N. Shimomura, H. Yoda, and Y. Suzuki, "Unified understanding of both thermally-assisted and precessional spin-transfer switching in perpendicularly magnetized giant magnetoresistive nanopillars," *Appl. Phys. Lett.* **102**, 042409 (2013); doi: http://dx.doi.org/10.1063/1.4789879.
10. L. Savtchenko, B. Engel, N. Rizzo, M. DeHerrera, and J. A. Janesky, "Method of writing to scalable magnetoresistance random access memory element," U.S. Patent 6,545,906 (2001).
11. B. Dieny, J. P. Gavigan, and J. P. Rebouillat, "Magnetisation processes, hysteresis and finite-size effects in model multilayer systems of cubic or uniaxial anisotropy with antiferromagnetic coupling between adjacent ferromagnetic layers," *J. Phys. Condens. Matter* **2**, pp. 159–185 (1990); doi: 10.1088/0953-8984/2/1/013.
12. B. N. Engel, J. Åkerman, B. Butcher, R. W. Dave, M. DeHerrera, M. Durlam, G. Grynkewich, J. Janesky, S. V. Pietambaram, N. D. Rizzo, J. M. Slaughter, K. Smith, J. J. Sun, and S. Tehrani, "A 4-Mb toggle MRAM based on a novel bit and switching method," *IEEE Trans. Magn.* **41**, pp. 132–136 (2005); doi: 10.1109/TMAG.2004.840847.
13. N. D. Rizzo, "Narrow gap cladding field enhancement for low power programming of a MRAM device," U.S. Patent 6,559,511B1 (2001).
14. D. C. Worledge, "Spin–flop switching for magnetic random access memory," *Appl. Phys. Lett.* **84**, pp. 4559–4561 (2004); doi: 10.1063/1.1759376.

15. Everspin Technologies, Chandler, Arizona; www.everspin.com.
16. J. Slonczewski, "Currents and torques in metallic magnetic multilayers," *J. Magn. Magn. Mater.* **159**, pp. L1–L7 (1996); doi: 10.1016/0304-8853(96)00062-5.
17. J. Slonczewski, "Excitation of spin waves by an electric current," *J. Magn. Magn. Mater.* **195**, pp. L261–L268 (1999); doi: 10.1016/S0304-8853(99)00043-8.
18. D. E. Nikonov, G. I. Bourianoff, G. Rowland, and I. N. Krivorotov, "Strategies and tolerances of spin-transfer torque switching," *J. Appl. Phys.* **107**, 113910 (2010); doi: 10.1063/1.3429250.
19. W. H. Rippard, M. R. Pufall, S. Kaka, S. E. Russek, and T. J. Silva, "Direct-current induced dynamics in $Co_{90}Fe_{10}/Ni_{80}Fe_{20}$ point contacts," *Phys. Rev. Lett.* **92**, 027201 (2004); doi: 10.1103/PhysRevLett.92.027201.
20. S. Amara-Dababi, R. C. Sousa, M. Chshiev, H. Béa, J. Alvarez-Hérault, L. Lombard, I. L. Prejbeanu, K. Mackay, and B. Dieny, "Charge trapping–detrapping mechanism of barrier breakdown in MgO magnetic tunnel junctions," *Appl. Phys. Lett.* **99**, 083501 (2011); doi: 10.1063/1.3615654.
21. Z. Diao, Z. Li, S. Wang, Y. Ding, A. Panchula, E. Chen, L.-C. Wang, and Y. Huai, "Spin-transfer torque switching in magnetic tunnel junctions and spin-transfer-torque random access memory," *J. Phys. Condens. Matter* **19**, 165209 (2007); doi: 10.1088/0953-8984/19/16/165209.
22. J. Z. Sun, "Spin angular momentum transfer in current–perpendicular nanomagnetic junctions," *IBM J. Res. Dev.* **50**, pp. 81–100 (2006); doi: 10.1147/rd.501.0081.
23. M. Hosomi, H. Yamagishi, T. Yamamoto, K. Bessho, Y. Higo, K. Yamane, H. Yamada, M. Shoji, H. Hachino, C. Fukumoto, H. Nagao, and H. Kano, "A novel nonvolatile memory with spin-torque transfer magnetization switching: spin-RAM," *IEEE International Electron Devices Meeting, IEDM Technical Digest*, pp. 459–462 (2005); doi: 10.1109/IEDM.2005.1609379.
24. J. Z. Sun, "Spin–current interaction with a monodomain magnetic body: a model study," *Phys. Rev. B* **62**, pp. 570–578 (2000); doi: 10.1103/PhysRevB.62.570.
25. R. Heindl, W. H. Rippard, S. E. Russek, M. R. Pufall, and A. B. Kos, "Validity of the thermal activation model for spin-transfer torque switching in magnetic tunnel junctions," *J. Appl. Phys.* **109**, 073910 (2011); doi: 10.1063/1.3562136.
26. Y. Huai, "Spin-transfer-torque MRAM (STT-MRAM): Challenges and prospects," *AAPPS Bull.* **18** (6), pp. 33–40 (2008).
27. H. W. Schumacher, C. Chappert, P. Crozat, R. C. Sousa, P. P. Freitas, J. Miltat, J. Fassbender, and B. Hillebrands, "Phase coherent precessional magnetization reversal in microscopic spin-valve elements," *Phys. Rev. Lett.* **90**, 017201 (2003); doi: 10.1103/PhysRevLett.90.017201.
28. S. Kaka and S. E. Russek, "Precessional switching of submicrometer spin-valves," *Appl. Phys. Lett.* **80**, pp. 2958–2960 (2002); doi: 10.1063/1.1470704.
29. O. Redon, B. Dieny, and B. Rodmacq, "Magnetic device with polarization of spin incorporating a three-layer stack with two magnetic layers separated by a non-magnetic conducting layer for the realization of magnetic memory," France Patent 2,817,999A1 (2000).
30. A.D. Kent, B. Özyilmaz, and E. del Barco, "Spin-transfer-induced precessional magnetization reversal," Appl. Phys. Lett. **84**, pp. 3897–3899 (2004); doi: 10.1063/1.1739271.
31. D. Houssameddine, U. Ebels, B. Delaet, B. Rodmacq, I. Firastrau, F. Ponthenier, M. Brunet, C. Thirion, J.-P. Michel, L. Prejbeanu-Buda, M.-C. Cyrille, O. Redon, and B. Dieny, "Spin-torque oscillator using a perpendicular polarizer and a planar free layer," *Nat. Mater.* **6**, pp. 447–453 (2007); doi: 10.1038/nmat1905.
32. C. Papusoi, B. Delaët, B. Rodmacq, D. Houssameddine, J.-P. Michel, U. Ebels, R. C. Sousa, L. Buda-Prejbeanu, and B. Dieny, "100 ps precessional spin-transfer switching of a planar magnetic random access memory cell with perpendicular spin-polarizer," *Appl. Phys. Lett.* **95**, 072506 (2009); doi: 10.1063/1.3206919.
33. O.J. Lee, V.S. Pribiag, P.M. Braganca, P.G. Gowtham, D.C. Ralph, and R.A. Buhrman, "Ultrafast switching of a nanomagnet by a combined out-of-plane and in-plane polarized spin current pulse," Appl. Phys. Lett **95**, 012506 (2009); doi: 10.1063/1.3176938.
34. H. Liu, D. Bedau, D. Backes, J. A. Katine, J. Langer, and A. D. Kent, "Ultrafast switching in magnetic tunnel junction based orthogonal spin-transfer devices," *Appl. Phys. Lett.* **97**, 242510 (2010); doi: 10.1063/1.3527962.
35. H. Liu, D. Bedau, D. Backes, J. A. Katine, and A. D. Kent, "Precessional reversal in orthogonal spin-transfer magnetic random access memory devices," *Appl. Phys. Lett.* **101**, 032403 (2012); doi: 10.1063/1.4737010.

36. B. Lacoste, M. Marins de Castro, T. Devolder, R. C. Sousa, L. D. Buda-Prejbeanu, S. Auffret, U. Ebels, C. Ducruet, I. L. Prejbeanu, L. Vila, B. Rodmacq, and B. Dieny, "Modulating spin-transfer torque switching dynamics with two orthogonal spin-polarizers by varying the cell aspect ratio," *Phys. Rev. B.* **90**, 224404 (2014); doi: 10.1103/PhysRevB.90.224404.
37. M. T. Johnson, P. J. H. Bloemen, F. J. A. den Broeder, and J. J. de Vries, "Magnetic anisotropy in metallic multilayers," *Rep. Prog. Phys.* **59**, pp. 1409–1458 (1996); doi: 10.1088/0034-4885/59/11/002.
38. P. P. Nguyen and Y. Huai, "Spin transfer magnetic element with free layers having high perpendicular anisotropy and in-plane equilibrium magnetization," U.S. Patent 7,531,882B2 (2009).
39. B.Rodmacq, B.Dieny, "Thin-film magnetic device with strong spin polarization perpendicular to the plane of the layers, magnetic tunnel junction and spin-valve using such device," U.S. patent US 7,813,202 B2 (2010)
40. L. Liu, T. Moriyama, D. C. Ralph, and R. A. Buhrman, "Reduction of the spin-torque critical current by partially cancelling the free layer demagnetization field," *Appl. Phys. Lett.* **94**, 122508 (2009); doi: 10.1063/1.3107262.
41. P. K. Amiri, Z. M. Zeng, J. Langer, H. Zhao, G. Rowlands, Y.-J. Chen, I. N. Krivorotov, J.-P. Wang, H. W. Jiang, J. A. Katine, Y. Huai, K. Galatsis, and K. L. Wang, "Switching current reduction using perpendicular anisotropy in CoFeB–MgO magnetic tunnel junctions," *Appl. Phys. Lett.* **98**, 112507 (2011); doi: 10.1063/1.3567780.
42. L. E. Nistor, B. Rodmacq, S. Auffret, and B. Dieny, "Low effective demagnetizing field in magnetic tunnel junctions," *INTERMAG Conference, Taipei (Taiwan)*, April 25–29, HE-07 (2011).
43. T. Moriyama, T. J. Gudmundsen, P. Y. Huang, L. Liu, D. A. Muller, D. C. Ralph, and R. A. Buhrman, "Tunnel magnetoresistance and spin-torque-switching in MgO-based magnetic tunnel junctions with a Co/Ni multilayer electrode," *Appl. Phys. Lett.* **97**, 072513 (2010); doi: 10.1063/1.3481798.
44. S. Monso, B. Rodmacq, S. Auffret, G. Casali, F. Fettar, B. Gilles, B. Dieny, and P. Boyer, "Crossover from in-plane to perpendicular anisotropy in Pt/CoFe/AlO_x sandwiches as a function of Al oxidation: a very accurate control of the oxidation of tunnel barriers," *Appl. Phys. Lett.* **80**, pp. 4157–4159 (2002); doi: 10.1063/1.1483122.
45. B. Rodmacq, A. Manchon, C. Ducruet, S. Auffret, and B. Dieny, "Influence of thermal annealing on the perpendicular magnetic anisotropy of Pt/Co/AlO_x trilayers," *Phys. Rev. B*, **79**, 024423 (2009); doi: 10.1103/PhysRevB.79.024423.
46. L. E. Nistor, B. Rodmacq, S. Auffret, and B. Dieny, "Pt/Co/oxide and oxide/Co/Pt electrodes for perpendicular magnetic tunnel junctions," *Appl. Phys. Lett.* **94**, 012512 (2009); doi: 10.1063/1.3064162.
47. S. Ikeda, K. Miura, H. Yamamoto, K. Mizunuma, H. D. Gan, M. Endo, S. Kanai, J. Hayakawa, F. Matsukura, and H. Ohno, "A perpendicular-anisotropy CoFeB–MgO magnetic tunnel junction," *Nat. Mater.* **9**, pp. 721–724 (2010); doi: 10.1038/nmat2804.
48. N. D. Rizzo, D. Houssameddine, J. Janesky, R. Whig, F. B. Mancoff, M. L. Schneider, M. DeHerrera, J. J. Sun, K. Nagel, S. Deshpande, H.-J. Chia, S. M. Alam, T. Andre, S. Aggarwal, and J. M. Slaughter, "A Fully Functional 64 Mb DDR3 ST-MRAM Built on 90 nm CMOS Technology," *IEEE Trans. Mag.* **49**, 4441–4446 (2013); doi: 10.1109/TMAG.2013.2243133.
49. S. Mangin, D. Ravelosona, J. A. Katine, M. J. Carey, B. D. Terris, and E. E. Fullerton, "Currentinduced magnetization reversal in nanopillars with perpendicular anisotropy," *Nat. Mater.* **5**, pp. 210–215 (2006); doi: 10.1038/nmat1595.
50. M. Nakayama, T. Kai, N. Shimomura, M. Amano, E. Kitagawa, T. Nagase, M. Yoshikawa, T. Kishi, S. Ikegawa, and H. Yoda, "Spin-transfer-switching in TbCoFe/CoFeB/MgO/CoFeB/TbCoFe magnetic tunnel junctions with perpendicular magnetic anisotropy," *J. Appl. Phys.* **103**, 07A710 (2008); doi: 10.1063/1.2838335.
51. N. Nishimura, T. Hirai, A. Koganei, T. Ikeda, K. Okano, Y. Sekiguchi, and Y. Osada, "Magnetic tunnel junction device with perpendicular magnetization films for high-density magnetic random access memory," *J. Appl. Phys.* **91**, pp. 5246–5249 (2002); doi: 10.1063/1.1459605.
52. H. Ohmori, T. Hatori, and S. Nakagawa, "Fabrication of MgO barrier for a magnetic tunnel junction in as-deposited state using amorphous RE–TM alloy," *J. Magn. Magn. Mater.* **320**, pp. 2963–2966 (2008); doi: 10.1016/j.jmmm.2008.08.005.
53. B. Carvello, C. Ducruet, B. Rodmacq, S. Auffret, E. Gautier, G. Gaudin, and B. Dieny, "Sizable roomtemperature magnetoresistance in cobalt based magnetic tunnel junctions with out-of-plane anisotropy," *Appl. Phys. Lett.* **92**, 102508 (2008); doi: 10.1063/1.2894198.

54. H. Yoda, T. Kishi, T. Nagase, M. Yoshikawa, K. Nishiyama, E. Kitagawa, T. Daibou, M. Amano, N. Shimomura, S. Takahashi, T. Kai, M. Nakayama, H. Aikawa, S. Ikegawa, M. Nagamine, J. Ozeki, S. Mizukami, M. Oogane, Y. Ando, S. Yuasa, K. Yakushiji, H. Kubota, Y. Suzuki, Y. Nakatani, T. Miyazaki, and K. Ando, "High efficient spin-transfer-torque writing of perpendicular magnetic tunnel junctions for high density MRAMs," *Curr. Appl. Phys.* **10** (Suppl. 1), pp. e87–e89 (2010); doi: 10.1016/j.cap.2009.12.021.
55. O. G. Heinonen and D. V. Dimitrov, "Switching-current reduction in perpendicular-anisotropy spin-torque magnetic tunnel junctions," *J. Appl. Phys.* **108**, 014305 (2010); doi: 10.1063/1.3457327.
56. S. A. Wolf, J. Lu, M. R. Stan, E. Chen and D. M. Treger, "The promise of nanomagnetics and spintronics for future logic and universal memory," *Proc. IEEE* **98**, pp. 2155–2168 (2010); doi: 10.1109/JPROC.2010.2064150.
57. A. V Khvalkovskiy, D. Apalkov, S. Watts, R. Chepulskii, R. S. Beach,A. Ong, X. Tang, A. Driskill-Smith, W.H. Butler, P. B. Visscher, D. Lottis, E. Chen, V. Nikitin and M. Krounbi, "Basic principles of STT-MRAM cell operation in memory arrays," *J. Phys. D: Appl. Phys.* **46**, 074001 (2013); doi: 10.1088/0022-3727/46/7/074001.
58. T. Kishi, H. Yoda, T. Kai, T. Nagase, E. Kitagawa, M. Yoshikawa, K. Nishiyama, T. Daibou, M. Nagamine, M. Amano, S. Takahashi, M. Nakayama, N. Shimomura, H. Aikawa, S. Ikegawa, S. Yuasa, K. Yakushiji, H. Kubota, A. Fukushima, M. Oogane, T. Miyazaki, and K. Ando, "Lower-current and fast switching of a perpendicular TMR for high speed and high density spin-transfer-torque MRAM," *IEEE International Electron Devices Meeting, IEDM Technical Digest*, vol. 4, pp. (2008), doi: 10.1109/IEDM.2008.4796680.
59. J. M. Shaw, H. T. Nembach, and T. J. Silva, "Measurement of orbital asymmetry and strain in $Co_{90}Fe_{10}$/Ni multilayers and alloys: origins of perpendicular anisotropy," *Phys. Rev. B* **87**, 054416 (2013); doi: 10.1103/PhysRevB.87.054416.
60. L. Néel, "Anisotropie magnétique superficielle et surstructures d'orientation," *J. Phys. Radium* **15**, pp. 225–239 (1954); doi: 10.1051/jphysrad:01954001504022500.
61. U. Gradmann and J. Müller, "Flat ferromagnetic epitaxial 48Ni/52Fe(1 1 1) films of few atomic layers," *Phys. Stat. Solidi* **27**, pp. 313–321 (1968); doi: 10.1002/pssb.19680270133.
62. A. Manchon, C. Ducruet, L. Lombard, S. Auffret, B. Rodmacq, B. Dieny, S. Pizzini, J. Vogel, V. Uhlir, M. Hochstrasser, and G. Panaccione, "Analysis of oxygen induced anisotropy crossover in Pt/Co/MO_x trilayers," *J. Appl. Phys.* **104**, 043914 (2008); doi: 10.1063/1.2969711.
63. A. Manchon, S. Pizzini, J. Vogel, V. Uhlir, L. Lombard, C. Ducruet, S. Auffret, B. Rodmacq, B. Dieny, M. Hochstrasser, and G. Panaccione, "X-ray analysis of oxygen-induced perpendicular magnetic anisotropy in Pt/Co/AlO_x trilayers," *J. Magn. Magn. Mater.* **320**, pp. 1889–1892 (2008); doi: 10.1063/1.2969711.
64. H. X. Yang, M. Chshiev, B. Dieny, J. H. Lee, A. Manchon, and K. H. Shin, "First-principles investigation of the very large perpendicular magnetic anisotropy at Fe|MgO interfaces," *Phys. Rev. B* **84**, 054401 (2011); doi: 10.1103/PhysRevB.84.054401.
65. T. Devolder, P.-H. Ducrot, J.-P. Adam, I. Barisic, N. Vernier, Joo-Von Kim, B. Ockert, and D. Ravelosona, "Damping of CoxFe80-2xB20 ultrathin films with perpendicular magnetic anisotropy," *Appl. Phys. Lett.* **102**, 022407 (2013); doi: 10.1063/1.4775684.
66. L. E. Nistor, B. Rodmacq, S. Auffret, A. Schuhl, M. Chshiev, and B. Dieny, "Oscillatory interlayer exchange coupling in MgO tunnel junctions with perpendicular magnetic anisotropy," *Phys. Rev. B* **81**, 220407 (2010); doi: 10.1103/PhysRevB.81.220407.
67. T. Kawahara, K. Ito, R. Takemura, H. Ohno, "Spin-transfer torque RAM technology: Review and prospect," *Microelectron. Reliab.* **52**, 613–627 (2012); DOI: 10.1016/j.microrel.2011.09.028
68. J. J. Nowak, R. P. Robertazzi, J. Z. Sun, G. Hu, David W. Abraham, P. L. Trouilloud, S. Brown, M. C. Gaidis, E. J. O'Sullivan, W. J. Gallagher, and D. C. Worledge, "Demonstration of ultralow bit error rates for spin-torque magnetic random-access memory with perpendicular magnetic anisotropy," *IEEE Magn. Lett.* **2**, 2–5 (2011); DOI: 10.1109/LMAG.2011.2155625
69. J.J. Nowak, R.P. Robertazzi, J.Z. Sun, G. Hu, J.H. Park, J. Lee, A.J. Annunziata, G.P. Lauer, R. Kothandaraman, E.J. O'Sullivan, P.L. Trouilloud, Y. Kim, and D.C. Worledge, "Dependence of Voltage and Size on Write Error Rates in Spin-Transfer Torque Magnetic Random-Access Memory", *IEEE Mag. Lett.* **7**, 3102604 (2016); DOI: 10.1109/LMAG.2016.2539256

70. B. Rodmacq, S. Auffret, B. Dieny, L. Nistor,"Three-layer magnetic element, magnetic field sensor, magnetic memory and magnetic logic gate using such an element," U.S. Patent 8,513,944 B2 (2008).
71. H. Sato, M. Yamanouchi, S. Ikeda, S. Fukami, F. Matsukura, and H. Ohno, "Perpendicular-anisotropy CoFeB-MgO magnetic tunnel junctions with a MgO/CoFeB/Ta/CoFeB/MgO recording structure," *Appl. Phys. Lett.* **101**, 022414 (2012); DOI: 10.1063/1.4736727
72. G. Jan, L. Thomas, S. Le, Y.J. Lee, H. Liu, J. Zhu, R.Y. Tong, K. Pi, Y.J. Wang, D. Shen, R. He, J. Haq, J. Teng, V. Lam, K. Huang, T. Zhong, T. Torng, and P.K. Wang,"Demonstration of fully functional 8 Mb perpendicular STT-MRAM chips with sub-5 ns writing for non-volatile embedded memories," *Proc. Symp. VLSI Technol. Dig. Tech. Papers*, 42–43 (2014).
73. L. Thomas, G. Jan, J. Zhu, H. Liu, Y.J. Lee, S. Le, R.Y. Tong, K. Pi, Y.J. Wang, D. Shen, R. He, J. Haq, J. Teng, V. Lam, K. Huang, T. Zhong, T. Torng and P.K. Wang, "Perpendicular spin-transfer torque magnetic random access memories with high spin-torque efficiency and thermal stability for embedded applications (invited)," *J. Appl. Phys.* **115**, 172615 (2014); DOI: 10.1063/1.4870917
74. C. Park, J.J. Kan, C. Ching, J. Ahn, L. Xue, R. Wang, A. Kontos, S. Liang, M. Bangar, H. Chen, S. Hassan, M. Gottwald, X. Zhu, M. Pakala, and S. H. Kang,"Systematic Optimization of 1 Gbit Perpendicular Magnetic Tunnel Junction Arrays for 28 nm Embedded STT- MRAM and Beyond," *Proc. Int. Elect. Devices Meet (IEDM 2015)*, 26.2.1 (2015); DOI: 10.1109/IEDM.2015.7409771.
75. B. Dieny and O. Redon, "Magnetic tunnel junction magnetic device, memory and writing and reading methods using said device," U.S. Patent 6,950,335B2 (2005), Fig. 8.
76. K. Yakushiji, T. Saruya, H. Kubota, A. Fukushima, T. Nayahama, S. Yuasa, and K. Ando, "Ultrathin Co/Pt and Co/Pd superlattice films for MgO-based perpendicular magnetic tunnel junctions," *Appl. Phys. Lett.* **97**, 232508 (2010); doi: 10.1063/1.3524230.
77. G. Hu, J. H. Lee, J. J. Nowak, J. Z. Sun, J. Harms, A. Annunziata, S. Brown, W. Chen, Y. H. Kim, G. Lauer, L. Liu, N. Marchack, S. Murthy, E. J. O'Sullivan, J. H. Park, M. Reuter, R. P. Robertazzi, P. L. Trouilloud, Y. Zhu and D. C. Worledge,"STT-MRAM with double magnetic tunnel junctions," *Proc. Int. Electron. Devices Meeting Dig., 2015*, 26.3.1–26.3.4. (2015).
78. L. Cuchet, B. Rodmacq, S. Auffret, R.C. Sousa, I.L. Prejbeanu, B. Dieny, "Perpendicular magnetic tunnel junctions with a synthetic storage or reference layer: A new route towards Pt- and Pd-free junctions", *Scient. Rep.* **6**, 21246 (2016); DOI: 10.1038/srep21246.
79. D. Apalkov, A. Khvalkovskiy, and S. Watts, "Spin-transfer torque magnetic random access memory (stt-mram)," *ACM J. Emerg. Technol. Comput. Syst.*, **9**, 13:1–13:35 (2013); DOI: 10.1145/2463585.2463589
80. S. Amara, R. C. Sousa, H. Bea, C. Baraduc, and B. Dieny, "Barrier breakdown mechanisms in MgO-based magnetic tunnel junctions and correlation with low-frequency noise," *IEEE Trans. Magn.* **48**, pp. 4340–4343 (2012); doi: 10.1109/TMAG.2012.2200243.
81. H. Zhao, A. Lyle, Y. Zhang, P. K. Amiri, G. Rowlands, Z. Zeng, J. Katine, H. Jiang, K. Galatsis, K. L. Wang, I. N. Krivorotov, and J.-P. Wang, "Low writing energy and sub-nanosecond spin-torque transfer switching of in-plane magnetic tunnel junction for spin-torque transfer random access memory," *J. Appl. Phys.* **109**, 07C720 (2011); doi: 10.1063/1.3556784.
82. M. Carpentieri, M. Ricci, P. Burrascano, and L. Torres, "Spreading sequences for fast switching process in spin-valve nanopillars," *Appl. Phys. Lett.* **98**, 122504 (2011); doi: 10.1063/1.3569947.
83. T. Nozaki, Y. Shiota, M. Shiraishi, T. Shinjo, and Y. Suzuki, "Voltage-induced perpendicular magnetic anisotropy change in magnetic tunnel junctions," *Appl. Phys. Lett.* **96**, 022506 (2010); doi: 10.1063/1.3279157.
84. Y. Shiota, T. Maruyama, T. Nozaki, T. Shinjo, M. Shiraishi, and Y. Suzuki, "Voltage-assisted magnetization switching in ultrathin $Fe_{80}Co_{20}$ alloy layers," *Appl. Phys. Express* **2**, 063001 (2009); doi: 10.1143/APEX.2.063001.
85. F. Bonell, S. Murakami, Y. Shiota, T. Nozaki, T. Shinjo, and Y. Suzuki, "Large change in perpendicular magnetic anisotropy induced by an electric field in FePd ultrathin films," *Appl. Phys. Lett.* **98**, 232510 (2011); doi: 10.1063/1.3599492.
86. S. Bandiera, R. C. Sousa, M. Marins de Castro, C. Ducruet, C. Portemont, S. Auffret, L. Vila, I. L. Prejbeanu, B. Rodmacq, and B. Dieny, "Spin-transfer-torque switching assisted by thermally induced

anisotropy reorientation in perpendicular magnetic tunnel junctions," *Appl. Phys. Lett.* **99**, 202507 (2011); doi: 10.1063/1.3662971.

87. W. Zhao, J. Duval, J.-O. Klein, and C. Chappert, "A compact model for magnetic tunnel junction (MTJ) switched by thermally-assisted spin-transfer-torque (TAS + STT)," *Nanoscale Res. Lett.* **6**, p. 368 (2011); doi: 10.1186/1556-276X-6-368.
88. R. C. Sousa and I. L. Prejbeanu, "Non-volatile magnetic random access memories (MRAM)," *Compt. Rend. Phys.* **6**, pp. 1013–1021 (2005); doi: 10.1016/j.crhy.2005.10.007.
89. B. Dieny and O. Redon, "Magnetic device with magnetic tunnel junction, memory array and read/write methods using same," Patent US6950335 (B2) (2003).
90. S. S. Cardoso, R. Ferreira, F. Silva, P. P. Freitas, L. V. Melo, R. C. Sousa, O. Redon, M. MacKenzie, and J. N. Chapman, "Double-barrier magnetic tunnel junctions with GeSbTe thermal barriers for improved thermally-assisted magnetoresistive random access memory cells," *J. Appl. Phys.* **99**, 08N901 (2006); doi: 10.1063/1.2162813.
91. C. Papusoi, R. Sousa, J. Herault, I. L. Prejbeanu, and B. Dieny, "Probing fast heating in magnetic tunnel junction structures with exchange bias," *New J. Phys.* **10**, 103006 (2008); doi: 10.1088/1367-2630/10/10/103006.
92. I. L. Prejbeanu, M. Kerekes, R. C. Sousa, H. Sibuet, O. Redon, B. Dieny, and J.-P. Nozières, "Thermally-assisted MRAM," *J. Phys. Cond. Mater* **19**, 165218 (2007); doi: 10.1088/0953-8984/19/16/165218.
93. I. L. Prejbeanu, W. Kula, K. Ounadjela, R. C. Sousa, O. Redon, B. Dieny, and J.-P. Nozières, "Thermally-assisted switching in exchange-biased storage layer magnetic tunnel junctions," *IEEE Trans. Magn.* **40**, pp. 2625–2627 (2004); doi: 10.1109/TMAG.2004.830395.
94. J. Wang and P. P. Freitas, "Low-current blocking temperature writing of double barrier magnetic random access memory cells," *Appl. Phys. Lett.* **84**, pp. 945–947 (2004); doi: 10.1063/1.1646211.
95. B. F. Cambou, "Multilevel magnetic element," U.S. Patent 8,630,112 (2014). B. F. Cambou, D. J. Lee, and K. Mackay, "Magnetic logic units configured as an amplifier," U.S. Patent 8,933,750 (2015).
96. B. F. Cambou, *Match-In-Place™: A Novel Way to Perform Secure and Fast User's Authentication*, Crocus Technology, Santa Clara, CA (2012).
97. E. Gapihan, R. C. Sousa, J. Herault, C. Papusoi, M. T. Delaye, B. Dieny, I. L. Prejbeanu, C. Ducruet, C. Portemont, K. Mackay, and J.-P. Nozières, "FeMn exchange biased storage layer for thermally-assisted MRAM," *IEEE Trans. Magn.* **46**, pp. 2486–2488 (2010); doi: 10.1109/TMAG.2010.2041198.
98. E. R. Callen and H. B. Callen, "Anisotropic magnetization," *J. Phys. Chem. Solids* **16**, pp. 310–328 (1960); doi: 10.1016/0022-3697(60)90161-X.
99. H. W. van Kesteren and W. B. Zeper, "Controlling the Curie temperature of magneto-optical recording media of Co/Pt multilayer," *J. Magn. Magn. Mater.* **120**, pp. 271–273 (1993); doi: 10.1016/0304-8853(93)91339-9.
100. P. Poulopoulos and K. Baberschke, "Magnetism in thin films," *J. Phys. Condens. Matter* **11**, pp. 9495–9515 (1999); doi: 10.1088/0953-8984/11/48/310.
101. Q. Meng, W. P. van Drent, J. C. Lodder, and Th. J. A. Popma, "Curie temperature dependence of magnetic properties of CoNi/Pt multilayer films," *J. Magn. Magn. Mater.*, **156**, pp. 296–298 (1999); doi: 10.1016/0304-8853(95)00875-6.
102. S. Yuasa and D. D. Djayaprawira, "Giant tunnel magnetoresistance in magnetic tunnel junctions with a crystalline MgO(0 0 1) barrier," *J. Phys. D Appl. Phys.* **40**, pp. R337–R354 (2007); doi: 10.1088/0022-3727/40/21/R01.
103. K. Mizunuma, S. Ikeda, J. H. Park, H. Yamamoto, H. Gan, K. Miura, H. Hasegawa, J. Hayakawa, F. Matsukura, and H. Ohno, "MgO barrier-perpendicular magnetic tunnel junctions with CoFe/Pd multilayers and ferromagnetic insertion layers," *Appl. Phys. Lett.* **95**, 232516 (2009); doi: 10.1063/1.3265740.
104. S. Bandiera, R. C. Sousa, Y. Dahmane, C. Ducruet, C. Portemont, V. Baltz, S. Auffret, I. L. Prejbeanu, and B. Dieny, "Comparison of synthetic antiferromagnets and hard ferromagnets as reference layer in magnetic tunnel junctions with perpendicular magnetic anisotropy," *IEEE Magn. Lett.* **1**, 3000204 (2010); doi: 10.1109/LMAG.2010.2052238.
105. K. Mizunuma, S. Ikeda, H. Sato, M. Yamanouchi, H. D. Gan, K. Miura, H. Yamamoto, J. Hayakawa, F. Matsukura, and H. Ohno, "Tunnel magnetoresistance properties and annealing stability in

perpendicular anisotropy MgO-based magnetic tunnel junctions with different stack structures," *J. Appl. Phys.* **109**, 07C711 (2011); doi: +10.1063/1.3554092.

106. M. Gajek, J. J. Nowak, J. Z. Sun, P. L. Trouilloud, E. J. O'Sullivan, D. W. Abraham, M. C. Gaidis, G. Hu, S. Brown, Y. Zhu, R. P. Robertazzi, W. J. Gallagher, and D. C. Worledge, "Spin-torque switching of 20 nm magnetic tunnel junctions with perpendicular anisotropy," *Appl. Phys. Lett.* **100**, 132408 (2012); doi: 10.1063/1.3694270.
107. E. P. Sajitha, J. Walowski, D. Watanabe, S. Mizukami, F. Wu, H. Naganuma, M. Oogane, Y. Ando, and T. Miyazaki, "Magnetization dynamics in CoFeB buffered perpendicularly magnetized Co/Pd multilayer," *IEEE Trans. Magn.* **46**, pp. 2056–2059 (2010); doi: 10.1109/TMAG.2009.2038929.
108. L. Lombard, E. Gapihan, R. C. Sousa, Y. Dahmane, Y. Conraux, C. Portemont, C. Ducruet, C. Papusoi, I. L. Prejbeanu, J. P. Nozières, B. Dieny, and A. Schuhl, "IrMn and FeMn blocking temperature dependence on heating pulse width," *J. Appl. Phys.* **107**, 09D728 *(2100);* doi: 10.1063/1.3340452.
109. P. M. Braganca, J. A. Katine, N. C. Emley, D. Mauri, J. R. Childress, P. M. Rice, E. Delenia, D. C. Ralph, and R. A. Buhrman, "A three-terminal approach to developing spin-torque written magnetic random access memory cells," *IEEE Trans. Nanotech.* **8**, pp. 190–195 (2009); doi: 10.1109/TNANO.20A08.2005187.
110. I. M. Miron, K. Garello, G. Gaudin, P.-J. Zermatten, M. V. Costache, S. Auffret, S. Bandiera, B. Rodmacq, A. Schuhl, and P. Gambardella, "Perpendicular switching of a single ferromagnetic layer induced by in-plane current injection," *Nature* **476**, pp. 189–194 (2011); doi: 10.1038/nature10309.
111. I. M. Miron, G. Gaudin, S. Auffret, B. Rodmacq, A. Schuhl, S. Pizzini, J. Vogel, and P. Gambardella, "Current-driven spin-torque induced by the Rashba effect in a ferromagnetic metal layer," *Nat. Mater.* **9**, pp. 230–234 (2010); doi: 10.1038/nmat2613.
112. L.Q. Liu, O.J. Lee, T.J. Gudmundsen, D.C. Ralph, R.A. Buhrman, "Current-Induced Switching of Perpendicularly Magnetized Magnetic Layers Using Spin Torque from the Spin Hall Effect", *Phys. Rev. Lett.* **109**, 096602 (2012); DOI: 10.1103/PhysRevLett.109.096602
113. S. Fukami, T. Anekawa, C. Zhang, H. Ohno, "A spin–orbit torque switching scheme with collinear magnetic easy axis and current configuration," *Nat. Nanotec.* **11**, 621 (2016); DOI: 10.1038/NNANO.2016.29
114. K. Garello, C.O. Avci, I.M. Miron, M. Baumgartner, A. Ghosh, S. Auffret, O. Boulle, G. Gaudin and P. Gambardella, "Ultrafast magnetization switching by spin–orbit torques," *Appl. Phys. Lett.* **105**, 212402 (2014); DOI: 10.1063/1.4902443
115. T. Endoh, "Spin-transfer torque MRAM (SPRAM) and its applications," ITRS Emerging Research Memory Technologies Workshop (2010).
116. J. Hutchby and M. Garner, Assessment of the Potential & Maturity of Selected Emerging Research Memory Technologies Workshop & ERD/ERM Working Group Meeting (April 6–7, 2010).
117. Y. de Charentenay, *Emerging Non Volatile Memory (NVM) Technology & Market Trends Report*, Yole Dévelopement, Lyon-Villeurbanne, France (2015). www.i-micronews.com/component/hikashop/product/emerging-non-volatile-memory-nvm-technology-market-trends-report-launch-offer.html
118. Y. Shiota, T. Nozaki, F. Bonell, S. Murakami, T. Shinjo and Y. Suzuki, "Induction of coherent magnetization switching in a few atomic layers of FeCo using voltage pulses," *Nat. Mater.*, **11**, 39 (2012); DOI: 10.1038/nmat3172
119. S. Kanai, M. Yamanouchi, S. Ikeda, Y. Nakatani, F. Matsukura, and H. Ohno, "Electric field-induced magnetization reversal in a perpendicular anisotropy CoFeB–MgO magnetic tunnel junction," *Appl. Phys. Lett.* **101**, 122403 (2012); DOI: 10.1063/1.4753816

CHAPTER 6

MAGNETIC BACK-END TECHNOLOGY

Michael C. Gaidis
IBM T. J. Watson Research Center, Yorktown Heights, NY, USA

6.1 MAGNETORESISTIVE RANDOM-ACCESS MEMORY (MRAM) BASICS

By combining magnetics (spin) and CMOS (electronics), the field of "spintronics" has created a solid-state memory touting nonvolatility, high-density, high-endurance, radiation hardness, high-speed operation, and inexpensive CMOS integration: *magnetoresistive random-access memory* (MRAM). A semiconductor device manufacturer that holds MRAM as an available option for its customers can command a larger share of the market and can enable novel applications that generate new markets. This chapter will review the fabrication and test methods that will be required by a manufacturing line to create their own MRAM offerings.

The commercialization of MRAM hinges on the advantages offered by the magnetic tunnel junction (MTJ). The MTJ tunnel barrier thickness can be tuned by just a fraction of a nanometer to set its resistance to be an ideal match to inexpensive CMOS drive transistors. The resistance change between magnetic states is large enough for fast detection by amplifiers, yet small enough for simple write circuitry with balanced drive currents. Everything involved in the back-end-of-line (BEOL) processing of MRAM is chosen with careful consideration to yield the best MTJ. One must simultaneously optimize the MTJ device resistance, the magnetoresistance, the write current, the read current, the tunnel barrier breakdown voltage, and the stability against thermal fluctuations.

Currently, the most glaring change forced upon standard CMOS BEOL integration is the added requirement for low process temperatures. To ensure good MTJ properties, one must substitute lower temperature versions of the conventional 350–400 °C dielectric and metal depositions, film curing and packaging, and high-temperature solders. Aside from the qualification of lower temperature process flows, relatively little additional expense is needed to fabricate MRAM. Tooling

Introduction to Magnetic Random-Access Memory, First Edition. Edited by Bernard Dieny, Ronald B. Goldfarb, and Kyung-Jin Lee.

requirements are modest: a precision magnetic film deposition system, an optional magnetic-field annealing tool, and a suitable etcher for patterning the magnetic metals.

6.2 MRAM BACK-END-OF-LINE STRUCTURES

Competing in the crowded field of random-access memory, MRAM is fabricated with careful attention to cost. Expense is driven in large part by the number of photomask levels required to fabricate the memory, as photolithography tools tend to be the most expensive tools to use in a fab line. Multiple varieties of MRAM exist with requirement for only one or a few extra photomask levels. MRAM is also flexible for incorporation into the CMOS wiring hierarchy because it is not locked-in to the front-end silicon. The integration of MRAM into a CMOS back end offers lower fabrication cost than memory located in the front-end-of-line (FEOL) because of relaxed ground rules, but comes at the expense of lower memory density. Thermal degradation of MRAM devices prevents incorporation of MRAM into the high-temperature FEOL. For the densest memory, MRAM is located at or near the lowest back-end level. The two most prominent types of MRAM each hold niches, where one type offers greater cost or performance benefits depending on the application being considered. The structures of these MRAMs are described in the following sections.

6.2.1 Field-MRAM

For applications with modest memory size requirements, field-MRAM is at present the less-expensive alternative. With relatively small numbers of bits in the array, larger MTJ devices can be used without incurring prohibitive cost. Large devices (>100 nm linear dimension) can be effectively and reliably field-switched with proven technology that can be purchased on the market today. Bit cell size for field-MRAM is set by the size of the large MTJ devices and driving wires, so older, cheaper, and more mature CMOS FEOL technology can be used without limiting the array density. As semiconductor manufacturers upgrade to fabrication facilities using 300 or 450 mm technology, the displaced older fabs can be effectively filled with inexpensive MRAM technology operating from 65 nm node (or larger) CMOS. Figure 6.1 shows one possible topology for the elements present in typical field-MRAM. Figure 6.2 is a SEM cross-sectional image of three CMOS-driven MRAM cell elements, each corresponding to the structure in Fig. 6.1. The MTJ elements in Fig. 6.2 have been fabricated with an additional copper wiring level below the MTJs for flexibility in circuit wiring.

Different versions of field-MRAM include older Stoner–Wohlfarth switching, toggle-MRAM with modified timing of write word and bit lines, and the newer thermal-assist (TAS) MRAM, where devices require only one magnetic field-generating wire to operate. Stoner–Wohlfarth and toggle-MRAM both require a field generated by a bit line and by a word line. The MRAM circuit can benefit from ferromagnetic cladding around the field-generating wires to focus the field on the desired MTJ. The cladding is a further deviation from standard CMOS and will be discussed in detail later in this chapter. TAS MRAM is somewhat simplified by the

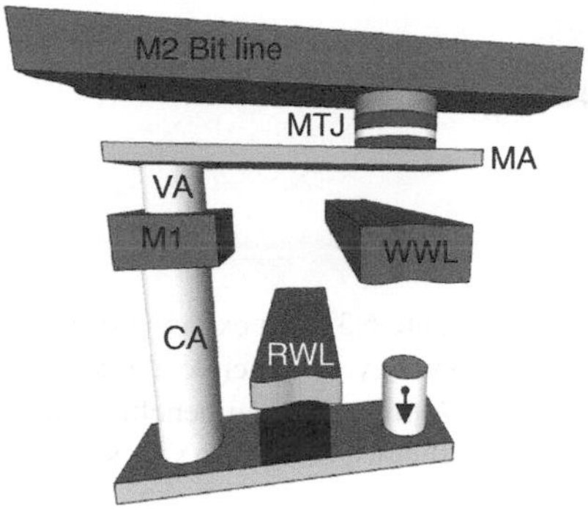

Figure 6.1 An example of field-MRAM cell circuit topology, showing individual word lines for reading (RWL) and writing (WWL). The read word line controls MTJ read current through a field-effect transistor (FET) in the silicon beneath the MTJ. The FET isolates the one MTJ being read, preventing leakage of read currents through nearby MTJ devices in the array. Thermal-assisted (TAS) MRAM also uses the RWL to control a heating current that passes through the MTJ during writing [1,2]. MRAM additions to standard CMOS include the following conductor elements shown in the figure: a shallow VA via and a MA metal strap to isolate the MTJ from the copper WWL. The MTJ is located between the MA metal and the top bit line. The bit line is used for both read and write operations. Distance between the M1 and M2 wiring is determined primarily by the thicknesses of VA, MA, and the MTJ top contact (the active magnetic films can be quite thin).

need for only one wiring layer to be clad, but the trade-off is the need for careful control of the thermal environment near the MTJ.

In summary, field-MRAM is a reliable, proven technology that integrates well with standard CMOS. Additional process steps to incorporate the MRAM are modest: roughly three additional photomask levels (VA, MA, and MTJ), and the potential need for ferromagnetic cladding around field-generating wires. Scaling properties of this type of memory are limited by near-neighbor disturbances from fringing fields, and by increasingly difficult switching field requirements as the MTJs shrink.

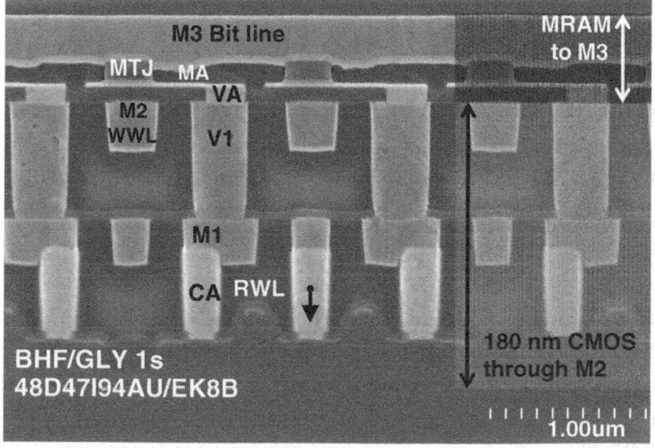

Figure 6.2 SEM cross section of three adjacent field-MRAM cells fabricated atop 180 nm node CMOS. The read word line is formed from gate polysilicon, and one side of the drive transistor connects to a M1 wiring grid at ground potential. The circuit elements germane to MRAM are highlighted in red on the right side of the image, with the standard CMOS elements highlighted in green.

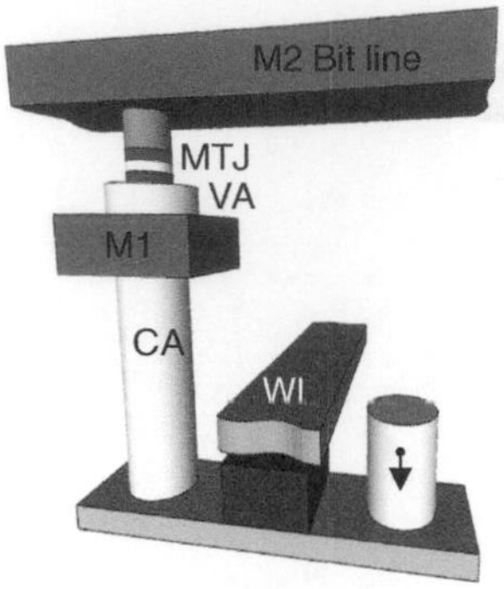

Figure 6.3 An example of STT-MRAM cell circuit topology. In principle, the MTJ can be placed directly on the CA pillar, but can benefit from an additional VA element fashioned to ensure a very smooth substrate on which to grow the MTJ.

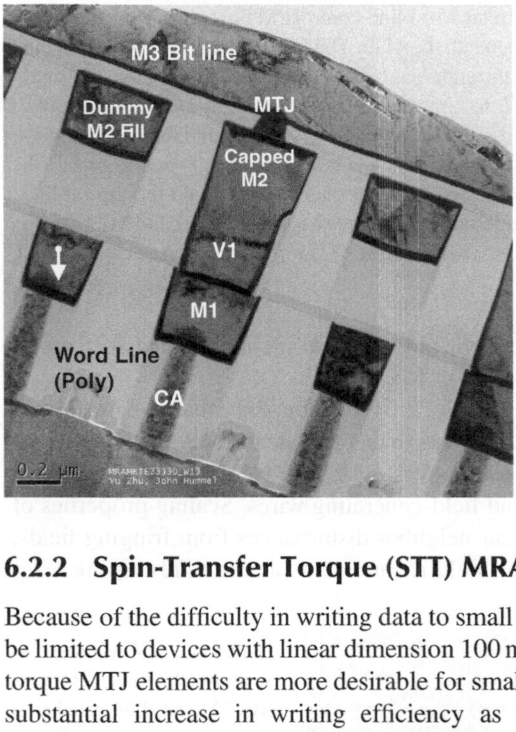

Figure 6.4 STT-MRAM cell TEM cross-sectional image showing the MTJ element placed atop the CMOS M2 wiring level. The word line for the array is formed from gate polysilicon, and the bit line is a copper wire contacting a self-aligned conductive hard mask atop the magnetics.

6.2.2 Spin-Transfer Torque (STT) MRAM

Because of the difficulty in writing data to small devices, field-MRAM is expected to be limited to devices with linear dimension 100 nm or larger. Alternative spin-transfer torque MTJ elements are more desirable for smaller sizes, and are expected to show a substantial increase in writing efficiency as they scale below 40 nm diameter. Therefore, most research for high-density MRAM memory applications is focused on STT-MRAM. The smaller footprint of STT-MRAM devices translates to a cell size that is presently limited by the size of the drive transistor. A major challenge in the fabrication of STT-MRAM is reducing the necessary write current so that smaller transistors can be employed for a smaller overall cell size. Figure 6.3 schematically illustrates the simplicity of the STT-MRAM cell and its potential for very high density if the drive transistor can be scaled as well. A corresponding transmission electron microscopy (TEM) cross section through a STT-MRAM cell in a CMOS-driven array is presented in Fig. 6.4.

STT-MRAM is a notably simple structure, with only one additional photomask required above standard CMOS. Depending on circuit size requirements, one can use

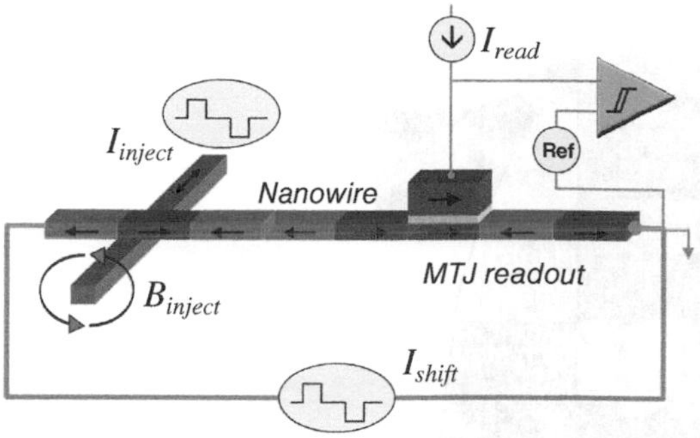

Figure 6.5 Racetrack memory circuit schematic. The fabrication is substantially more challenging than STT- or field-MRAM, but the benefit could be an extremely dense memory. Future development aims to create these magnetic nanowire shift registers in a vertical orientation. With a space requirement of only two or three transistors per racetrack nanowire capable of holding over a hundred bits, the vertical racetrack memory can be as dense as flash, but with superior speed and endurance properties.

large-area or intentionally shunted MTJs as high-conductivity vias in parallel with the MTJ layer (3). This can obviate the need for a separate via mask, making the total number of mask adders zero. STT-MRAM performs better when it is smaller, does not require field lines or ferromagnetically clad wires, and is therefore an inexpensive addition to CMOS.

6.2.3 Other Magnetic Memory Device Structures

Field-MRAM and STT-MRAM are not the only two players in the magnetics–CMOS partnership. Adding slightly more complexity to the fabrication can generate three-terminal MRAM devices that relax requirements on device spreads and tunnel barrier stress (4). Another major development in the field of spintronics comes from new research on the motion of domain walls within nanowires. Schematically described in Fig. 6.5, "racetrack" memory was proposed in the previous decade with potential as a very high-density shift-register memory (5,6). The fabrication is quite challenging, but the payoff could be a storage-class memory with features preferable to flash or phase-change memory (PCM).

6.3 MRAM PROCESS INTEGRATION

The following sections highlight the differences involved in BEOL MRAM processing compared with standard CMOS processing techniques. Tooling requirements for MRAM production in a CMOS fab are outlined, as well as potential pitfalls in the processing of the magnetic structures.

6.3.1 The Magnetic Tunnel Junction

At the heart of all competitive MRAM concepts is the MTJ. As discussed in other chapters in this book, much effort is expended in optimizing the magnetic layers for

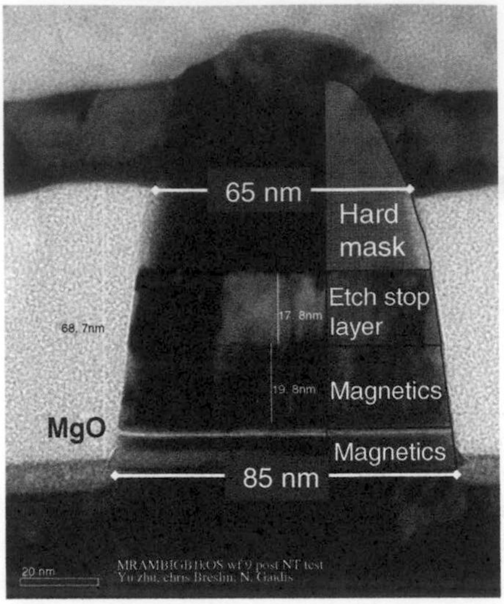

Figure 6.6 The components of a practical BEOL MRAM MTJ, from the bottom wiring contact up through to the top wiring contact. Note the moderate taper angle to the structure sidewalls, employed to reduce redeposits on the device sidewall during etching.

ideal switching and readout behavior. It is the process integrator's job to ensure the processing of structures surrounding the MTJ does little to degrade the performance one would expect from unprocessed sheet-film magnetic stacks. Figure 6.6 shows a high-magnification image of a representative MTJ device after undergoing a CMOS-compatible process flow. Key items of note in the figure are the use of MgO tunnel barriers for high magnetoresistance and reasonably good thermal resilience, and the use of a conductive hard mask. Subtractive (etch) patterning is used for MTJ formation, necessitating a hard mask able to withstand the etchants used for the magnetic materials. By employing a conductive hard mask, one realizes a self-aligned top contact to the MTJ device.

The most substantial change to CMOS BEOL needed for MTJ fabrication is the use of lower temperature processes. Although progress is being made in the fabrication of magnetic stacks capable of withstanding 400 °C BEOL processing, in general, processes must remain below 350 °C when performed after patterning of the MTJ. Sheet-film magnetic stacks are able to withstand higher temperatures, but the presence of interfaces in the patterned devices enhances material diffusion that can degrade the magnetics and alter the resistance of the tunnel barrier. Some conventional integration techniques do not lend themselves well to low-temperature processing and should be avoided. Tooling options are available for high-reliability low-temperature BEOL processing, and proven-reliable low-temperature-processed MRAM is available on the market today (7). Tooling changes can often be simple adaptations of recipes in high-temperature tools, but film quality can suffer. Therefore, reliability qualification must be part of the development of new low-temperature BEOL process flows. It is also best if tools are not shared with high-temperature CMOS processes because thermal cycling of tools can result in particulate generation and consequently on-wafer defect formation.

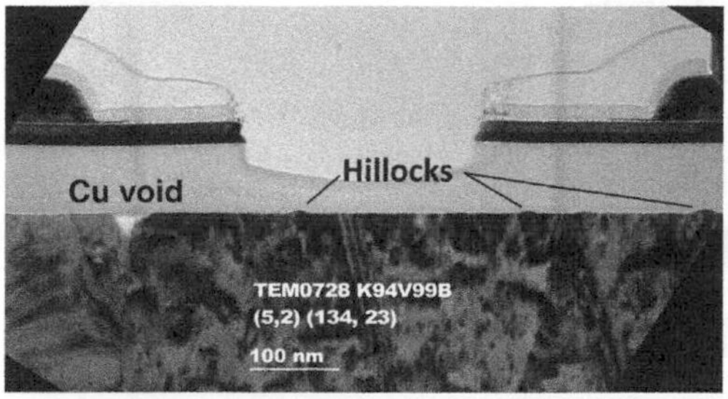

Figure 6.7 TEM cross-sectional image of a smooth VA dielectric layer (refer to Fig. 6.1) used to insulate and buffer the MTJ devices from defects in underlying copper.

6.3.1.1 Substrate Preparation Even small defects in the approximately 1 nm thick MTJ barrier can cause major problems for the reliability of the structure. Underlying substrate roughness should be held to the 0.1 nm rms level for MTJs with high breakdown voltage, tight spreads in device resistance, and minimal exchange coupling. Film choices for substrate must keep this in mind: no porous dielectrics, no large-grain-boundary metals such as copper, and, for many materials, the as-deposited roughness is unacceptable and requires chemical–mechanical polishing (CMP) or other surface treatment to smooth the surface prior to magnetic stack deposition.

It is risky to build the device on copper without a smooth cap layer to decouple defects in the copper from the MTJ. Copper voids and hillocks of size <1 nm are often ignored in standard CMOS fabrication flows, but can have severe effects on MTJ performance. Figure 6.7 shows an example of capping underlying copper with a CMP-smoothed dielectric film to provide an optimal MTJ substrate.

Process-induced surface nonplanarity is a concern when using Damascene polish of wiring beneath the MTJ layer. Excessive dishing of CMP surfaces on relatively large length scales can translate directly to topography in the MTJ films. One option for field-MRAM is to include an extra smoothing polish on a dielectric or metal layer above the wiring layer. Compare the relatively well-planarized devices in Fig. 6.2 with the device in Fig. 6.8. The dielectric beneath the MTJ in Fig. 6.8 did not receive an extra smoothing polish before MTJ stack deposition and large-scale

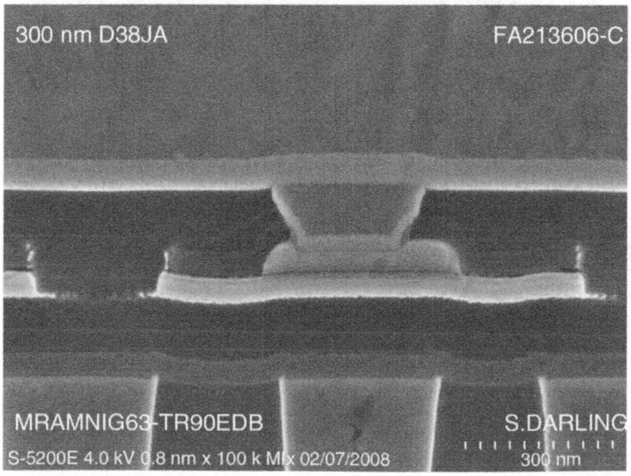

Figure 6.8 SEM cross-sectional image of an MTJ deposited on unpolished dielectric layer above the polish-induced topography of the underlying copper wiring. The MTJ stack films are noticeably nonplanar due to dishing from the copper polish.

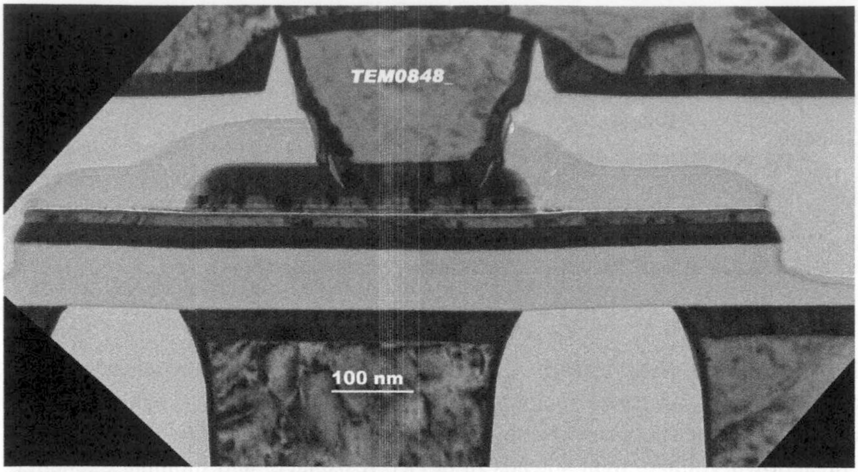

Figure 6.9 TEM cross-sectional image of an MTJ above a copper wire capped with self-aligned smooth metal. The planarity of the MTJ stack films is similar to that of Fig. 6.7, but did not require dielectric polish.

dishing mirroring the underlying wiring is apparent. Regarding STT-MRAM, an additional metal via layer such as the VA feature shown in Fig. 6.3 can be used to ensure an ideal MTJ substrate. This can be a simple shallow via with smoothing Damascene CMP of metal fill in the via.

Another option for both field- and STT-MRAM is a self-aligned copper cap such as that shown in Figs. 6.4 and 6.9 (8). After a standard copper Damascene polish, a gentle etch is employed to remove a small amount of the copper. Backfilling with suitable metal and then employing a smoothing CMP process leaves behind a substrate ideal for MTJ growth–and self-aligned to minimize cell size and process cost.

If abrasive CMP is used to create the smooth MTJ substrate, one must exercise care to clean the substrate thoroughly of all residual slurry particles. A 20 nm slurry particle may be a minor yield concern for 90 nm CMOS, but as shown in Fig. 6.10, it is a device killer if it is located beneath an MTJ. *In situ* megasonic cleaning and the use of abrasive-free slurries are options to enhance yield from CMP processes.

A final consideration in substrate preparation involves alignment of the MTJ-patterned photomask to the underlying circuit. Because magnetic film stacks typically include opaque metal layers, one must rely on topography or additional process modifications to ensure fine alignment of the MTJ pattern to underlying circuitry. The use of an ultrasmooth substrate can be at odds with alignment of the lithography used to pattern the magnetic stack. Generation of topography in the kerf, coarse block-out masks with etch-through opaque stack before fine alignment, and/or masking of alignment marks during stack deposition can enable alignment to the underlying layers (9).

6.3.1.2 *Film Deposition and Anneal* Among the most critical technology advances enabling integration of MRAM with 200 mm and larger CMOS wafers

Figure 6.10 TEM cross-sectional image showing the devastation to the MTJ device grown atop a residual CMP slurry particle.

is the tooling now available for large-area deposition of highly uniform films with thickness control below 1/10th of a nanometer (10). Most conventional physical vapor deposition (PVD) tools are poorly suited to MTJ film deposition because they lack the necessary control of film thickness, film uniformity, and surface roughness. Additional chambers are typically needed for finely controlled oxidation of tunnel barriers without breaking vacuum. The development of such precision film deposition tooling for high-dielectric-constant (high-K)/metal gate CMOS applications is synergistic with the MRAM requirements, and has created a larger market to drive more tooling choices. Throughput is a concern because the precision deposition of up to dozens of layers in the magnetic stack requiring small deposition rates and multiple target exposures can make for 30 min or longer depositions on each wafer. Seed/underlayer and magnetic films must be chosen to minimize MTJ barrier roughness so as to minimize the magnetostatic coupling across the barrier (the so-called Néel or "orange peel" coupling) and avoid pinhole formation through the barrier. Figure 6.11 shows an

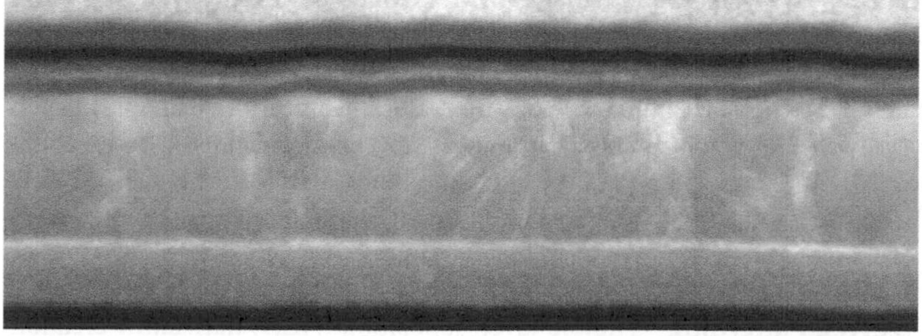

Figure 6.11 High-resolution dark-field TEM image of a magnetic stack with flux-closed trilayer beneath the MgO tunnel barrier (in black). The smooth substrate does not translate to a smooth MgO barrier due to the granularity of the base layer film.

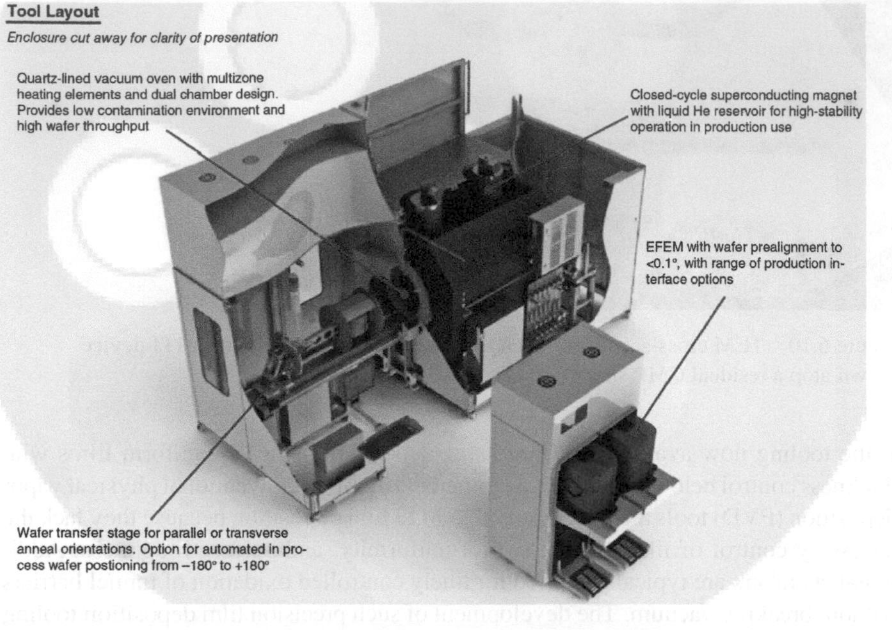
Tool Layout
Enclosure cut away for clarity of presentation
Quartz-lined vacuum oven with multizone heating elements and dual chamber design. Provides low contamination environment and high wafer throughput
Closed-cycle superconducting magnet with liquid He reservoir for high-stability operation in production use
EFEM with wafer prealignment to <0.1°, with range of production interface options
Wafer transfer stage for parallel or transverse anneal orientations. Option for automated in process wafer postioning from −180° to +180°

Figure 6.12 Magnetic Solutions MRT/300 magnetic annealing tool. The tool offers precision alignment of wafer orientation with respect to the field generated by a superconducting magnet around the anneal furnace. (Courtesy of Magnetic Solutions.)

example of a magnetic stack where the granularity of the base layer film is substantial compared with the thickness of the tunnel barrier.

The magnetic film stack is generally grown entirely without breaking vacuum as interface properties are of high importance in magnetics. The sheet film deposition is typically followed by a 250–450 °C anneal of the blanket magnetic stack. This anneal is used to optimize crystal structure and set pinning layers with desired magnetization direction. During annealing, the MgO barrier undergoes some crystallization and magnetic films can change stoichiometry dramatically (e.g., CoFeB can exhibit substantial loss of boron during annealing, with concomitant crystallization of the as-deposited amorphous stoichiometric CoFeB). For MRAM designs that require film magnetization to be anisotropically set at high temperatures, the anneal oven must have a relatively large magnet capable of applying a uniform field of approximately 1 T across the entire batch of wafers during anneal. An example of such a commercially available tool is shown in Fig. 6.12. This is considered an essential part of most MRAM fabrication toolkits, but some innovative MRAM stack designs do not rely on high-temperature magnetization manipulation and thus do not require this type of magnetic oven.

6.3.1.3 Device Patterning Atomic layer deposition (ALD) may allow creation of MTJ stacks that conform to substantial topography in the substrate. At this time, however, the most common method of MTJ patterning relies on subtractive (etch) processes to define pillars in a planar multilayer magnetic stack. For decades, ion

Figure 6.13 IBE used to pattern an MTJ through the top layers, landing on the tunnel barrier (the horizontal white line). Redeposits on the sidewall of the upper portion of the device have minimal effect on device performance. The choice of IBE incidence angle was made for optimal cleanliness and minimal damage to the films immediately above the tunnel barrier. (Reproduced from Ref. 11 with permission from Wiley/IEEE Press.)

beam etching (IBE) has been the standard method of patterning magnetic devices in the disk drive and tape head industries. IBE has advantages such as low reactivity with the films being etched, control over incident angle of etching atoms, and similar etch rates across a wide range of different materials. Figure 6.13 gives an example of the control one can achieve with IBE when etching low-density MTJ arrays.

IBE difficulties in high-volume MRAM production revolve around the desire for short process times, tight-pitch device arrays, and high aspect ratio structures for easy contact to the top electrode wiring. The tight pitch negates the flexibility in IBE beam angle and can result in conductive sidewall redeposits that are not easily removed. Formation of high aspect ratio structures demands good selectivity between mask etch rate and the magnetic films' etch rates. While field-MRAM device density is certainly within the realm of IBE patterning, there is little experience with IBE processing of 300 mm wafers, and even small beam divergence can be problematic for device uniformity across such large wafers. The high-inertia CMOS industry has instead driven most field-MRAM and STT-MRAM processing toward the use of reactive ion etch (RIE) methods to satisfy patterning needs. RIE of magnetic materials is nontrivial because it is difficult to form volatile by-products through moderate-temperature chemical reactions with magnetic materials. Chlorine has been used with moderate success as a magnetic metal etchant, offering mask selectivity slightly more favorable than a simple argon sputter etch (12). The vapor pressure of magnetic metal chlorides is quite low when processing at temperatures below 300 °C, and a heavy

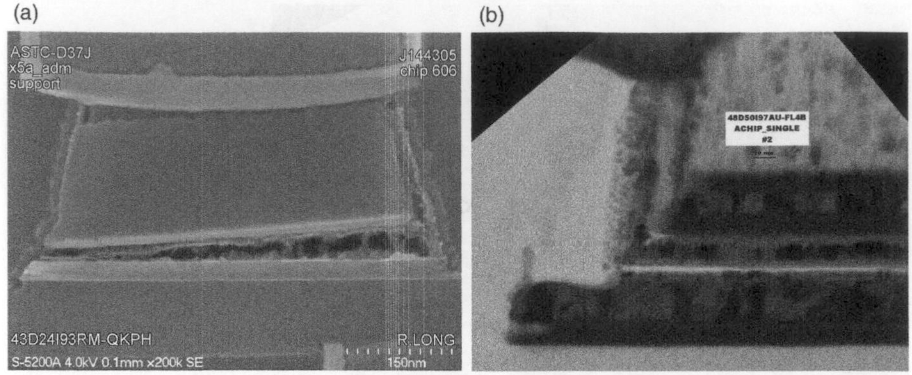

Figure 6.14 Chlorine attack of sensitive magnetic films (a) and poor reactant volatility seen as sidewall redeposits (b).

reliance on ion bombardment is a characteristic of the chlorine etches. Certain magnetic films can be quite reactive and will corrode in the presence of chlorine, particularly when adjacent to the noble metals such as iridium and platinum that are used in antiferromagnet pinning layers. A disturbing example of such chlorine-induced corrosion is shown in Fig. 6.14. Note in this figure as well the existence of metallic sidewall redeposits that effectively short-circuit the MTJ.

Magnetic properties of films etched in chlorine have also been shown to degrade due to halogen exposure (13). Searching for a solution to the problem of poor chlorine etch performance, the MRAM community has largely turned to carbonyl-based RIE. Through the use of carbonyl-forming etchants such as methanol or carbon monoxide/ammonia, one can create slightly more volatile by-products in reactions with the magnetic metals. Perhaps more importantly, one can achieve quite high selectivity between etching of metallic mask materials (e.g., titanium or tantalum) and the etching of magnetic materials. Most magnetic films show low propensity toward corrosion in these chemistries as well.

Both capacitively coupled plasmas (CCP) and inductively coupled plasmas (ICP) have been used with success in RIE of magnetic films, so RIE tooling choices are driven primarily by availability of existing equipment or by the flexibility needed to improve the immature existing etch methods. Irrespective of tool choice, a source of carbon monoxide/ammonia or methanol is a valuable addition to the RIE toolkit. ICP tooling has gained some ground on CCP options because it allows for independent control of the plasma density and impinging ion energy and thus a more anisotropic etching profile. The ultimate goal is development of a process with etch profile compatible with tight-pitch arrays, high mask selectivity, and etch rates fast enough for adequate wafer throughput (14).

The use of a conducting hard mask for MTJ patterning offers the potential for self-aligned contact between the top of the MTJ and the upper wiring layer. This is arguably the least expensive process choice with the fewest processing steps. An alternative method involves using a short hard mask so as not to shadow the MTJ during etch. This can be beneficial in reducing sidewall redeposits, but requires

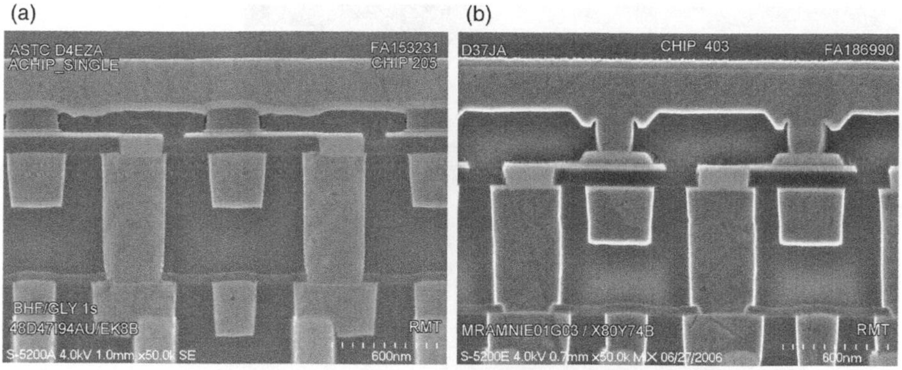

Figure 6.15 (a) Example of MTJ conductive hard mask forming a self-aligned contact to the upper layer wiring. (b) Example of MTJ with short hard mask for improved magnetic etch performance, but requiring additional via formation for contact to the top of the MTJ.

additional processing (e.g., inclusion of a via) to bridge the top of the MTJ to the upper wiring level. Figure 6.15 shows SEM cross-section images of the two different processes for field-MRAM. The short hard mask/additional via approach is practical only for MTJ devices with relatively large lateral dimensions (as in field-MRAM) due to the overlay tolerances in aligning the via to the MTJ. STT-MRAM development is targeting sub-40 nm MTJ diameters, and is incompatible with standard via processing to form such an upper contact. Thus, STT-MRAM is most often produced with the self-aligned tall conducting hard mask approach. As shown in Fig. 6.16, a gentle taper angle to the sidewall of the conducting hard mask can reduce etch shadowing effects and prevent redeposits from building up on the sidewalls and shunting the MTJ.

If sidewall redeposits are a particularly difficult problem, one possible solution is to adjust the MTJ etch to stop in or on the tunnel barrier. Since the tunnel barrier is not fully breeched, one avoids the mechanism of conductive redeposits shunting the tunnel barrier edges. It can be difficult to obtain enough etch selectivity to stop reliably in the tunnel barrier. As shown in Fig. 6.17, a slight underetch can result in a residual foot adjacent to the etched pillar. An overetch to ensure removal of the foot can result in damage to the layers beneath the tunnel barrier. A thinner hard mask can alleviate this selectivity/shadowing problem, as in the field-MRAM process flow with separate via for top contact from Fig. 6.15. With the "stop on barrier" etch process flow, a second masking/etch step is needed to isolate devices from each other. This isolation etch can be done with a spacer technique, or with a separate mask with openings at a relatively large distance from the edge of the tunnel barrier. For structures with the free layer above the tunnel barrier, this can be beneficial because it separates the free layer from a potentially large demagnetizing field emanating from the reference layer edges. Reliable device operation requires the demagnetizing offset field to be smaller than the free-layer coercive field. For free layers with relatively small coercive field, the stop on barrier etch with far removed reference layer etch can be an effective solution. A drawback to this approach is the associated growth in cell size coming from the stepped magnetic profile, so it is not practical for the densest of MRAM arrays.

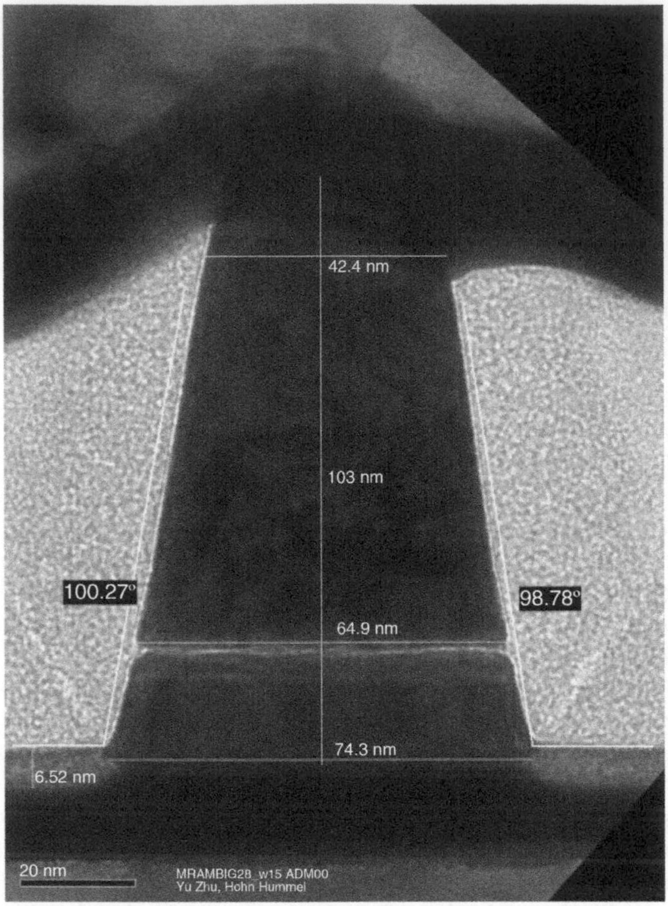

Figure 6.16 Example of gentle sidewall taper angle to assist with physical removal of sidewall redeposits. The sidewalls are roughly 10° from normal to the surface.

Continuing the MTJ etch below the tunnel barrier in a continuous manner will eventually expose the nonmagnetic base layer beneath. An advantage to this approach is that the etch can be made highly selective against etching the nonmagnetic layer, and one can thus scrub the MTJ sidewalls clean of residues with a healthy overetch. Figure 6.18 shows an example of a clean sidewall etch that uses a base layer selective against etching in methanol. To reduce demagnetizing offset fields, one can utilize flux-closed synthetic antiferromagnet structures or additional field-compensating layers in the magnetic stack. This can be difficult to control in practice, particularly for materials with perpendicular magnetic anisotropy (PMA).

The difficulty in patterning complex MTJ stacks for high-density memory arrays can be avoided by using a planar racetrack memory structure. In the racetrack memory, the densely packed data storage is done with domain walls in thin nanowires of relatively simple composition. The challenging feature size patterning is done on

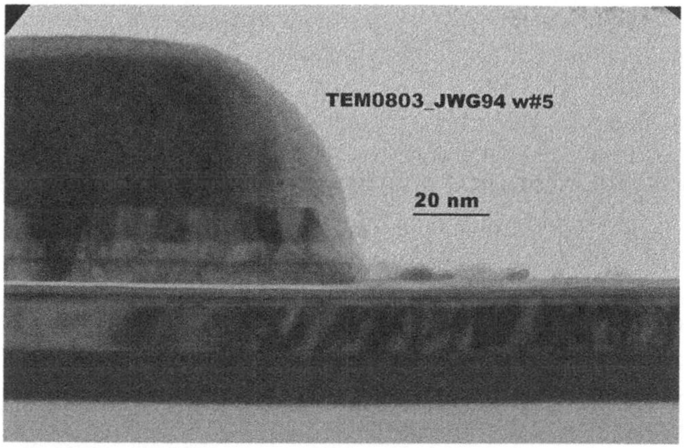

Figure 6.17 TEM cross section showing residual material adjacent to the MTJ pillar. This is a result of shadowing of the etch by the hard mask.

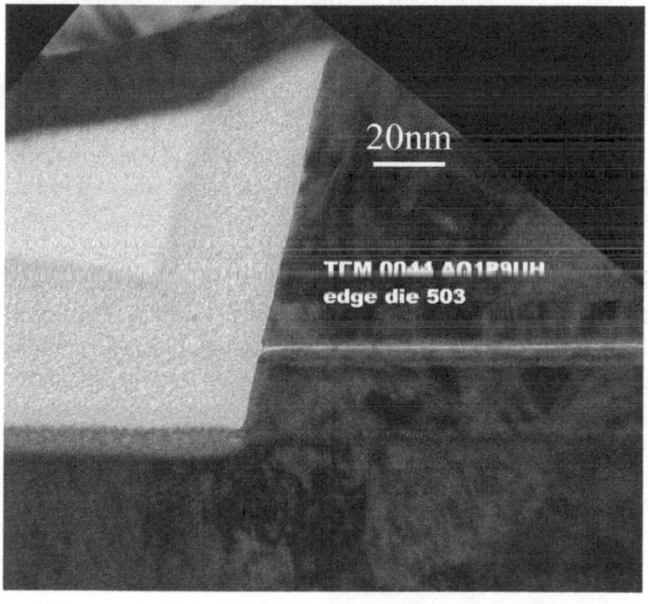

Figure 6.18 TEM cross section of an MTJ device formed with an etch that stopped on a base layer with high selectivity.

simple structures, whereas peripheral MTJs used to drive the nanowires can be patterned at relaxed pitch.

6.3.1.4 Dielectric Encapsulation Patterned MTJs must be adequately encapsulated for protection during further wafer and chip processing. Diffusion of materials along interfaces is one of the more prominent causes of device degradation during elevated-temperature processing, so the best choices of dielectric encapsulants will inhibit atomic migration. Additional limits on temperature exposure for patterned MTJs typically dictate that one must employ low-temperature-deposited dielectrics instead of conventional 400 °C CMOS BEOL dielectrics as MTJ encapsulation and ensuing insulation layers. One must take care with deposition conditions to ensure good dielectric adhesion, and conventional plasma surface treatments may need to be adjusted to prevent damage to the MTJ edges. Plasma or other pretreatments prior to encapsulant deposition are particularly important for partial (e.g., stop-on-barrier) MTJ etches. The post-etch surface may be composed of relatively weak material with properties that may inhibit good nucleation of void-free, adherent encapsulant. Noble metals containing antiferromagnets and reference layers are notable sources of poor adhesion and often drive further optimization of the dielectric encapsulation process. Magnetostriction effects can be present in MTJs, so the thickness and stress of dielectrics used should be explored in the course of optimizing the process flow. There is some concern about the need for *in situ* MTJ encapsulation immediately after etch, but presently there does not seem to be a large amount of degradation coming from exposure to atmosphere between the etch and encapsulation steps.

MTJ encapsulation with silicon nitride and similar compounds is desirable for film adhesion and for strong interfacial bonds that reduce troublesome diffusion of metal atoms along the dielectric–metal interfaces. Unfortunately, the deposition of silicon nitrides with industry standard plasma-enhanced chemical vapor deposition (PECVD) techniques requires the use of expensive high-density plasma tooling and process temperatures of 400 °C or higher to achieve a semblance of conformal fill around the protruding MTJ studs. Worse for higher aspect ratio structures and lower temperature deposition, seams and voids like those shown in Fig. 6.19 are often found in the silicon nitride dielectric at the juncture of films growing from two nonplanar surfaces. Mitsubishi Heavy Industries (MHI) offers a new design of CVD deposition tool that reduces heat and damage on the surface of the wafer, yet affords deposition of low-temperature (200 °C) silicon nitride and silicon oxide with improved conformality (15). Low substrate temperature is ensured by a pseudo-remote plasma and a high thermal conductivity wafer chucking mechanism. Leakage current of SiO_x and SiN films deposited with such tooling is adequate for high-reliability dielectrics.

Figure 6.19 Cross sectional SEM image of an MTJ device with a buffered hydrofluoric acid (BHF) + glycerine dip utilized to highlight weaknesses in the dielectric and reveal the poor conformality of the low-temperature-grown CVD SiN adjacent to the MTJ's metal films.

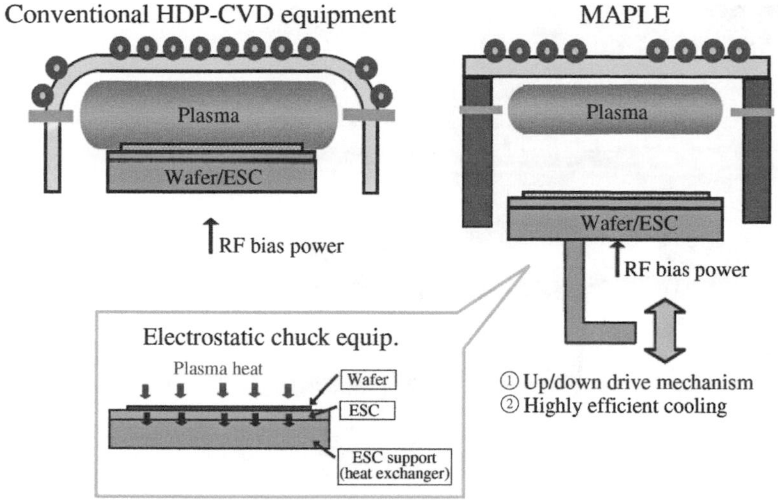

Figure 6.20 Comparison of conventional high-density plasma (HDP) CVD tool with the pseudo-remote plasma control of the MAPLE tool. (Courtesy of MHI (15).)

Further details are presented in Figs. 6.20 and 6.21. Another option for MTJ encapsulation is ALD aluminum oxide. This is commonly used in the hard disk drive and tape head industries after an IBE MTJ patterning step, but is not commonplace in CMOS BEOL tool sets at present.

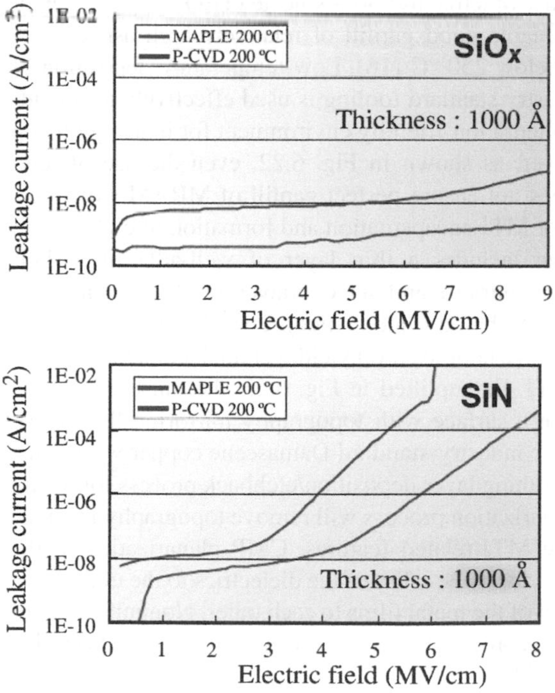

Figure 6.21 Comparison of leakage currents for conventional low-temperature plasma CVD and the MAPLE pseudo-remote plasma deposition with lower hydrogen and water content in the low-temperature deposited films. (Courtesy of MHI (15).)

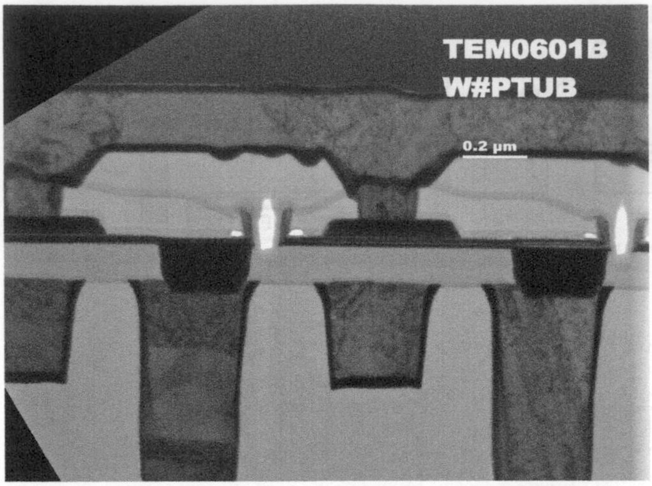

Figure 6.22 The TEM image above shows that even TEOS gapfill is not perfect, and for high aspect ratio features, the low-temperature TEOS is unable to completely fill the voids. A redesigned process flow with reduced aspect ratio topography was needed to achieve acceptable gapfill.

Silicon oxide films offer more options for low-temperature deposition than nitrides. As an alternative to the conventional silane-based CVD and the MHI pseudo-remote CVD plasma tooling, the use of tetraethyl orthosilicate (TEOS) as a precursor in silicon oxide deposition can enable good gapfill of relatively high aspect ratio structures, even at temperatures below 250 °C (16). Low-temperature deposition of TEOS-based silicon oxide in industry standard tooling is used effectively as a void-free MTJ encapsulant and as an integration-friendly environment for housing the top electrode wiring features. However, as shown in Fig. 6.22, even the use of low-temperature deposition TEOS does not ensure perfect gapfill of MRAM structures.

An example process flow for MTJ encapsulation and formation of dielectric to house the top-layer metal wiring includes a thin layer of well-adhering, MTJ-compatible dielectric (e.g., silicon nitride) and an overlayer of thick, conformal CVD-deposited dielectric (e.g., TEOS-based silicon oxide). The silicon nitride is kept thin enough that seams and voids have no impact on yield, yet thick enough to inhibit undesirable diffusion near the MTJ. Exemplified in Fig. 6.23, the thick, conformal dielectric deposition will generate a surface with topography mirroring that of the MTJ layer underneath. To facilitate industry-standard Damascene copper wiring, the wafers will require a CMP or smoothing layer deposition/etchback process following the dielectric deposition. This planarization process will remove topography from the surface caused by the underlying MTJ-related features. CMP planarization of the dielectric is an initial check of the adhesive quality of the dielectrics to the underlying metal films, as well as the cohesion of the metal films to each other. Nonuniformity of this planarization step can reduce the process window of ensuing process steps used to contact the top electrodes of the MTJs.

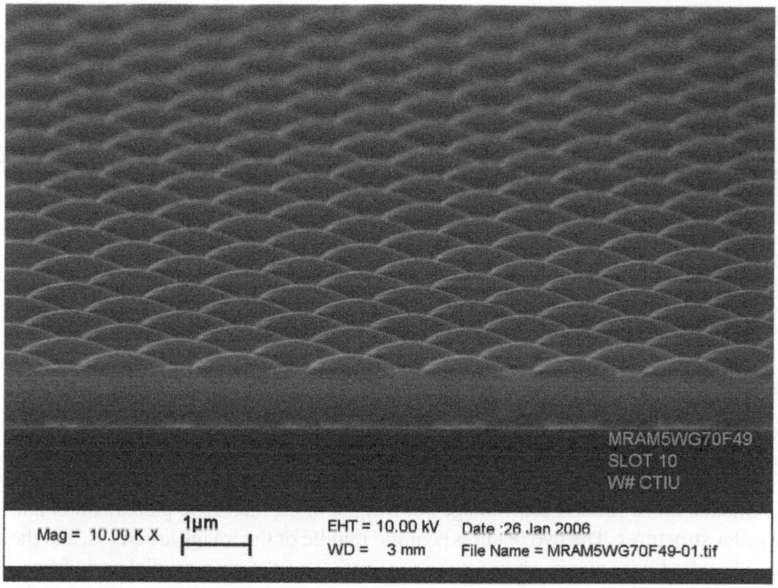

Figure 6.23 TEOS encapsulation of an array of MTJs propagates underlying topography and will require CMP dielectric planarization before next metal level formation.

If the encapsulating dielectrics are suitably planarizing in their deposition, this post-dielectric-deposition planarization step can be eliminated for faster turn-around time and potentially higher yield. With recent advances in the field of low-K dielectrics, the use of low-temperature-cured spin-on dielectrics for film uniformity, gapfill, and planarization has become an attractive option for this dielectric layer formation. JSR Corporation developed several grades of spin-on dielectric with cure temperatures as low as 200 °C (17). Gapfill is superior to CVD TEOS oxide, film uniformity is outstanding (<1% 3-sigma thickness across 200 mm wafers), and planarization obviates the need for additional steps before Damascene copper processing can take place. Figure 6.24 shows an example of the use of a planarizing spin-on dielectric to form the upper wiring layer dielectric.

6.3.2 Wiring and Packaging

MRAM wiring differs from standard CMOS BEOL wiring in a few nontrivial ways. First, wire-to-wire vertical spacing is set largely by the process flow used for creating and contacting the MTJ devices. Interwire capacitance is relegated to a secondary concern, as the focus of the MRAM processing is on yielding the best MTJs. The use of low-K dielectrics may be limited by the need to carefully encapsulate the MTJs and simplify the process flow. In principle, embedded MRAM can share wiring layers with surrounding logic circuitry so long as design kits are modified for any alterations in wire conductance and capacitance. The insertion of MRAM into a standard BEOL CMOS process flow will require requalification of the CMOS circuit reliability at substantial expense.

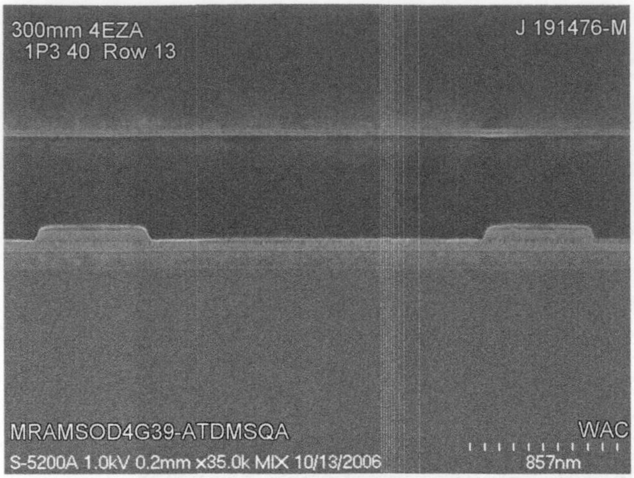

Figure 6.24 Polished cross-section SEM image showing spin-on dielectric planarization and gapfill of MTJ pillar structures. The two features near the middle of the image are MTJs, and the bright line a few hundred nanometers above is the top surface of the spin-on dielectric film.

Reliability is one of MRAM's strong attributes. Write endurance is expected to be effectively infinite if write schemes are such that tunnel barrier breakdown is avoided. However, electromigration in field-generating wires is a threat to reliability, since current density can approach 10 MA/cm^2. Electromigration can be mitigated through circuit designs that use bidirectional switching currents. Ferromagnetic liners around field-generating wires can further improve the situation by reducing required current densities.

6.3.2.1 Ferromagnetic Cladding Ferromagnetic flux guides to direct the magnetic fields generated by currents in the field-MRAM structures serve a dual purpose: concentrate the magnetic field on desired device regions for lower write currents and reduce near-neighbor interaction. Write power is relatively large for field-MRAM, so the possibility of a factor of 2 or more reduction in write current by focusing the field is of importance. Near-neighbor disturbs are a leading source of soft errors in field-MRAM arrays, and can be effectively quenched through the use of these ferromagnetic focusing elements. Figure 6.25 illustrates the use of ferromagnetic liners as a cladding around the bit line in a field-MRAM circuit.

Cladded wires offer the added benefit of alleviating electromigration failures. Not only do ferromagnetic liners offer reduction in current density, but they can also improve electromigration performance relative to conventional copper processes. By reducing the interface diffusion of copper atoms, ferromagnetic material surrounding the copper will enhance electromigration reliability to an extent similar to that seen in the industry by advanced Ta/TaN or CoWP capping processes (18).[1]

[1] J. Walls, M. Durlam, D. Gajewski, T. Kim, M. Martin, and D. Qualls, "Improved electromigration resistance of copper interconnects using multiple cladding layers," unpublished; http://www.IP.com, disclosure No. IPCOM000009315D.

Figure 6.25 SEM cross-sectional image of three cells in an array, with ferromagnetic cladding in an inverted U shape around the bit line above the MTJ device. Bit line currents will generate a field to switch the state of the MTJ. Yellow arrow overlays represent the extent of the field if ferromagnetic cladding had not been utilized: the field would spread to affect neighboring devices, and will have only moderate effect on the device under the bit line. The field from a clad wire (represented by green arrow, with cladding highlighted by the peach colored lines) will be focused with greater strength on those devices immediately under the bit line.

Formation of the cladding structure for wires beneath the MTJ can be a relatively simple procedure. If the chosen ferromagnetic cladding material is CMP-tolerant, the process flow involves simple replacement of the existing copper liner material with some variant including the ferromagnetic material. Figure 6.26 shows an example of a wire structure resulting from such a process flow. The efficiency of the liner in focusing the field is dependent on the permeability of the material surrounding the copper wire. One can improve performance with thicker sidewall coverage through the use of techniques such as collimation during deposition, self-ionized plasma (SIP) deposition, or atomic layer deposition (ALD). The chosen process must deposit a sufficiently high-quality, damage-free film so as to maintain the necessary permeability.

For variants of MRAM requiring field generation from the bit line wiring atop the MTJ, the cladding fabrication process can be substantially more difficult. The

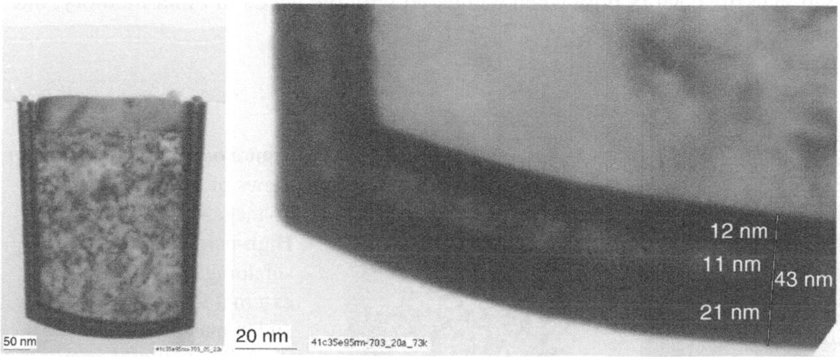

Figure 6.26 TEM cross-sectional images of a high-permeability ferromagnetic material sandwiched between conventional liner materials in a Damascene copper wire.

desirable inverted U shape of cladding around the bit line is formed in a multistep process to prevent the presence of ferromagnetic material at the bottom of the bit line. Coverage on the sidewalls and atop the bit line with seam-free interfaces is desirable. Figure 6.25 shows the structure created by a self-aligned approach described in the patent literature (19).

The use of ferromagnetic cladding in the CMOS BEOL wiring will unfortunately displace high-conductivity copper with lower conductivity magnetic material. In addition to the design modifications required for wire size and capacitance, field-MRAM necessitates the consideration of wire resistivity and inductance changes driven by the use of the ferromagnetic cladding. Designers can reserve certain layers for MRAM use only, or can modify their wiring models to conform to MRAM layer requirements. It is worth noting that the ferromagnetic cladding has the added benefit of acting as a shield against external fields, protecting the device somewhat to extend the range of possible operating environments.

6.3.2.2 Packaging Intended operating environments for MRAM include high-temperature automotive and military applications. In addition, operation in fluctuating external magnetic fields is to be expected. The properties of the MRAM element must be tuned to maintain stable storage through the fluctuations in temperature and field. It can be difficult to design the MTJ for suitable operating margins for read and write while at the same time making it robust against expected external magnetic field environments. One can rely on the packaging of the device to include additional magnetic shielding to ease the requirements on the MTJ devices. Inclusion of thick, plated layers or separate foils of high-permeability material in the packaging can increase external field operating margins by up to a factor of 10. It is best if the chip can be encased in the high permeability material, but even a single layer of such shielding can provide substantial benefit and allow use of the MRAM in harsh environments. A packaged 16 Mbit MRAM circuit is shown in Fig. 6.27.

6.3.3 Processing Cost Considerations

Fabrication of MRAM in the back end of a CMOS process flow is a relatively small perturbation to the CMOS flow when compared with the needs of other memory, like

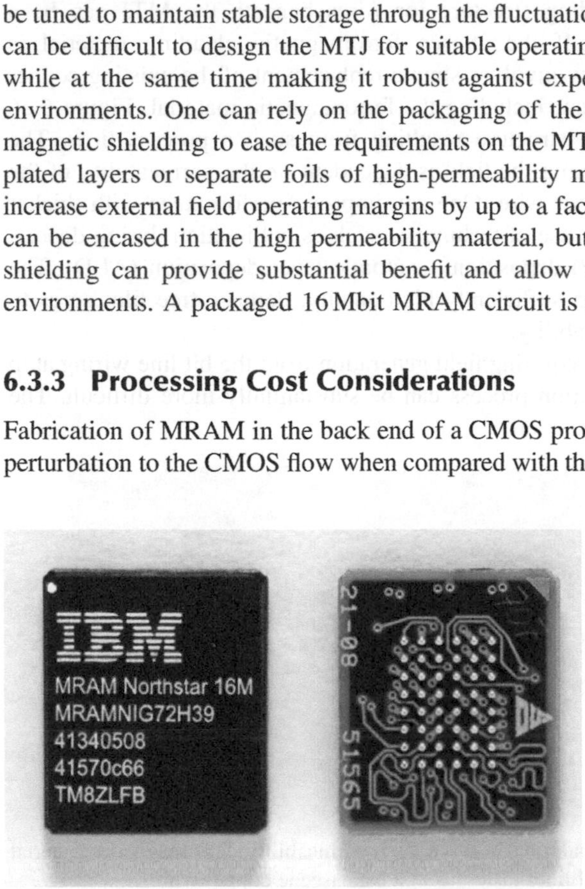

Figure 6.27 Top and bottom views of a packaged 16 megabit MRAM memory. High-permeability foil shielding is used to extend external field operating range, and connections are made through a ball grid array suitable for low-temperature soldering.

embedded Flash or ferroelectric RAM. MRAM can be tucked neatly between two CMOS wiring layers with only a small number of additional photomask levels. Particularly straightforward for STT-MRAM, wiring can be shared with surrounding CMOS logic circuitry for most efficient use of chip area in embedded applications. Specific adjustments to standard CMOS process techniques are driven mainly by the following:

- The temperature sensitivity of the MTJ devices (at present nearly 300 °C, but progress is being made toward 400 °C compatibility)
- The need for smooth substrates for MTJ film growth
- Ferromagnetically clad wiring in field-MRAM structures

The tooling requirements are modest as well, since minor adaptations of existing CMOS processes can suffice in many of the MRAM processes. Major exceptions include the following:

- A precision magnetic stack deposition tool (perhaps the largest expense and greatest limitation to wafer throughput)
- The potential need for an anneal oven capable of 400 °C temperatures in 1 T fields uniformly distributed over, for example, 300 mm wafers
- Specialized etch equipment capable of carbonyl-based RIE

Low-temperature BEOL dielectrics and ferromagnetically clad wires can drive the need for requalification of CMOS circuit reliability and necessitate process recipe development or new tooling considerations. In addition to the specific tooling requirements for fabrication of the MRAM, a suite of new process characterization methods is required for process development and monitoring.

6.4 PROCESS CHARACTERIZATION

6.4.1 200–300 mm Wafer Blanket Magnetic Films

The need for rapid in-line process monitoring is critical for a high-yield fab line. In addition to the standard CMOS test macros, one must include measurements of the MRAM-specific layers. Standard serpentine-comb and via chain structures can give high-speed readout of electrical process yield for the aforementioned VA vias, MA local strap (Fig. 6.3), and MTJ devices (20). Magnetic characterization is generally a much slower process because in the MRAM development phase, one generally must utilize high-inductance (slow response time) external magnets. After achieving reasonable yield on the MRAM devices, test and characterization teams can move to faster on-chip testing of arrays without external magnetic fields. Several methods have been developed to provide rapid feedback to the magnetic materials teams to monitor the health of their magnetic stacks (21). The following sections cover the use of rapid characterization methods to verify the sensitive magnetic stack deposition and anneal process is performing as expected. This is followed by a discussion of representative test data one may utilize from a process-centered viewpoint to increase device yield.

6.4.1.1 Current-in-Plane Tunneling (CIPT) Similar to the way a CMOS engineer might check RIE tool etch rate or metal tool deposition parameters with a standard recipe daily for process monitoring, the MRAM fabrication engineer will regularly monitor the quality of the magnetic stack being deposited and annealed for the MRAM product. Blanket films deposited on 200 or 300 mm wafers can be rapidly characterized for tunnel barrier resistance and magnetoresistance using a modified four-point probe measurement technique called current-in-plane tunneling (CIPT) (22). At the heart of this method is the use of tightly spaced probes wherein the current spreading is controlled to sample the tunnel barrier and adjacent films rather than films farther below the surface (Fig. 6.28). One generally adjusts the metal cap film deposition atop the magnetic films to optimize match of the CIPT tool to the conductivity of the layers above the tunnel barrier. Delicate microfabricated probes touch down on the surface of the blanket film with care not to puncture the tunnel barrier beneath. The use of a cap film such as a thin layer of ruthenium can ensure

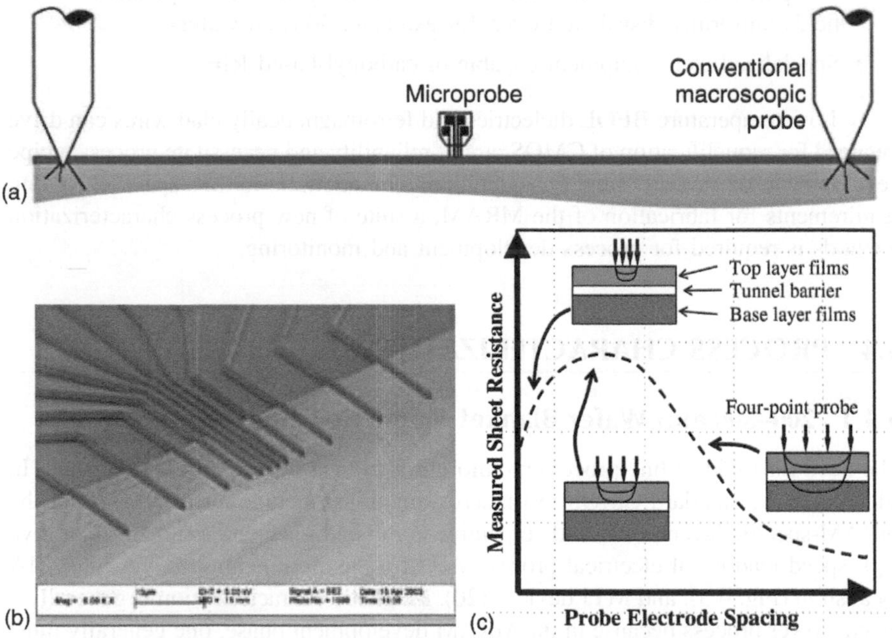

Figure 6.28 Examples of (a) the microprobe geometry, (b) commercially available finely spaced microprobes, and (c) experimental sheet resistance data as a function of probe spacing. In part (c), the path of current flow is represented by the thin lines connecting the outer probes. The experimentally measured sheet resistance versus probe spacing curve is fit mathematically to give an *RA* value for the blanket (unpatterned) magnetic stack beneath the probes. Collecting data from more combinations of probe spacings will give better accuracy to the curve fit. (Courtesy of Ref. 23.)

good probe-to-stack conduction even with little downforce from the probes. The resistance is measured between various combinations of pairs of probes spaced apart by several different distances. A mathematical fit is performed on the multiple measurements to extract the resistance–area (RA) product of the tunnel barrier. Ideal probe spacing depends on film resistance atop the tunnel barrier, film resistance below the tunnel barrier, and tunnel barrier *RA* product. The process can be repeated with applied magnetic fields to switch the state of the device and extract a magnetoresistance value. The different resistance values for the high-resistance state (R_{high}) and the low-resistance state (R_{low}) can be used to define a magnetoresistance ratio (*MR*):

$$MR = \frac{\left(R_{high} - R_{low}\right)}{R_{low}}. \tag{6.1}$$

One can substitute the CIPT-calculated *RA* product into this equation in place of the resistance values without changing the ultimate result.

6.4.1.2 Kerr Magnetometry Magneto-optical Kerr effect (MOKE) measurements are among the most widely used in the industry to characterize sheet films of magnetic materials. Kerr magnetometry relies on the physical effect that polarized light reflecting from a magnetic film obtains a small rotation of polarization. The measurement setup is relatively simple and inexpensive: a laser beam with well-known polarization is reflected off a sample and the change in polarization is measured. Its simplicity means one can easily build one's own in the laboratory, although commercial equipment is also available (24,25). The equipment is friendly to 200 and 300 mm wafers, but the bulk of the equipment is the magnet that is used to bias the film, with the magnet size being driven by the size of the sample that is placed inside. The *absolute* value of magnetic moment in the measured films cannot be accurately determined, but *relative* changes are straightforward to detect. The Kerr signal is proportional to the sample magnetization component parallel to the optical beam polarization. Many fundamental characteristics of the films can be inferred from Kerr magnetometry by sweeping the applied field and measuring the laser beam's polarization response. Accurate knowledge of the applied field can give a measurement of film coercivity, anisotropy, and exchange coupling. Polarization rotation versus applied field loops can be analyzed for squareness and multiple transitions corresponding to the behavior of multilayer magnetic films. Kerr magnetometry can also be performed on patterned films to examine the effects of shape anisotropy and demagnetization fields on the switching behavior of the devices. The low cost and great flexibility of a Kerr system makes it indispensable for rapid development of magnetic film stacks with good switching behavior (see footnote 1).

6.4.2 Parametric Test of Integrated Magnetic Devices

Once a fab has advanced to the point of producing patterned MTJs, one can make large strides in yield learning with some relatively simple techniques. Arguably, cycle

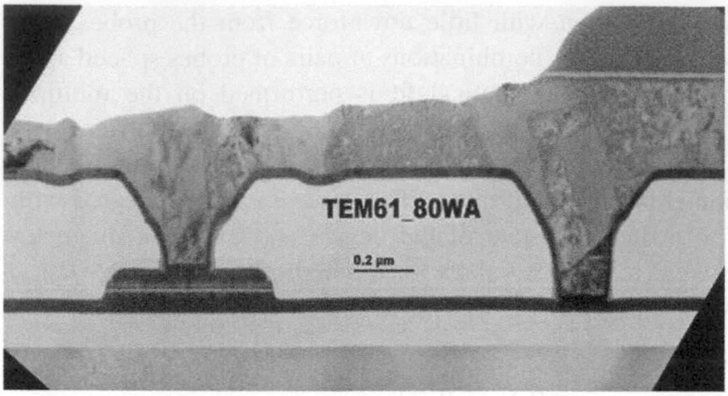

Figure 6.29 TEM cross-sectional image of a simplified structure for rapid turn-around time magnetics learning. The cross section exemplifies a rapid-build integration scheme for investigating the important magnetic elements without the expense and delay of building a fully integrated CMOS structure.

time of wafer processing and test is the key indicator of how quickly a fab can bring up a working MRAM product offering. Rapid turnaround learning can be achieved with little loss of functionality through the use of so-called "short loop" process flows that are subsets of fully integrated CMOS array fabrication. As illustrated in Fig. 6.29, by building only the MTJ and accessible contacts to the top and bottom, one can learn about the critical elements of magnetic stack quality and tunnel junction patterning with a 1/10th the turnaround time of a fully integrated CMOS-driven array (26). If one uses a tall self-aligned hard mask to contact the MTJ, and makes use of large MTJs as "vias" to a bottom blanket layer metal conductor, this short loop process can be performed with just one mask for the MTJ and one mask to define the top-layer wiring. Among other tests, such wafers can be measured for magnetoresistance, resistance, resistance versus field loops for coercivity and offsets, breakdown voltage, and thermal activation energy barrier. It cannot be emphasized enough that learning from simplified structures is key to bringing up the eventual product process flow.

6.4.2.1 *Magnetoresistance versus Resistance and Resistance versus Reciprocal Area* One of the most useful tools in diagnosing problems is a simple plot of magnetoresistance versus device resistance. Figure 6.30 shows data from a particularly interesting array of devices that show many different characteristic signatures of imperfect MTJs.

The different resistance values for the MTJ high-resistance state (R_{high}) and the MTJ low-resistance state (R_{low}) can be used to define a magnetoresistance ratio (MR) as in Eq. (6.1). Addition of a series resistance R_{series} to the MTJ modifies the equation for MR to look like

$$measured\,MR = \frac{(R_{high} - R_{low})}{(R_{low} + R_{series})} = \frac{(R_{high} - R_{low})}{R_{meas}}, \qquad (6.2)$$

which has a hyperbolic character for *measured MR* versus the *measured* value of the low-resistance state R_{meas} when a large series resistance is present. This situation

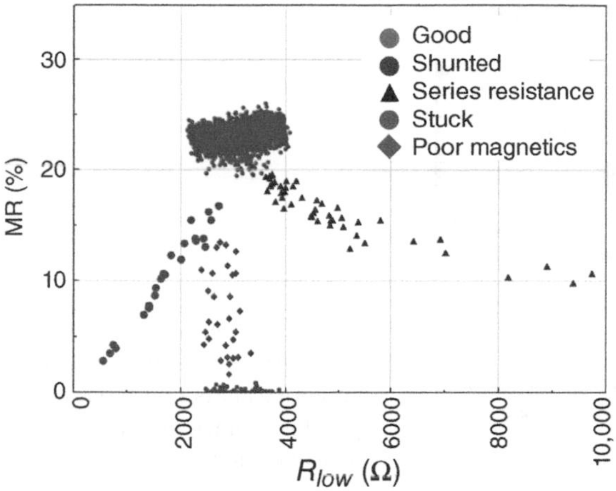

Figure 6.30 A plot of magnetoresistance versus low-state device resistance for several thousand devices in an array. Each point corresponds to a different MTJ. Magnetoresistance and low-state resistance were determined by applying an external magnetic field to set the device in one state, then measuring resistance in zero field, and then using the external magnet to switch the device to the opposite state, and finally measuring resistance again in zero field.

corresponds to the tail of black triangles below the desirable devices in the cluster of green circles in Fig. 6.30. Addition of a shunt resistance modifies Eq. (6.1) to look like

$$\textit{measured MR} = \frac{R_{meas}\left(R_{high} - R_{low}\right)}{R_{high}R_{low} - R_{high}R_{meas} + R_{low}R_{meas}}, \tag{6.3}$$

which has a linear character for *measured MR* versus the *measured* value of the low-resistance state R_{meas} when a large shunt conductance is present:

$$\textit{measured MR} = \frac{R_{meas}\left(R_{high} - R_{low}\right)}{R_{high}R_{low}}. \tag{6.4}$$

This situation corresponds to the red circles in Fig. 6.30 in a line from the origin to the desired cluster location. Two other markers in Fig. 6.30 represent situations where the magnetic films do not behave as desired. Devices with no *MR*, but having reasonable resistance are represented by the orange circles as "stuck" devices where the magnetic free layer does not switch between two stable, hysteretic states. This can arise from defective films where there may be a defective exchange-coupled interface, a weakly pinned reference layer, or large offset field at the device. If the offset from, for example, demagnetizing field is larger than the coercivity of the device, only one state is stable at zero field, and the measured *MR* will be zero. The blue diamond markers in Fig. 6.30 represent devices with imperfect magnetics such as those with free layer magnetization incompletely parallel or antiparallel to the reference layer.

It is also useful to have a test pattern on the wafer whereby one can measure many different-sized devices to obtain a resistance versus reciprocal area plot like that shown in Fig. 6.31. For good tunnel-barrier-dominated conduction, one wishes to have a linear relation between the device resistance and the inverse of device area. A linear fit to the resistance data in Fig. 6.31 is shown as a green line with a positive y-axis intercept. While the linear character of the fit is desirable, the positive y-axis intercept is indicative of either a parasitic resistance in series with the MTJ or a

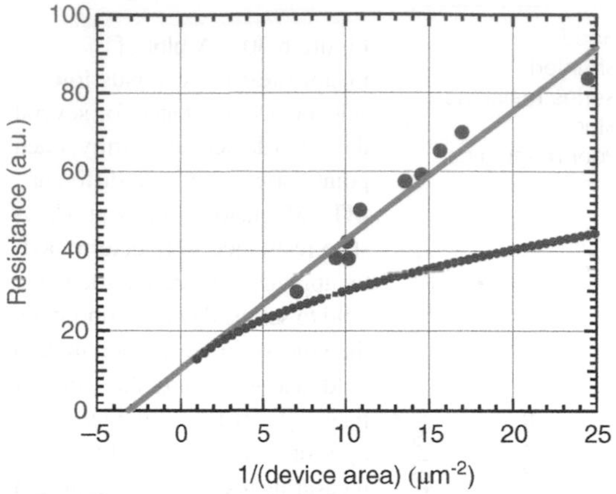

Figure 6.31 Blue dots are data points, green line is linear fit to the data, and red line is a curve one might expect for perimeter-dominated conduction (sublinear).

nonconducting portion of the tunnel barrier. The latter effect can be caused by overoxidation of the tunnel barrier during etch or encapsulation whereby the device now has a ring of slightly thicker tunnel barrier around the edges. Because tunnel barrier resistance is exponential in the thickness of the barrier, even small changes in barrier oxidation can have large effects. If one were to find a fit with negative *y*-axis intercept, this would be indicative of poor device size calibration. The optimal condition would have a zero intercept.

It is interesting to note that one of the major failure mechanisms in MTJ fabrication can be diagnosed through the use of a plot like that in Fig. 6.31. As mentioned earlier, sidewall redeposits during MTJ etching can be a troublesome result of the poor volatility of magnetic film etch by-products. Additionally, defects or excessive stress in the region near the edge of the MTJ can cause a shunting effect at the perimeter of the device. If conduction is dominated by perimeter paths, one will find a sublinear fit to the resistance versus reciprocal area data. One then knows to focus on improving yield through better device patterning and encapsulation rather than spending effort to improve the tunnel barrier characteristics.

Poor tunnel barrier formation can result in pinholes of high conductivity that dominate the current path through the device. The characteristic resistance versus reciprocal area plot will then be supralinear, with exact shape dependent on the density of the pinholes relative to the size of the devices being measured. If unsure about the quality of the tunnel barrier, one can further rely on breakdown voltage measurements to compare the efficacy of various tunnel barrier formation methods.

6.4.2.2 Breakdown Voltage Breakdown voltage of MTJ barriers is a useful tool for the diagnosis of problems regarding formation of the magnetic stacks and subsequent patterning of the devices. Devices with pinholes in the tunnel barrier will have a different breakdown voltage signature than "perfect" devices or devices with shunting along the device edges. An example of softer breakdown characteristics for devices with tunnel barrier pinholes is shown in Fig. 6.32. The sharper nature of breakdown in properly working devices can be seen in contrast to the more gradual, lower voltage breakdown of devices with pinholes in the tunnel barrier. The lower

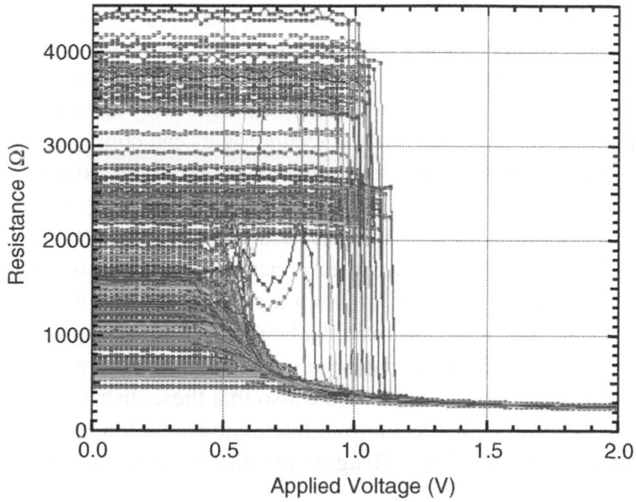

Figure 6.32 Breakdown voltage for several dozen MTJ devices. Sharp breakdown is characteristic of the higher resistance devices (note that the two states parallel and antiparallel give resistance groupings near 3.7 and 2.3 kΩ). More gradual, lower voltage breakdown is characteristic of the lower resistance devices. This can result from tunnel barriers with pinholes or similar defects.

initial resistance of the soft-breakdown devices is another clue to the origin of poor device behavior. The sharp breakdown in the higher resistance devices can be attributed to the high current densities present in STT devices. Rather than gradual pinhole formation, electromigration and heating leads to displacement of the MgO junction material and sharp decrease of barrier resistance (27,28).

MRAM reliability concerns are generally focused on the tunnel barrier dielectrics, as the extreme thinness makes them susceptible to pinhole formation. These tunnel barriers are artificially grown layers atop differing materials, rather than the conventional CMOS gate dielectrics where the silicon surface is *converted* to an insulator through an oxidation process that is quite uniform and reliable. Newer CMOS technology involving high-K and metal gate processes is more like the situation with MRAM, but MRAM will generally have stricter temperature limits during the tunnel barrier growth and has less flexibility in improving materials properties through heated deposition or high-temperature post-anneal. The moderate-temperature (<450 °C) postmagnetic stack deposition anneal is used in part to crystallize the MgO tunnel barrier, and is critical to achieving highly transmissive barriers with high breakdown voltage (29).

Tunnel barrier breakdown is a major concern for the commercialization of STT-MRAM. Because the switching mechanism for STT-MRAM relies on high current density through the tunnel barrier, switching voltage can be dangerously close to the breakdown voltage. Read-then-write operations can improve device longevity by eliminating the need for high-power device writing if the device is already in the desired state. This comes at the expense of a longer cycle time for memory operation.

6.4.2.3 Device Spreads A nagging concern for large memory arrays is the need to create thousands, millions, or billions of devices with nearly identical characteristics so they can all be driven with the same peripheral circuitry. Device "outliers" that have characteristics vastly different from other devices in the memory array must be repaired by patching-in redundant circuitry. Mature technologies can compute the trade-offs between additional chip area required for redundancy versus the expected number of outliers. In addition to minimizing the number of outliers, the process integrator must also attempt to minimize spreads in characteristics for devices that are of nominally identical size. Figure 6.33 illustrates the concerns for device distributions in an STT-MRAM memory array. In the example in the figure, the distribution curves assume a Gaussian profile. The distributions must be separate from neighboring distributions by a certain number of standard deviations, determined by requirements such as 10 years of error-free operation (30). Note also that these distributions are not static, but will change over time due to thermal fluctuations and longer term physical aging of the circuit. It is particularly challenging to tailor the magnetic stack

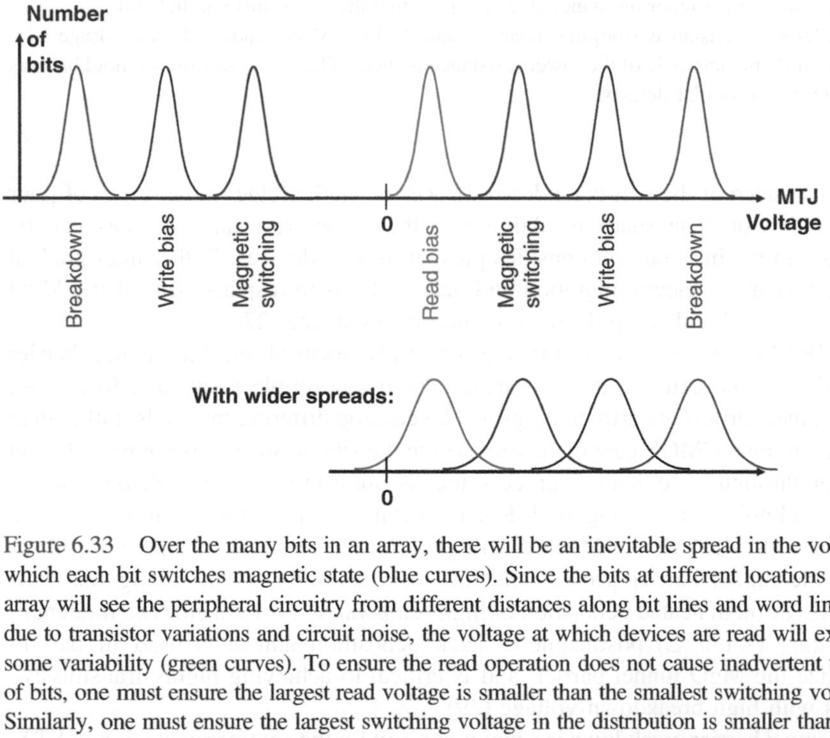

Figure 6.33 Over the many bits in an array, there will be an inevitable spread in the voltage at which each bit switches magnetic state (blue curves). Since the bits at different locations in the array will see the peripheral circuitry from different distances along bit lines and word lines, and due to transistor variations and circuit noise, the voltage at which devices are read will exhibit some variability (green curves). To ensure the read operation does not cause inadvertent writing of bits, one must ensure the largest read voltage is smaller than the smallest switching voltage. Similarly, one must ensure the largest switching voltage in the distribution is smaller than the smallest applied write bias voltage (black curves). Finally, the device breakdown voltage (red curves) must be larger than the highest write bias for all devices. The upper portion of the figure shows the situation for an array with good operating characteristics. The lower portion of the figure shows the positive voltage portion of a situation where device and circuit spreads are too large, resulting in soft errors or device breakdown.

to adequately resist thermal fluctuations at elevated operating temperatures such as those found in automotive or military applications. Some relief can be found with error correction code (ECC) and operation modes that allow multiple attempts at writing—including a write-then-read cycle that is repeated until the bit lands in the proper state. Measurement of and attention to the device spreads is a never-ending task in MRAM product development.

ACKNOWLEDGMENTS

I gratefully acknowledge the contribution of CIPT images and diagrams from Peter F. Nielsen of Capres A/S. Tadashi Shimazu and Mitsubishi Heavy Industries kindly provided figures related to low-temperature dielectric deposition with the MAPLE tool. David Hurley from Magnetic Solutions Ltd. graciously provided the figure describing their state-of-the-art magnetic anneal tool. W. J. Gallagher provided editorial assistance and advice. I also thank IBM's Microelectronics Research Laboratory (MRL) for process development and fabrication; E.O'Sullivan for process integration and guidance; J. Nowak for test and analysis; P. Rice, T. Topuria, E. Delenia, and Yu Zhu for TEM imaging; E. Galligan and C. Jessen for failure analysis. Additional significant contributions were from D. W. Abraham, J. Sun, E. Joseph, R. M. Martin, P. Trouilloud, S. L. Brown, G. Hu, and A. J. Annunziata.

REFERENCES

1. D. W. Abraham and P. L. Trouilloud,"Thermally assisted magnetic random access memory," U.S. Patent 6,385,082 (2002).
2. I. L. Prejbeanu, M. Kerekes, R. C. Sousa, H. Sibuet, O. Redon, B. Dieny, and J. P. Nozières, "Thermally assisted MRAM," *J. Phys. Condens. Mater.* **19**, 165218 (2007); doi: 10.1088/0953-8984/19/16/165218.
3. M. C. Gaidis,"Tunnel junction via," U.S. Patent 8,124,426 (2012).
4. J. Z. Sun, M. C. Gaidis, E. J. O'Sullivan, E. A. Joseph, G. Hu, D. W. Abraham, J. J. Nowak, P. L. Trouilloud, Yu Lu, S. L. Brown, D. C. Worledge, and W. J. Gallagher, "A three-terminal spin-torque-driven magnetic switch," *Appl. Phys. Lett.* **95**, 083506 (2009); doi: 10.1063/1.3237025.
5. S. S. P. Parkin,"Shiftable magnetic shift register and method of using the same," U.S. Patent 6,834,005 (2004); "System and method for writing to a magnetic shift register," U.S. Patent 6,898,132 (2005).
6. A. J. Annunziata, M. C. Gaidis, L. Thomas, C. W. Chien, C. C. Hung, P. Chevalier, E. J. O'Sullivan, J. P. Hummel, E. A. Joseph, Y. Zhu, T. Topuria, E. Delenia, P. M. Rice, S. S. P. Parkin, and W. J. Gallagher,"Racetrack memory cell array with integrated magnetic tunnel junction readout," in *IEEE International Electron Devices Meeting (IEDM), Technical Digest*, Washington, DC. (2011); doi: 10.1109/IEDM.2011.6131604.
7. See, for example, the MR4A16B 16 Mb field-MRAM available from Everspin, http://www.everspin.com (August 2012).
8. M. C. Gaidis, J. Nuetzel, W. Glashauser, E. O'Sullivan, G. Costrini, S. L. Brown, F. Findeis, and C. Park,"Recessed metal lines for protective enclosure in integrated circuits," U.S. Patent 6,812,141 (2004).
9. C. Sarma, S. Kanakasabapathy, I. Kasko, G. Costrini, J. Hummel, and M. C. Gaidis,"Method for improved alignment of magnetic tunnel junction elements," U.S. Patent 6,933,204 (2005).

10. C-7100EX sputter deposition tool from Canon Anelva Corp., Tokyo (http://www.canon-anelva.co.jp/english), and Timaris sputter deposition tool from Singulus Technologies, Kahl, Germany (http://www.singulus.de) (August 2012).
11. M. C. Gaidis, "Magnetoresistive random access memory," in *Nanotechnology: Information Technology I*, Vol. 3, ed. R. Waser, Wiley-VCH Verlag GmbH, Weinheim, Germany, 2010, pp. 419–446; doi: 10.1002/9783527628155.nanotech033.
12. S. J. Pearton, H. Cho, K. B. Jung, J. R. Childress, F. Sharifi, and J. Marburger, "Dry etching of MRAM structures," *Mater. Res. Soc. Symp. Proc.* **614**, pp. F10.2.1–F10.2.11 (2000); doi: 10.1557/PROC-614-F10.2.1.
13. E. A. Joseph, R. M. Martin, J. S. Washington, D. W. Abraham, S. Raoux, J. L. Jordan-Sweet, D. Miller, H.-Y. Cheng, M. C. Gaidis, M. Gajek, M. Breitwisch, S.-C. Lai, Y. Zhu, R. Dasaka, R. Sawant, D. Neumeyer, R. M. Shelby, H.-L. Lung, C. H. Lam, and N. C. M. Fuller,"Characterizing the effects of processing on materials for PCM and STT-based NVM technologies," *58th International Symposium on American Vacuum Society*, Nashville, TN (2011).
14. M. T. Moneck and J.-G. Zhu,"Challenges and promises in the fabrication of bit patterned media," *Proceedings of SPIE* 7823, Photomask Technology 2010, 78232U, September 2010; doi: 10.1117/12.875542.
15. Mitsubishi Heavy Industries (MHI) plasma CVD Multi-Application Plasma Equipment (MAPLE) tool (http://www.mhimaple.com) (August 2012).
16. J. Crowell, L. Tedder, H. Cho, F. Cascarano, and M. Logan, "Model studies of dielectric thin film growth: chemical vapor deposition of SiO_2," *J. Vac. Sci. Technol.* **A8**, pp. 1864–1870 (1990); doi: 10.1116/1.576817.
17. M. Yoshioka, E. Hayashi, K. Sumiya, and A. Shiota,"Insulation film," U.S. Patent 7,297,360 (2007); http://www.jsr.co.jp
18. D. A. Gajewski, T. Meixner, B. Feil, M. Lien, and J. Walls,"Electromigration of MRAM-customized Cu interconnects with cladding barriers and top cap," IEEE Integrated Reliability Workshop Final Report, pp. 90–92 (2004); doi: 10.1109/IRWS.2004.1422746.
19. S. K. Kanakasabapathy, E. O'Sullivan, M. C. Gaidis, and M. Lofaro,"Structure and method for formation of cladded interconnects for MRAMs," U.S. Patent 7,442,647 (2008).
20. M. Bhushan and M. B. Ketchen, *Microelectronic Test Structures for CMOS Technology*, Springer, New York, 2011.
21. D.W. Abraham, P. L. Trouilloud, and D. C. Worledge, "Rapid-turnaround characterization methods for MRAM development," *IBM J. Res. Dev.* **50**, pp. 55–67 (2006); doi: 10.1147/rd.501.0055.
22. D. C. Worledge and P. L. Trouilloud, "Magnetoresistance measurement of unpatterned magnetic tunnel junction wafers by current-in-plane tunneling," *Appl. Phys. Lett.* **83**, pp. 84–86 (2003); doi: 10.1063/1.1590740.
23. CIPTech tool, CAPRES A/S, with 12-probe cantilever sampling over distances from 1.5 μm to 59 μm; http://www.capres.com (accessed 2016).
24. M. Cormier, J. Ferré, A. Mougin, J. P. Cromières, and V. Klein, "High resolution polar Kerr magnetometer for nanomagnetism and nanospintronics," *Rev. Sci. Instrum.* **79**, 033706 (2008); doi: 10.1063/1.2890839.
25. J. M. Teixeira, R. Lusche, J. Ventura, R. Fermento, F. Carpinteiro, J. P. Araujo, J. B. Sousa, S. Cardoso, and P. P. Freitas, "Versatile, high sensitivity, and automatized angular dependent vectorial Kerr magnetometer for the analysis of nanostructured materials," *Rev. Sci. Instrum.* **82**, 043902 (2011); doi: 10.1063/1.3579497.
26. M. C. Gaidis, E. J. O'Sullivan, J. J. Nowak, Y. Lu, S. Kanakasabapathy, P. L. Trouilloud, D. C. Worledge, S. Assefa, K. R. Milkove, G. P. Wright, and W. J. Gallagher, "Two-Level BEOL processing for rapid iteration in MRAM development," *IBM J. Res. Dev.* **50**, pp. 41–54 (2006); doi: 10.1147/rd.501.0041.
27. M. Schäfers, V. Drewello, G. Reiss, A. Thomas, K. Thiel, G. Eilers, M. Münzenberg, H. Schuhmann, and M. Seibt, "Electric breakdown in ultrathin MgO tunnel barrier junctions for spin-transfer torque switching," *Appl. Phys. Lett.* **95**, 232119 (2009); doi: 10.1063/1.3272268.
28. R. C. Sousa, C. Papusoi, Y. Conraux, C. Maunoury, I. L. Prejbeanu, K. Mackay, B. Delaet, J. P. Nozieres, and B. Dieny, "Pulsewidth dependence of barrier breakdown in MgO magnetic tunnel junctions," *IEEE Trans. Magn.* **44**, pp. 2581–2584 (2008); doi: 10.1109/TMAG.2008.2003063.

29. J. Hayakawa, S. Ikeda, Y. M. Lee, F. Matsukura, and H. Ohno, "Effect of high annealing temperature on giant tunnel magnetoresistance ratio of CoFeB/MgO/CoFeB tunnel junctions," *Appl. Phys. Lett.* **89**, 232510 (2006); doi: 10.1063/1.2402904.
30. T. M. Maffitt, J. K. DeBrosse, J. A. Gabric, E. T. Gow, M. C. Lamorey, J. S. Parenteau, D. R. Willmott, M. A. Wood, and W. J. Gallagher, "Design considerations for MRAM," *IBM J. Res. Dev.* **50**, pp. 25–39 (2006), doi: 10.1147/rd.501.0025.

CHAPTER 7

BEYOND MRAM: NONVOLATILE LOGIC-IN-MEMORY VLSI

Takahiro Hanyu,[1,2,3] *Tetsuo Endoh,*[1,2,4] *Shoji Ikeda,*[1,2] *Tadahiko Sugibayashi,*[5] *Naoki Kasai,*[1] *Daisuke Suzuki,*[6] *Masanori Natsui,*[3] *Hiroki Koike,*[1] *and Hideo Ohno*[1,2,7]

[1]Center for Innovative Integrated Electronics Systems, Tohoku University, Sendai, Japan
[2]Center for Spintronics Integrated Systems, Tohoku University, Sendai, Japan
[3]Laboratory for Brainware Systems, RIEC, Tohoku University, Sendai, Japan
[4]Department of Electrical Engineering, Tohoku University, Sendai, Japan
[5]System Platform Research Laboratories, NEC Corporation, Tsukuba, Japan
[6]Frontier Research Institute for Interdisciplinary Sciences, Tohoku University, Sendai, Japan
[7]Laboratory for Nanoelectronics and Spintronics, RIEC, Tohoku University, Sendai, Japan

7.1 INTRODUCTION

This chapter discusses a strategy for solving key problems that face fine-scale very large-scale integrated (VLSI) circuits and systems by applying spintronic devices (1) to memory and general logic circuits. The use of nonvolatile, fast spintronic devices with virtually unlimited endurance, such as magnetic tunnel junction (MTJ) devices (1) and domain wall (DW) motion elements (2) in logic circuits, is the key to meeting these challenges.

7.1.1 Memory Hierarchy of Electronic Systems

Figure 7.1 illustrates current memory hierarchy, which is commonly used to illustrate the role of each memory device layer (3–5). The hierarchy consists of cache memory (static random-access memory (SRAM)), main memory (dynamic random-access memory (DRAM)), and storage (hard-disk drive (HDD) and NAND flash memory), in order of increasing distance from the processor core at the top of the hierarchy. SRAM and DRAM are volatile memory, whereas HDD and NAND are nonvolatile memory for the purpose of storage. The figure also indicates the memory speed by position in the vertical direction and the memory density by width in the horizontal

Introduction to Magnetic Random-Access Memory, First Edition. Edited by Bernard Dieny, Ronald B. Goldfarb, and Kyung-Jin Lee.

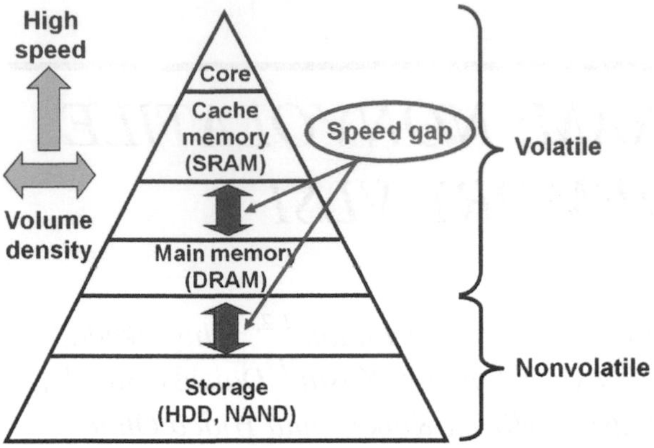

Figure 7.1 Current memory hierarchy as viewed from a processor core.

direction. The hierarchy is usually depicted as a pyramid, as shown in Fig. 7.1, because very high speed but relatively small density is required near the core at the top, whereas less speed but much higher density is required for storage distant from the core at the bottom. Data transfer is possible only between the neighboring memory layers, because the operation speeds of the devices must match.

Recently, several challenges have become apparent in the hierarchy structure. One is the slowdown in the rate of DRAM scaling. This has come about because of difficulties in ensuring sufficient capacitance and low leakage in memory cells with reduced cell size (6). This means that DRAM may not be able to keep its present position in the hierarchy. Another challenge is the increasing speed gap between individual memory layers. Compared to the increase in speed for processors and cache memory, which fully benefit from the performance improvement of the state-of-the-art complementary metal–oxide semiconductor (CMOS) process technology, the rate of improvement in DRAM speed is rather slow; hence, the speed gap between the two device layers is becoming wider. Also, there are speed gaps between DRAM and HDD, which is hardly affected by the performance improvements due to CMOS scaling, and between DRAM and NAND flash; the latter is manufactured using special process and circuit technologies that result in much slower operation than for DRAM. These speed gaps are becoming a major bottleneck limiting system performance.

An effective way to counter these challenges is to introduce spin-transfer torque random-access memory (STT-MRAM) and its variants such as DW motion elements into the memory hierarchy. STT-MRAM is nonvolatile and there is a great possibility to achieve high density comparable to DRAM with faster write/read operation (3,7). Thus, in addition to use as a substitute for DRAM, STT-MRAM can bridge the speed gaps between current SRAM and DRAM or DRAM and storage because of its high speed and high density.

Nonvolatility counters the increase in power consumption. Such a future memory system, consisting of conventional CMOS and newly introduced STT-MRAM technologies, is illustrated in Fig. 7.2 (8). Here, nonvolatile cache and main memory (NV-RAM) is introduced between SRAM and DRAM. Furthermore, high-speed storage memory (high-speed NVM) is introduced between DRAM and storage. Using NV-RAM and high-speed NVM together with a power-gating technique, as

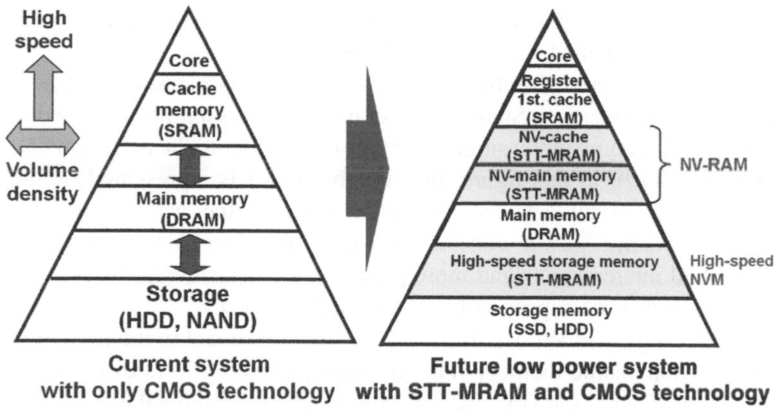

Figure 7.2 New low-power system enabled by STT-MRAM.

discussed later, would enable extremely low power systems. As discussed later, however, this approach alone is not capable of countering the challenges faced by current VLSI systems. This prompts us to introduce a fundamentally different approach, which is the subject of this final chapter.

7.1.2 Current Logic VLSI: The Challenge

In conventional VLSI, as shown in Fig. 7.3a, logic modules and memory modules are separately located. Since the amount and length of wiring between modules increases as the VLSI integration level progresses, chip performance will increasingly be restricted by interconnection delays. Reports from the International Technology Roadmap for Semiconductors (ITRSs) in recent years point out that global interconnects will especially have a large impact on the delay degradation compared to lower interconnect layers because of their long wiring length (9). This means that the

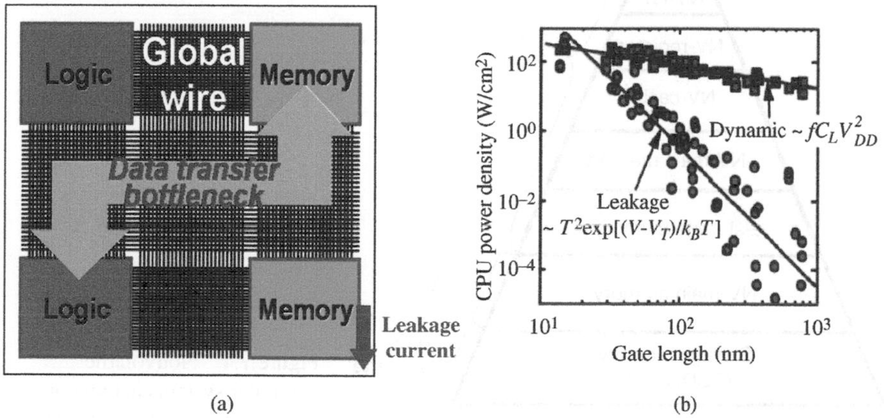

Figure 7.3 Increasing problems for future-scale CMOS VLSI. (a) Data transfer bottleneck in a VLSI chip. (b) Power density as a function of gate length. (from Pop (10).)

delay for global wiring will become one of the main factors restricting chip performance; this is the so-called input/output (I/O) bottleneck.

Another problem concerns standby power, that is, the increase in standby power as the VLSI integration level grows. Standby power used to be negligible compared to active power, but it has become increasingly important and closer to the active power, as shown in Fig. 7.3b (10). In this figure, the standby power is represented by the leakage current present in all CMOS transistors, whereas the active power is represented by the dynamic power consumption corresponding to the capacitive charging energy of all interconnects and more generally all capacities present in the chip and exposed to pulses of supply voltage V_{DD} at frequency f.

The nonvolatile logic-in-memory (LiM) architecture discussed in the following sections, in which a nonvolatile memory function implemented by MTJ devices and their variants and logic functions are merged, can simultaneously counter the two major challenges: the I/O bottleneck and the standby power increase. By distributing memory elements over the logic-circuit plane, LiM reduces the length of global interconnections, thereby suppressing the I/O bottleneck. Moreover, standby power is required only when there is a need to supply power to a circuit. Hence, a power-down mechanism called power gating, through which the power supply for inactive circuits is cut off, must be introduced. When nonvolatile memory elements are distributed over the logic-circuit plane, we can incorporate a mechanism to completely shut off power to inactive parts of the circuit, while at the same time having a much finer memory granularity (i.e., small memory blocks dispersed among the logic circuits) than in a conventional architecture. Therefore, a nonvolatile LiM architecture incorporating a power-down mechanism can facilitate a system like that shown in Fig. 7.4, which can achieve low-power, high-speed operation. In the following sections, various technologies related to the nonvolatile LiM

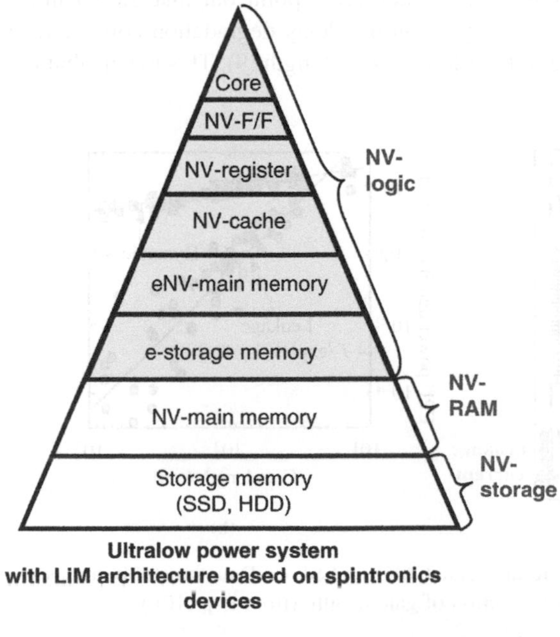

Figure 7.4 Nonvolatile computer systems achieving ultralow-power operation. In this figure, the "e-" prefix stands for "embedded."

architecture are described, together with circuits and device-level implementations based on this architecture.

The rest of this chapter is organized as follows. In Section 7.2, the benefits of nonvolatile LiM architecture combined with a power-gating technique is discussed to realize ultralow-power VLSI systems in a nanometer-scaled era. In order to realize such VLSI systems based on a nonvolatile LiM architecture, two types of hardware structures are proposed: first based on magnetic flip-flops (MFFs) and the second, MTJ/MOS-hybrid circuit structure. As examples of the former, two kinds of MFFs are presented in Section 7.3. First, a CMOS-compatible MFF with three-terminal MTJ devices is proposed and its advantage is pointed out in terms of robustness against read disturbance; the circuit block is designed and fabricated with 90 nm CMOS/MTJ technologies to demonstrate the advantages. Next, a novel MTJ/MOS-hybrid nonvolatile latch, where incubation time in MTJ switching is fully utilized, is presented. A 600 MHz stable switching operation is demonstrated by the fabricated test chip using 90 nm CMOS/MTJ technologies. Key logic-circuit components, where MTJ devices are compactly merged into CMOS logic gates, are presented in Sections 7.4–7.6. In Section 7.4, a nonvolatile full adder with MTJ devices combining MOS transistors is designed and fabricated using 180 nm CMOS/MTJ technologies. It demonstrates instant-on behavior with nonvolatile capability. This full adder makes it possible to greatly reduce the power dissipation in a motion-vector detection system.

As typical examples of special-purpose logic VLSIs, several kinds of (ternary) content-addressable memory (CAM) demonstrations are presented and their benefits are compared in terms of area density, search energy, and search speed in Section 7.5. A 16 kilobit nonvolatile CAM chip with domain-wall motion cells and a 2 kilobit nonvolatile ternary CAM chip with MTJ devices are successfully fabricated using 90 nm CMOS/MTJ technologies, and their basic behavior is confirmed by measured waveforms. Section 7.6 describes, as an example of general-purpose logic VLSIs, an MTJ/MOS hybrid nonvolatile lookup-table circuit chip that is designed and fabricated for a compact and standby power-free field-programmable gate array. Finally, the future prospects of the spintronics-based LiM architecture are summarized.

7.2 NONVOLATILE LOGIC-IN-MEMORY ARCHITECTURE

A power-down mechanism called power gating is one of the most effective methods for cutting off standby power in recent nanometer-scale VLSI chips. In power gating applied to conventional VLSI architecture with logic modules and a "volatile" memory, the power supply to logic modules can be completely cut off during the standby mode when no operation is performed. However, the power supply must be continuously provided to the volatile memory because the system loses critical data stored in the volatile memory when the power supply is turned off.

Assume two separate power supplies, V_{DDH} and V_{DDL}, where V_{DDH} is provided to the volatile memory during active mode and V_{DDH} is changed to V_{DDL} (the minimum level of power supply to retain contents of the memory) during the standby mode. In this case, there is no data access to an external nonvolatile memory for power

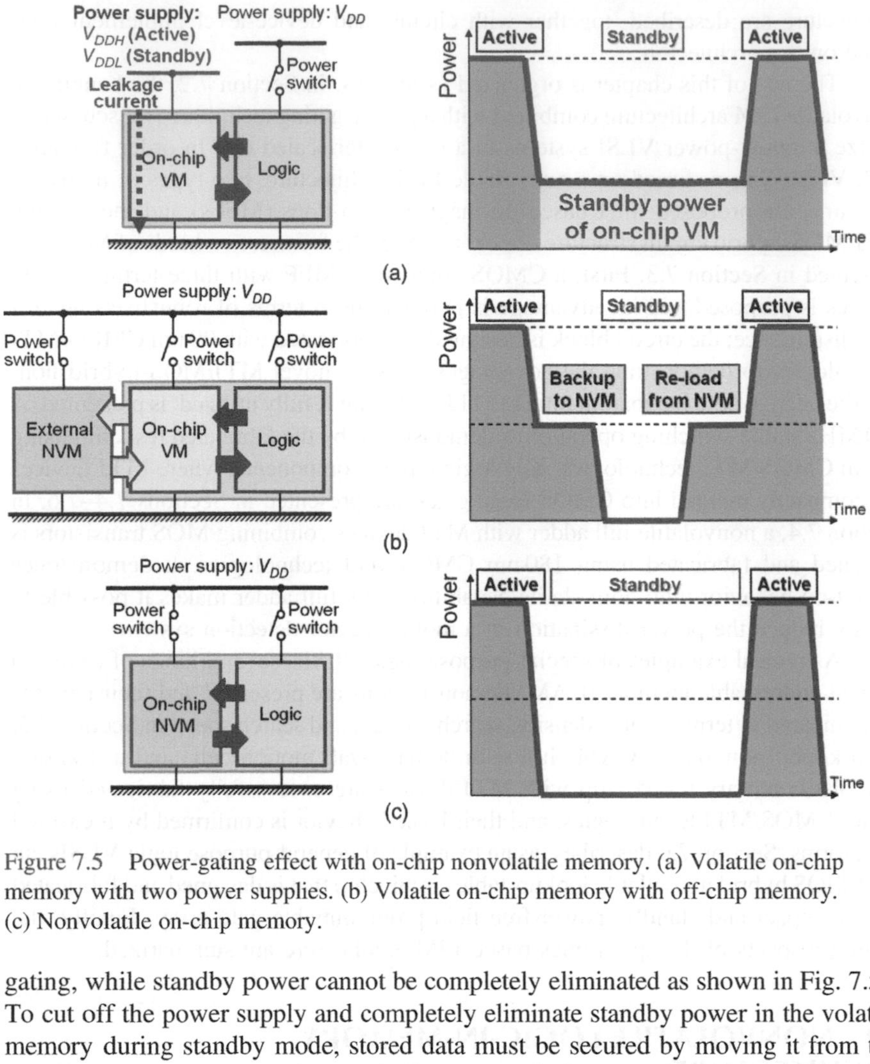

Figure 7.5 Power-gating effect with on-chip nonvolatile memory. (a) Volatile on-chip memory with two power supplies. (b) Volatile on-chip memory with off-chip memory. (c) Nonvolatile on-chip memory.

gating, while standby power cannot be completely eliminated as shown in Fig. 7.5a. To cut off the power supply and completely eliminate standby power in the volatile memory during standby mode, stored data must be secured by moving it from the volatile memory to external nonvolatile memory before entering the standby mode. When the standby mode changes to the active mode again, the stored data are reloaded from the external nonvolatile memory to the volatile memory, as shown in Fig. 7.5b. In order to completely eliminate standby power without any data access to external nonvolatile memory in power-gating technique, the unique and the best method is to replace "volatile" on-chip memory with "nonvolatile" memory, as shown in Fig. 7.5c.

The effectiveness of this approach depends strongly on hardware granularity (i.e., how finely the memory blocks are dispersed among the logic circuits). When a logic module is large, it is difficult to find a time interval in which none of the module's logic circuits is active, which means very little chance to use the power-gating technique to cut off the standby power. In addition, a large-scale hardware module generally consumes a large amount of power with a large stray capacitance on the power supply line, making power supply control difficult because this large flow of power must be turned off and on when the module enters and leaves the standby mode.

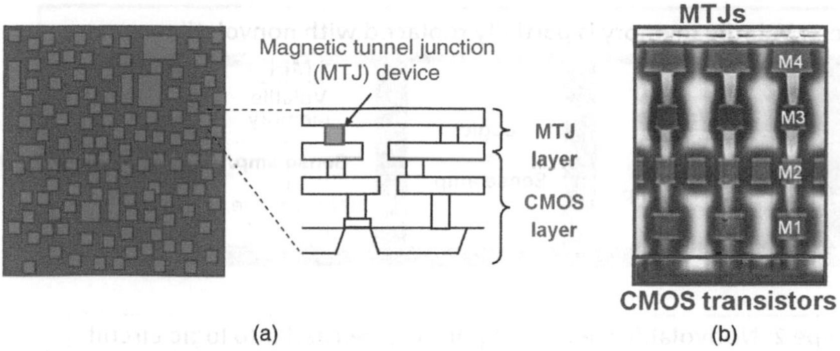

Figure 7.6 Conceptual illustration of a nonvolatile logic-in-memory (LiM) architecture: (a) concept and (b) MTJ on CMOS.

This difficulty associated with coarse-grained logic modules can be overcome by applying nonvolatile LiM. Figure 7.6a shows a conceptual block diagram of a nonvolatile LiM architecture, where nonvolatile storage elements are distributed over the logic-circuit plane. LiM was originally proposed in 1969 (11), but its implementation through CMOS alone provided no advantage because combining CMOS memory and logic circuits in the logic module led to a large device count and hence an area penalty. The recent development of "MTJ on CMOS" has made it possible to fully leverage the advantages of LiM architecture, along with nonvolatility, as shown in Fig. 7.6b. With nonvolatile storage elements, stored data in on-chip memory requires no backup. To successfully implement such a nonvolatile LiM architecture, the storage elements must have the following properties: nonvolatility, large resistance ratio, fast read/write accessibility, scalability, compatibility with a CMOS process, virtually unlimited endurance, and low resistance. At present, spintronics devices such as MTJ devices and domain-wall motion elements are the only available candidates that meet all of the above requirements.

In the following sections, we describe two kinds of nonvolatile LiM architectures: one using magnetic flip-flop circuits (called "type 1"), and the other using MTJ devices in combination with CMOS circuits (called "type 2"). As shown in Fig. 7.7, in the type 1 architecture, nonvolatile storage elements are inserted into conventional volatile flip-flops. The nonvolatile storage function is efficiently implemented because the sense amplifier inside the flip-flop is shared through this approach. Similarly, nonvolatile storage functions are merged into logic circuits in the type 2 architecture, which achieves compactness as well as high performance and low power consumption in nanometer-scale VLSI.

7.2.1 Nonvolatile Logic-in-Memory Architecture Using Magnetic Flip-Flops

As mentioned in the previous section, the power-gating technique is very effective for reducing the standby power when a logic module is in standby mode. If all logic gates connected to a power-isolation switch are nonvolatile, the LSI function block performance is also maintained because power gating causes no data loss. In

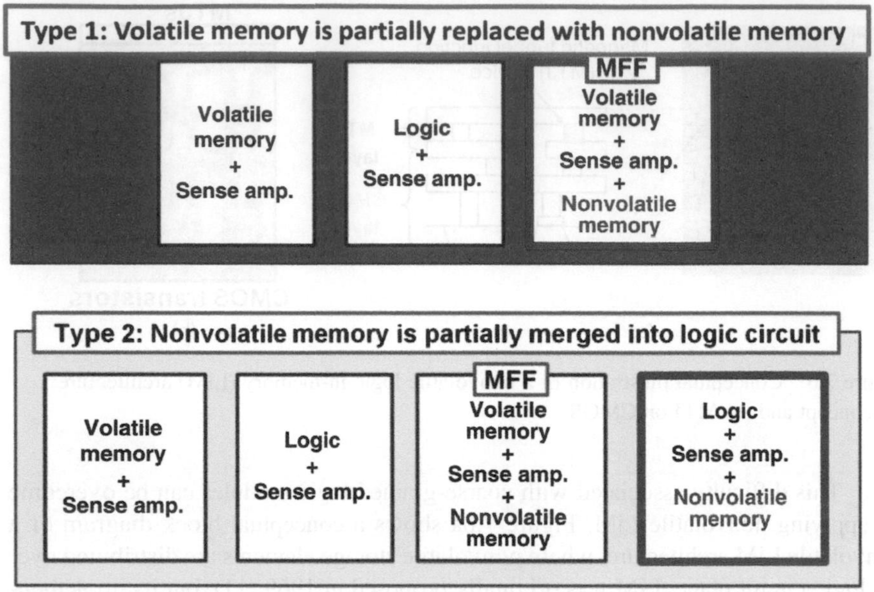

Figure 7.7 Classification of nonvolatile LiM architectures.

particular, when the function block circuitry is clock-synchronized, it is necessary only for all the flip-flops (F/Fs) to be nonvolatile to make the circuit nonvolatile.

Figure 7.8 shows a clock-synchronized function block. It consists of three parts: clock buffers, combinational-logic circuits, and F/Fs. The combinational-logic circuits operate with static logic gates such as NAND gates. The flip-flops retain intermediate data generated by the function block's operations. Many kinds of F/Fs for data retention have long been investigated for the benefits of standby power reduction. Table 7.1 compares candidates for the storage element of logic states. Retention F/Fs (12), for example, are advantageous in that they can be fabricated without using any optional memory process. Since they are volatile, however, they also require an additional power line for data retention and transistors with multiple threshold voltages. Although F/Fs fabricated by FeRAM (ferroelectric RAM) processes are nonvolatile (13), their market has not grown because of their high operating voltage compared to the state-of-the-art CMOS power supply voltage, and because of write endurance limitations. These devices also complicate the layout design flow because of their low compatibility with logic primitive libraries.

On the other hand, spintronics devices such as MTJs have great potential to facilitate high-speed, low-voltage, nonvolatile F/Fs with virtually unlimited endurance (14). The primitive cells do not include any additional power lines or special transistors, like normal F/Fs. This makes it possible to design zero standby power function blocks with conventional automatic layout design flow. The logic-in-memory architecture, however, has even greater potential. It can also reduce both the active power and the circuit area. These advantages are discussed further in the next section.

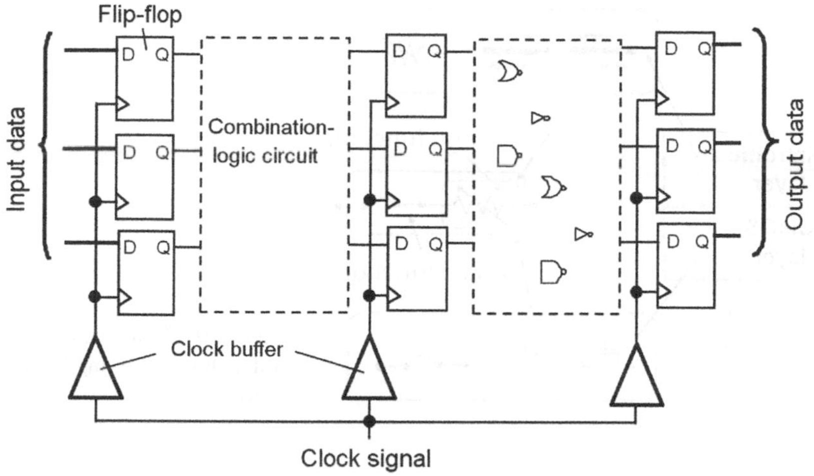

Figure 7.8 Clock-synchronized logic circuit.

7.2.2 Nonvolatile Logic-in-Memory Architecture Using MTJ Devices in Combination with CMOS Circuits

If MTJ devices are directly used for both storage and logic functions in basic logic gates, operation with zero standby power can be achieved with a smaller device count. Figure 7.9 shows a design concept for a nonvolatile LiM architecture using MTJ devices in combination with MOS transistors (15–17). Since MTJ devices are used as not only a 1-bit storage element but also as a "variable" resistor whose resistance value is programmed to a high- or low-resistance state, a nonvolatile logic circuit can be implemented by using MTJ devices in combination with conventional MOS transistors. In the MTJ/MOS hybrid circuit style, storage functions are merged into a logic function, so that the device count of the resulting logic circuit can be greatly reduced through appropriate design choices.

Figure 7.10 shows a basic design example for a logic circuit with a 1-bit storage element. This circuit has two logic functions: 2-input NOR and 2-input NAND. The

TABLE 7.1 Comparison of Flip-Flops (F/F) for Data Retention.

	Retention F/F	FeRAM F/F	MTJ F/F
Volatility	Volatile	Nonvolatile	Nonvolatile
Speed	Very high	Middle	High
Additional power line	Necessary	Necessary	Unnecessary
Endurance	Infinite	$<10^{14}$	Infinite
Transistor	Multithreshold	MOX	Core only
Process compatibility	High	Low	High
Standby current	Almost zero	Zero	Zero

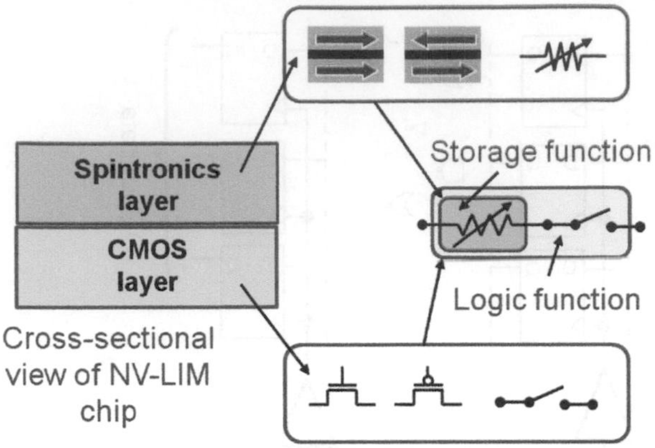

Figure 7.9 Nonvolatile LiM architecture using MTJ devices.

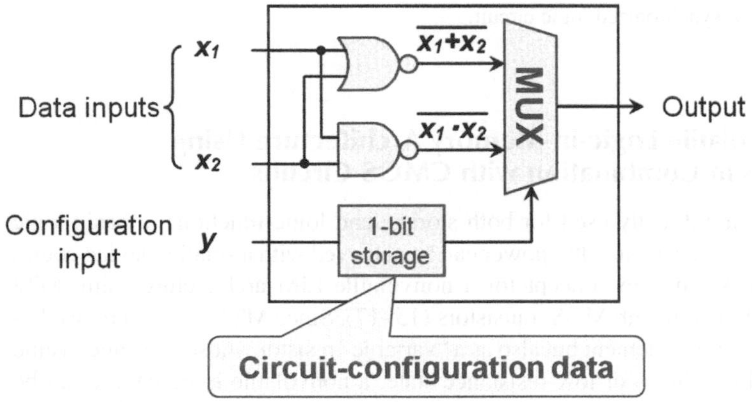

Figure 7.10 Design example of a simple logic circuit with a 1-bit storage element.

storage element is used to select one of the two logic operations via a 2-input multiplexer (MUX), where the input y is a "circuit-configuration input." Figure 7.11 shows a circuit diagram implementing this logic circuit example by using conventional CMOS gates. The NOR and NAND gates each requires 4 transistors, whereas the MUX and 1-bit SRAM cell each requires 6 transistors, giving a total transistor count of 20. This is because the logic and storage parts are separated in the CMOS implementation.

On the other hand, Fig. 7.12 shows a circuit diagram applying the proposed nonvolatile LiM circuitry. This circuit consists of three parts: the logic circuit, a current comparator, and an output driver. Since this circuit uses differential-pair logic, the result of a logic operation is generated by comparing two current levels, **I** and **I**′. The MTJ devices are utilized as both storage and logic elements, which reduces the number of MOS transistors. As a result, a compact LiM circuit can be implemented by merging the logic and storage elements.

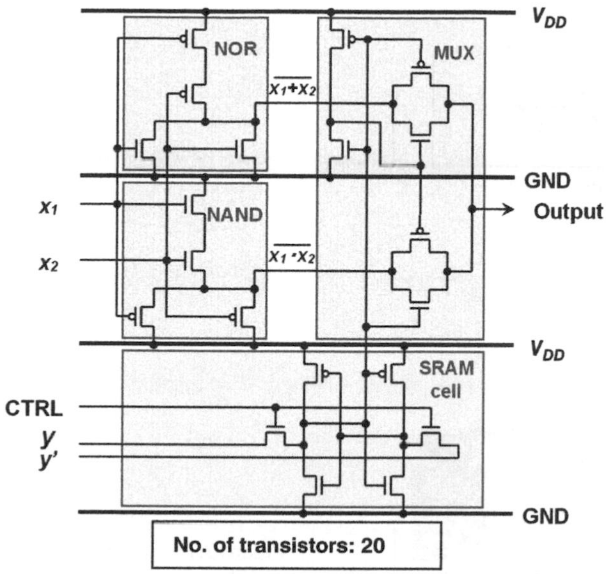

Figure 7.11 CMOS implementation of the circuit design shown in Fig. 7.10.

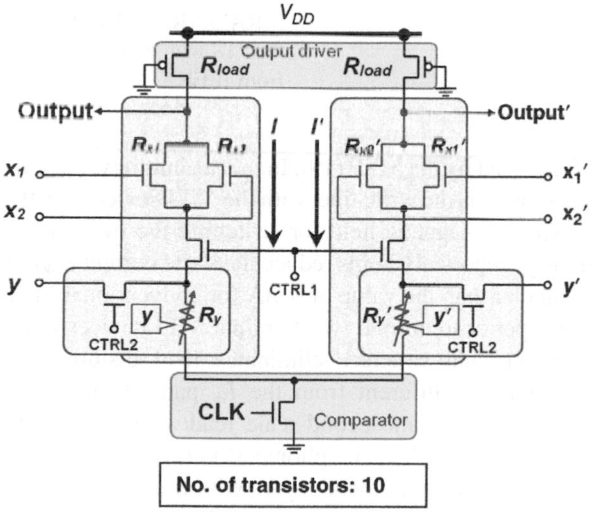

Figure 7.12 MTJ–MOS hybrid implementation of the circuit design shown in Fig. 7.10.

7.3 CIRCUIT SCHEME FOR LOGIC-IN-MEMORY ARCHITECTURE BASED ON MAGNETIC FLIP-FLOP CIRCUITS

7.3.1 Magnetic Flip-Flop Circuit

An MFF circuit constituting a primitive cell in a system-on-a-chip (SoC) design library was first demonstrated by Sakimura et al. (18). Figure 7.13 shows a top view

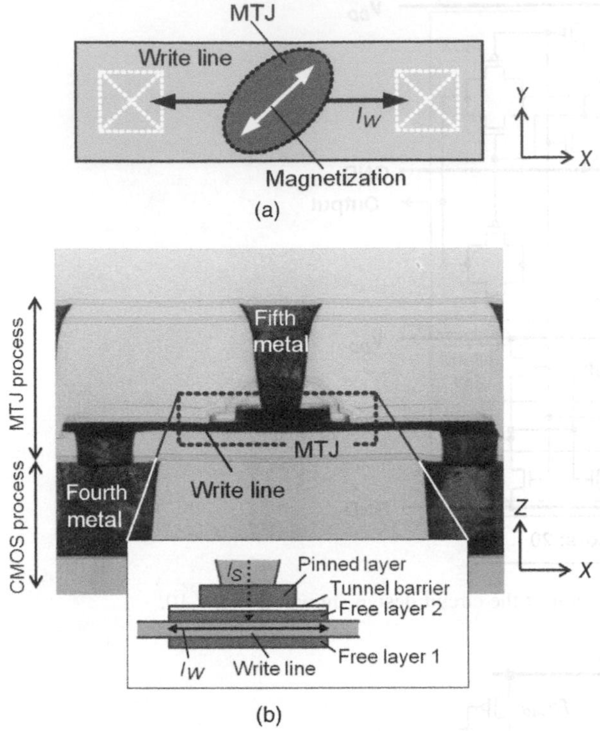

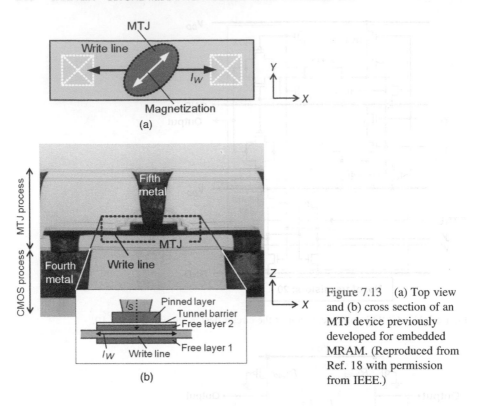

Figure 7.13 (a) Top view and (b) cross section of an MTJ device previously developed for embedded MRAM. (Reproduced from Ref. 18 with permission from IEEE.)

and cross section of the MTJ device used by the MFF (19). In this technology, the thin base electrode of the MTJ device is used as the write line, with the MTJ device directly placed on it to effectively enhance the magnetic field for switching the free layer's magnetization. This MTJ technology reportedly helps reduce the write current (I_W) to less than 1 mA, which is much smaller than the value of 5 mA for conventional MTJ devices. This technology is also superior to normal two-terminal MTJ devices in its high-speed operation. The three-terminal structure eliminates read disturbances because the sensing current (I_S) path is different from the I_W path. This disturbance-free circuit structure helps facilitate nanosecond-scale read/write operations and also makes it easy to design nonvolatile F/Fs. An optional MTJ process is applied only in the insulator between the fourth and fifth metal layers with four extra masks, which does not overly complicate the manufacturing process. The MTJ resistance (R_{LOW}) has been optimized to 5 kΩ through a previously developed MRAM macro design, while a TMR ratio of about 150% was obtained with an MgO barrier.

Figure 7.14 shows a circuit diagram of an MFF in which an MTJ-based latch is adopted as a slave latch. The latch consists of a six-transistor SRAM cell-based latch circuit (M1–M3 and M1′–M3′), connected to a pair of MTJ devices (J and J_b), two face-to-face NOR gates for the store operation, and M4 for the recall operation. A data bit and its complement are stored in a pair of MTJ devices (J and J_b) to expand the sensing signal. The write line under each MTJ device is connected to the NOR gates, which supply a bidirectional I_W according to the data bit. The MFF's operation is

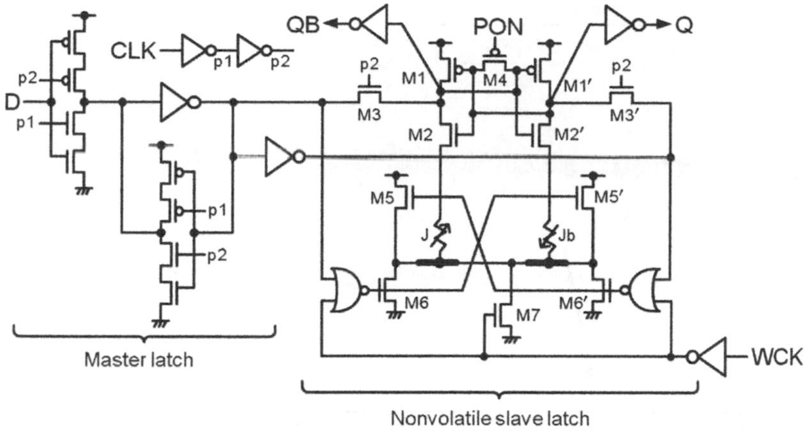

Figure 7.14 MFF circuit employing an MTJ-based slave latch, where the MTJ device is depicted as a variable resistor. (After Ref. 18.)

identical to that of a conventional F/F except for the store and recall operations. When power is supplied, the recall operation starts immediately on the condition that both the clock signal (CLK) and the power-on signal (PON) are low. Then, the 1-bit information saved in the MTJ devices is loaded to the MTJ-based slave latch according to the resistivity difference between J and J_b, and the MFF's output is determined. The store operation is executed by activating the write clock signal, WCK, at high level. Then, the NOR gates direct I_W into the write line of each MTJ device and back up the data stored in the latch circuit to the MTJ pair. The I_W paths are laid out so that write magnetic fields can be supplied to each MTJ device in opposite directions.

The MFF makes it possible to carry out both latch and store operations simultaneously during each cycle because the store operation does not disturb the stored voltage of the slave latch. The designer can freely determine whether to back up the MFF output every cycle. The store operation can be executed every clock cycle because of its virtually unlimited endurance.

Figure 7.15 shows a die photo of an MFF test chip, which was designed with 150 nm, 1.5 V CMOS, and MTJ process technologies. The design library includes the MFF primitive cell. The chip consists of four domains (D0 through D3) for evaluating various types of MFFs. A 16-stage, 8-bit shift register using MFFs is implemented in each domain.

7.3.2 M-Latch

For application to recent CMOS VLSI technology operating at a frequency of several hundred megahertz, MTJ/CMOS hybrid circuits must operate at an equivalent frequency. Writing data to an MTJ device at such a high frequency causes certain problems, however, such as requiring a large current and increasing the write error rate.

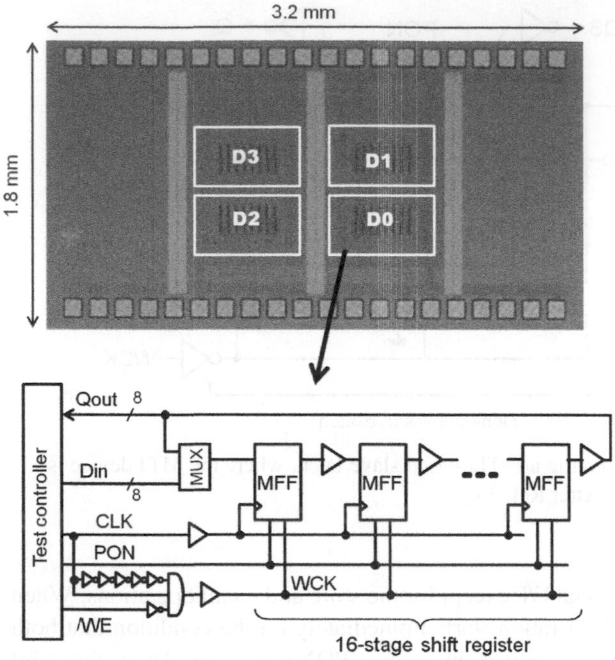

Figure 7.15 Die photo of an MFF test chip divided into four domains. A 16-stage, 8-bit shift register using various types of MFFs is fabricated in each domain. (After Ref. 18.)

Endoh et al. experimentally investigated the intrinsic characteristics of the write operation to an MTJ device, that is, the incubation time (t_A) and transit time (t_B) (20). The first, t_A, is the period required for applying the bias necessary to flip magnetization, while the second, t_B, is the time for actually flipping the magnetization. They discovered that t_A depends on the write voltage (V_p), whereas t_B is independent of V_p for about 10 ns for perpendicular MTJ devices with a 100 nm diameter.

Making use of these experimental results, a novel MTJ/MOS hybrid circuit has been proposed (20). Figure 7.16 illustrates the concept of the proposed circuit. Adjustment of V_p is used to control t_A to be more than $f/2$ and $t_A + t_B$ to be less than t_{PG}, where f is the operation frequency and t_{PG} is the interval time for power-gating technique. This timing setting by V_p ensures that only the CMOS circuit operates, while the MTJ device is not written to during high-frequency operation. In this case there is only a small write current to the MTJ device. On the other hand, a sufficiently long write pulse flips the MTJ's magnetization. This operation requires only a small amount of write current and achieves a nearly zero write error rate because of the sufficiently long write period. After the data in the CMOS circuit are written to the MTJ device, the power can be shut down. Thus, power gating can be executed with this circuit. Because the operation requires only setting the V_p level without any special control circuitry, it has the benefit of facilitating a simple circuit.

Figure 7.17a shows an M-latch, a latch circuit using this concept. It consists of a traditional CMOS latch circuit with two directly connected MTJ devices. By fabricating a test chip using 90 nm CMOS and 100 nm MTJ process technologies,

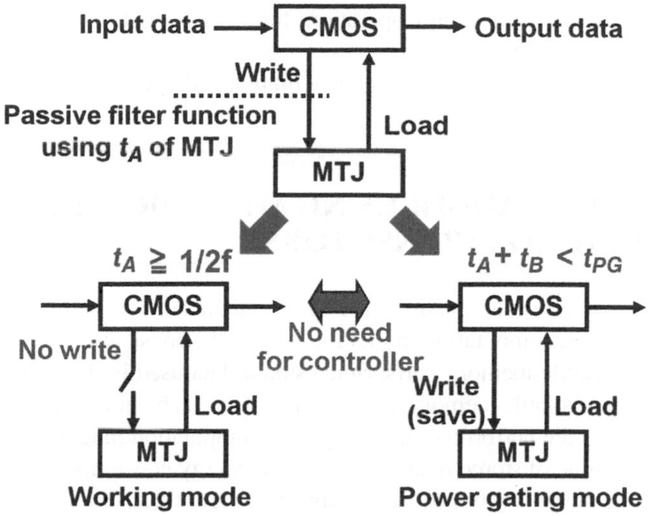

Figure 7.16 Proposed concept for an M-latch circuit. (From Ref. 20.)

600 MHz CMOS circuit operation and a 30 MHz MTJ write operation have been experimentally demonstrated. The 600 MHz value was the highest frequency reported for an MTJ/MOS hybrid circuit until 2011. Moreover, the write error rate of the MTJ device is practically negligible at the write frequency of 30 MHz. Figure 7.17b exhibits measurement waveforms demonstrating successful power gating, in which

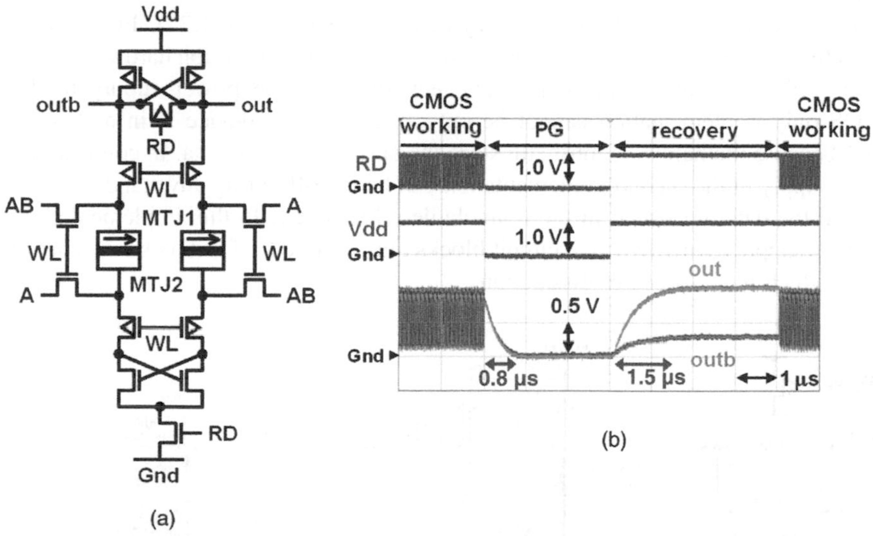

Figure 7.17 The proposed M-latch circuit and measured waveforms demonstrating its successful operation: (a) Circuit diagram and (b) measured operating waveforms. (After Ref. 20.)

the power was shut down after CMOS circuit operation, followed by data recovery and continued CMOS circuit operation.

Hence, this MTJ/CMOS hybrid circuit can achieve high-speed operation with a practically negligible error rate.

7.4 NONVOLATILE FULL ADDER USING MTJ DEVICES IN COMBINATION WITH MOS TRANSISTORS

In this section we discuss a concrete example of logic LSIs based on a nonvolatile LiM architecture. Figure 7.18 shows a computation model for logic LSIs, consisting of two components: logic and nonvolatile memory parts. Since stored data used in the logic circuit are backed up in the nonvolatile memory, the power supply can be immediately cut off whenever no operations are performed. In this type of computation model, it is important to consider the lifetime of stored data. Although the energy consumption of MTJ devices in the data-write phase is currently rather large, there are still advantages to using the nonvolatile LiM architecture with current MTJ devices in applications with long-lifetime storage capability. The reason is that in nonvolatile LiM circuitry, the logic and nonvolatile storage functions are compactly merged at the device level; hence, fine-grained power gating can be simply performed without backup/reload of data to/from an external nonvolatile memory. This greatly reduces both the write energy of MTJ devices and the standby power consumption of the circuit blocks.

One application of nonvolatile LiM architecture is the motion-vector detection system shown in Fig. 7.19, in which data from the current window are used as reference-window data and stored in storage elements in nonvolatile adders. The "sum of absolute differences" (SAD) operation, which is a basic operation for detecting the motion vector of the current window, is performed in parallel by many adders to compute the similarity between two adjacent image frames (21,22). Figure 7.20 shows time charts of the power dissipation in motion-vector detection hardware using CMOS-only and nonvolatile adder (23) technologies. The power supply to the CMOS-only implementation cannot be cut off even if the device is in a standby phase because the hardware must still store the current window data. In contrast, the power supply in the second implementation can be cut off during any standby phase because the data are stored in the nonvolatile adder array. In this implementation, power is supplied only to active circuit blocks and is turned off in blocks in standby mode, which reduces power dissipation.

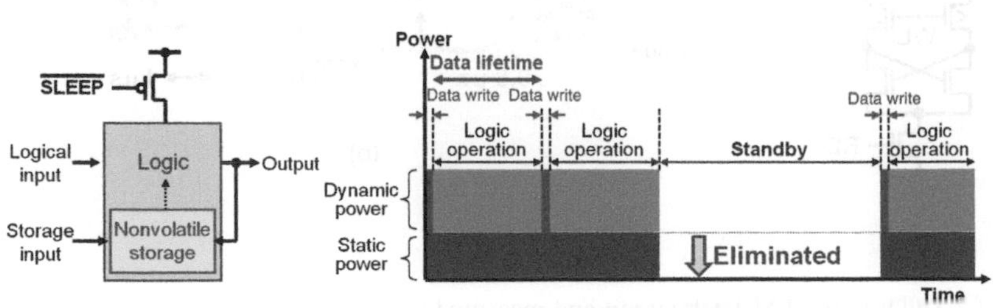

Figure 7.18 Nonvolatile LiM circuit model and a sample time chart of its operation.

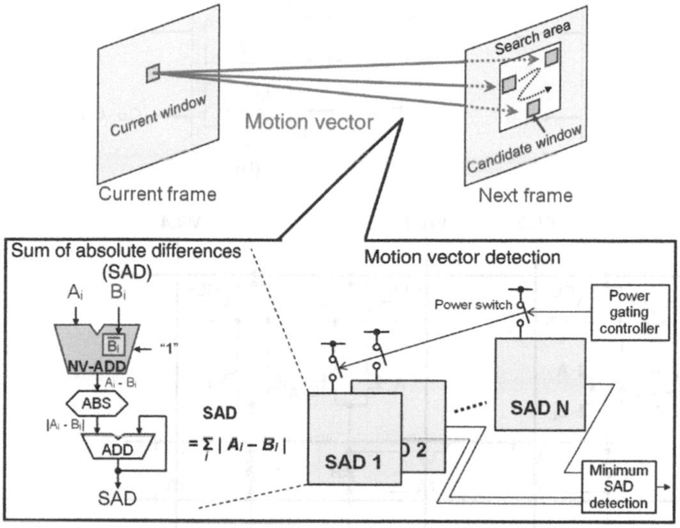

Figure 7.19 Motion-vector detection system using a nonvolatile adder.

Figure 7.21 shows full adders with 1-bit storage (21–25). The nonvolatile full adder is a basic component of a nonvolatile adder to calculate one-digit addition of an input, a stored input, and a carry input. A nonvolatile storage element and full adder can be compactly merged by applying the MTJ–MOS hybrid LiM circuitry as shown in Fig. 7.21a and b. Since MTJ devices are used as both nonvolatile storage elements and logic elements, as shown in Fig. 7.21c, the transistor count is reduced in comparison with a conventional CMOS-based implementation.

Figure 7.22a shows a die photo of a nonvolatile full-adder test chip fabricated with 0.18 mm CMOS and MTJ process technologies (23,24). The MTJ device is fabricated on top of the CMOS layer at the back end. Figure 7.22b shows measured input–output waveforms for the test chip. Because the circuit stores preprogrammed

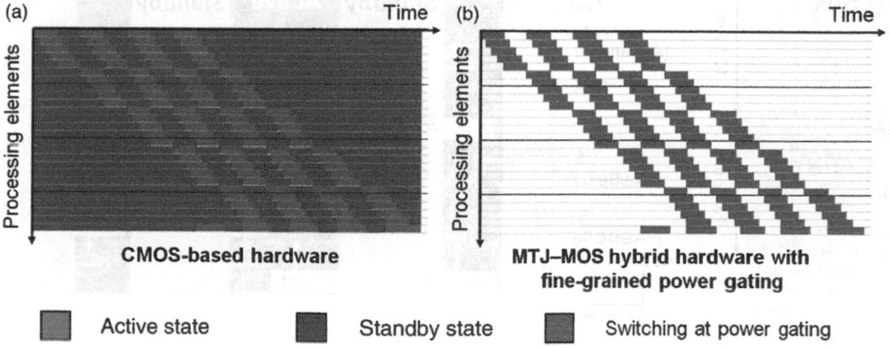

Figure 7.20 Activation states of SAD blocks in motion-vector detection hardware (Reproduced from Ref. 3 with permission from IEEE.)

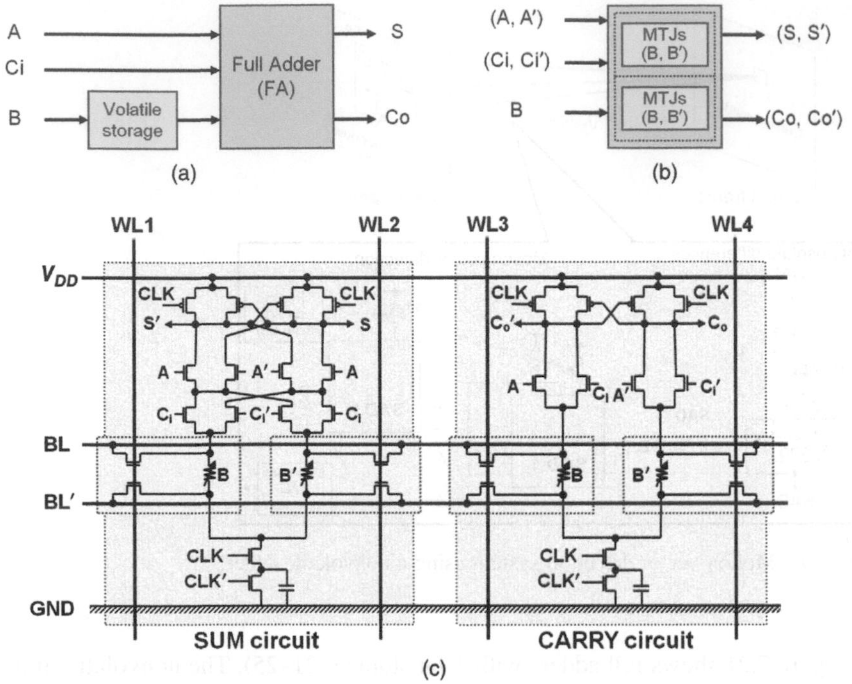

Figure 7.21 Full adder with 1-bit storage. (a) Block diagram of a conventional CMOS-based full adder with a 1-bit volatile storage element. (b) Block diagram of a proposed MOS/MTJ hybrid nonvolatile full adder, in which nonvolatile storage functions are merged. (c) Schematic of the proposed nonvolatile full adder. (From Ref. 23.)

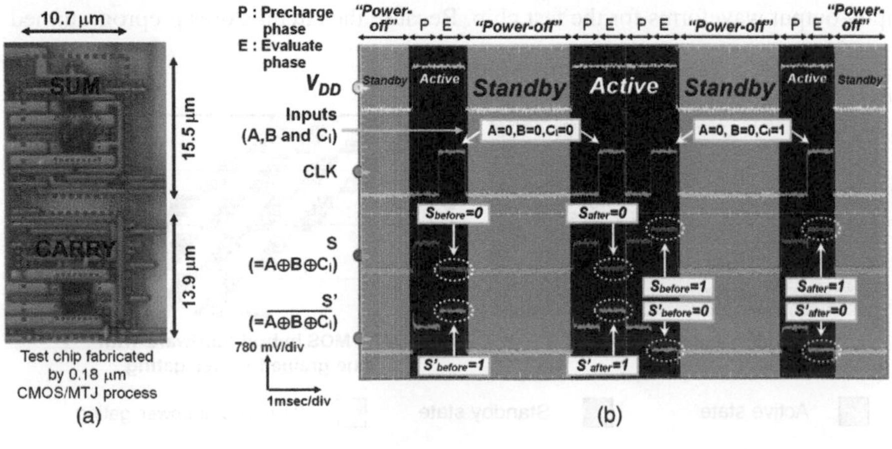

Figure 7.22 The fabricated nonvolatile full adder. (a) Die photo. (b) Measured waveforms demonstrating instant-on/off behavior. (From Ref. 23.)

configuration data in the nonvolatile MTJ devices, the power supply can be shut down and reapplied without storing and recalling data to and from external memory, thus facilitating continuous full-adder operation independent of the power supply. The output signal S_{after} is the same as the output signal S_{before}, indicating that the stored data remain even if the power supply V_{DD} is shut off and turned on again. As a result, the desired instant-on and instant-off behaviors have been confirmed for the fabricated nonvolatile full adder chip.

As a future prospect, if the write energy of an MTJ device is sufficiently small, the nonvolatile LiM architecture becomes applicable to a wider range of logic LSIs with shorter lifetime storage capability.

7.5 CONTENT-ADDRESSABLE MEMORY

7.5.1 Nonvolatile Content-Addressable Memory

In this section, we introduce a spintronics-based content-addressable memory (CAM), or spin-CAM, as another nonvolatile LiM architecture application. We chose this application because its demonstration almost proves the feasibility of a nonvolatile system-on-a-chip (SoC), because CAMs consist of both logic circuits and memory circuits. CAMs have been widely used in many recent SoCs requiring high search speeds, such as translation look-aside buffers (TLBs). CAMs can search for fixed-length data sets, called words. For example, a CAM can search for an input word among previously stored words. Hardware searches with a CAM enables high-speed and/or low-power searches as compared to software searches. In TLBs, CAMs transform virtual addresses into physical addresses. The CAM search speed directly influences memory latency, a significant factor in attaining high performance. In such applications, spintronics-based CAMs must be at least as fast as CMOS-based CAMs.

A spintronics-based CAM was demonstrated by Nebashi et al. (26). Since the data stored in the CAM are still retained even in the power-off state, the power for both retention and reloading was significantly reduced, while the access speed was at the same level as that of a CMOS-based CAM. To achieve these goals, a three-terminal spintronics device, the domain wall (DW) motion element (2) was used. This is the first example of using DW elements for a large-scale LSI function block.

Figure 7.23 shows a fabricated test chip for a 16-kbit spin-CAM, along with a cross-sectional micrograph of the DW element used in the CAM. A 5 ns search operation was successfully carried out, even under the worst-case condition of a 1-bit miss. Power of 9.4 mW was measured at a 200 MHz clock frequency and 1 V. This operation speed and power were nearly the same as those obtained for a CMOS-based CAM when the same 90 nm CMOS design rule was used. If the nonvolatile CAM was used with a CPU, the effective power could be reduced to 1.1 mW by power-gating technique with an assumed CPU operation ratio of 10%. The CAM area measured 330 μm × 460 μm with 6.6 μm^2 per bit cell. The array organization was 128 bits × 128 words.

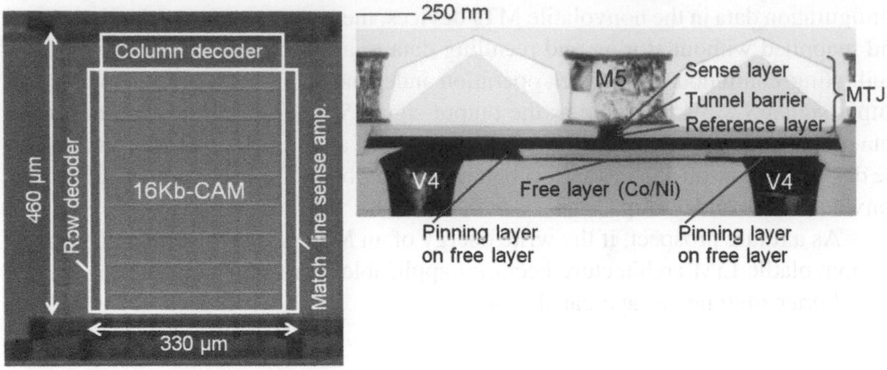

Figure 7.23 A 16-kbit CAM chip die photo and cross-sectional micrograph of a domain wall (DW) element. (Reproduced from Ref. 26 with permission from IEEE.)

Figure 7.24 illustrates the DW element used in the CAM. The free layer consists of a perpendicular magnetic thin film of Co/Ni. The data region is between the two fixed regions. The pinning layer is made of Co/Pt, which has greater magnetic coercive force than Co/Ni. The magnetizations of the two fixed regions are aligned antiparallel by the pinning layers. Each end of the data region becomes a pinning site for the DW. Thus, two stable states exist in this system. When the DW is on the right side, the element is in the "0" state. When the DW is on the left side, the element is in the "1" state. For writing, a bidirectional current is used. The DW is moved by the spin-torque of the polarized current. When the width of the free layer is 90 nm, the write current is 200 μA.

Figure 7.25 shows how the DW data are read. An MTJ device is added above the data region to read the data. This device consists of a sense layer, tunnel barrier, and reference layer. The cross section on the right side of the figure shows the data

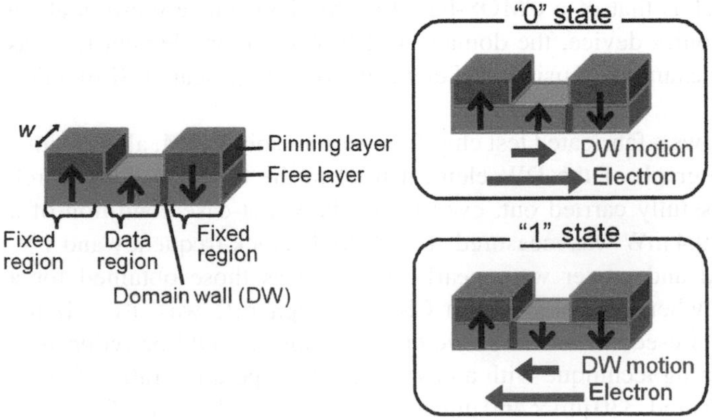

Figure 7.24 Sketch of the retention state of a DW element.

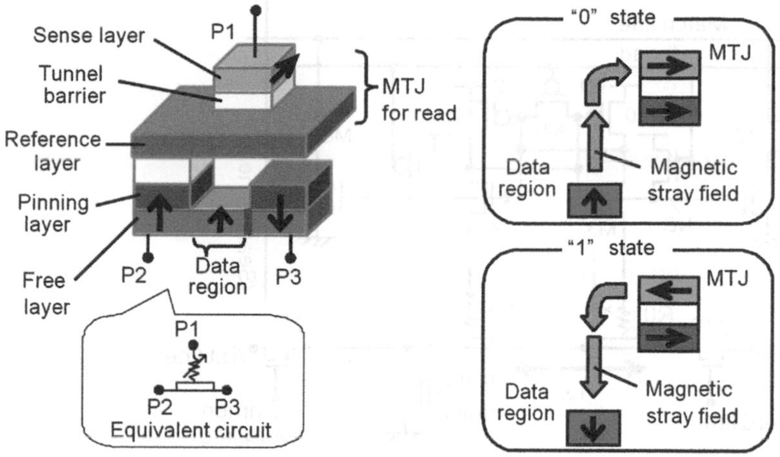

Figure 7.25 Sketch of a DW element reading data.

region as seen from the fixed layer. The magnetization of the sense layer is switched by the magnetic stray field generated from the data region. The level of MTJ resistance is low when the magnetizations of the sense layer and reference layer are parallel. In other words, MTJ resistance is high when the magnetizations of the two layers are antiparallel. In this structure, a high TMR ratio is obtained by using an in-plane magnetized sense layer. The sense layer for reading and the free layer for writing are optimized independently.

As mentioned earlier, the DW element has three terminals. The write current path and sense current path are separated. Therefore, the probability of read disturbance is sufficiently low. A memory array can be controlled by a simple circuit. Such simple control both reduces the peripheral circuit area of the CAM and enhances the operation speed.

Figure 7.26 shows a circuit diagram of the CAM cell designed to accelerate search operations. It comprises a memory cell and a comparison circuit. To reduce the cell area, the write transistor (M13) and pull-down transistor (M14) are shared with the adjacent memory cell. The CAM cell has three operation modes: load, program, and search. In the load operation for memory cell 0, the activated transistors M9-11 balance the cross-coupled inverters (M1-4). After the nodes ND and NDb are fully precharged, M14 is turned on, and M9-11 are turned off. Through the use of sense currents IS and ISb, the inverters output the stored data in the MTJ devices (R0 and R0b) according to the difference in their resistances. In the program operation, the activated M13 and M12 provide the write current I_W for R0 and R0b to write the complementary data. The write current paths are laid out so that a spin-torque can be applied to each DW element in opposite directions. In the search operation, M5-8 compare the search data on the search lines with the stored data in the CAM cell. In a mismatch case, the low-level voltage is output to match lines (MLs), whereas in other cases, the high-level voltage is output to the MLs.

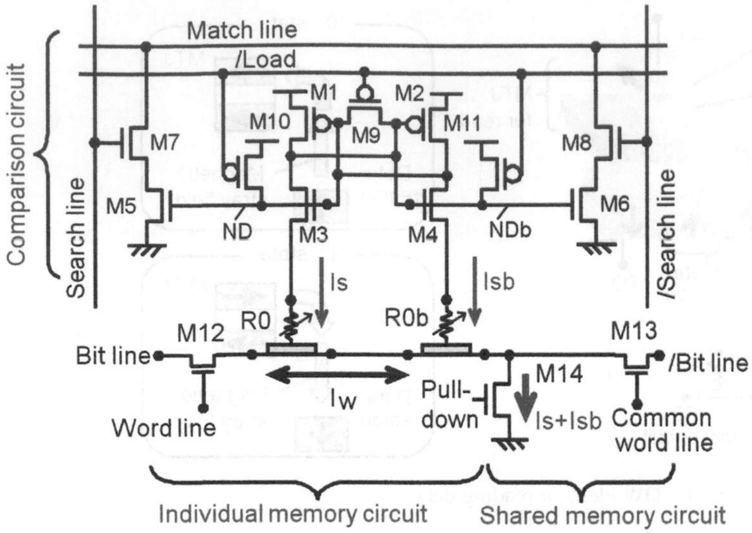

Figure 7.26 Circuit diagram of a CAM cell. (Reproduced from Ref. 26 with permission from IEEE.)

7.5.2 Nonvolatile Ternary CAM Using MTJ Devices in Combination with MOS Transistors

This section demonstrates nonvolatile ternary content-addressable memory (TCAMs) as another possible application of the nonvolatile (LiM) architecture. A TCAM is a rich functional memory that carries out fully parallel equality search between an input word, among stored words of variable length, by using masking bits stored in its cell array. Because of its rich functionality, a TCAM can be applied in various search applications, such as databases, virus checkers, and network routers. A TCAM consisting of a volatile CMOS-based 16-transistor (16T) or 12T cell circuit, however, has two major disadvantages. The first is increasing standby power due to leakage current in the nanometer-scale CMOS era. The second is high bit-cell cost, because a TCAM cell requires 2-bit storage elements and a comparison logic element. To extend the applications of TCAMs, they must be implemented with more compactness, lower power consumption, and higher search speed.

Figure 7.27 shows a design concept for a compact TCAM cell with zero standby power. By using an MTJ-based nonvolatile LiM architecture, 2-bit storage elements (SRAM cells) in a CMOS-based implementation can be replaced with 2 MTJ devices over 2 MOS devices, which drastically reduces the cell area required for nonvolatile storage capability (27,28). While this 2T-2MTJ circuit potentially has the cell function of a TCAM, it can barely perform bit-serial equality-search operations at the current TMR ratios of MTJ devices (about 150%). Therefore, additional circuit components are required to facilitate bit-parallel equality-search operation.

Figure 7.28 shows a proposed 6T-2MTJ nonvolatile TCAM cell circuit and a mechanism for a nonvolatile TCAM word circuit using a match-line voltage keeper in

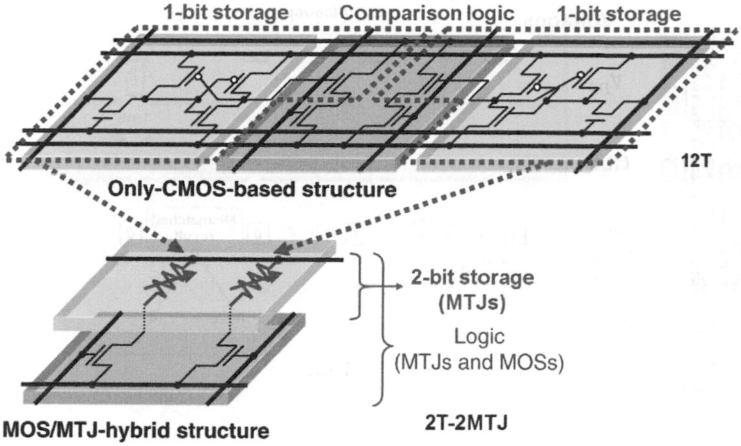

Figure 7.27 Design concept for a nonvolatile TCAM cell. (After Ref. 27.)

each cell (29). The match-line voltage keeper array operates as a minimum voltage conductor; that is, the output voltage V_{ML} is almost the same as the minimum voltage of each V_{CELL}. In the case of a full-bit match, V_{ML} becomes the high level, because every V_{CELL} becomes high. In case of a mismatch, V_{ML} becomes the low level,

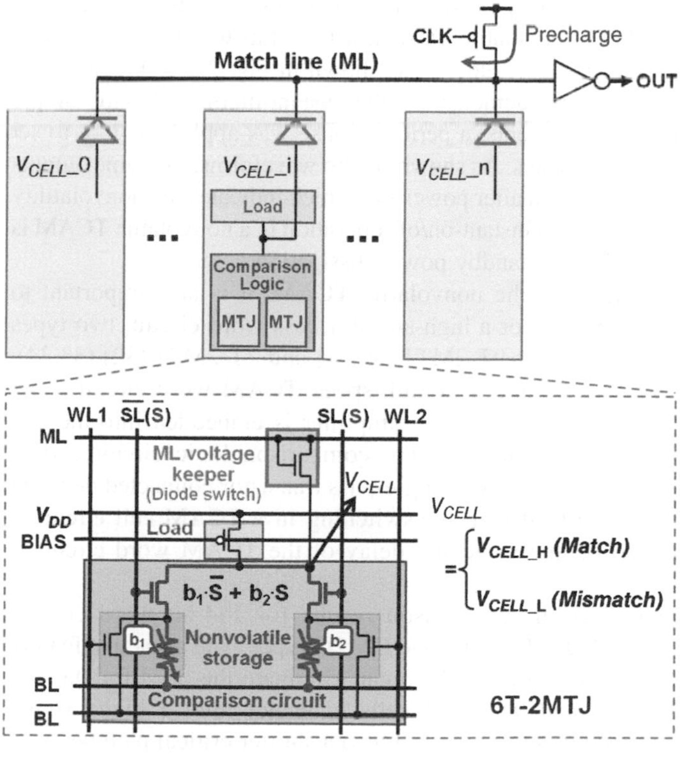

Figure 7.28 6T-2MTJ nonvolatile TCAM word circuit.

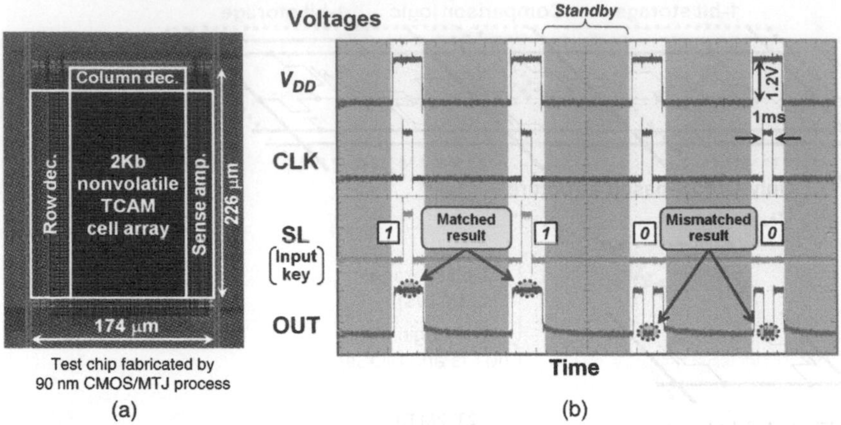

Figure 7.29 Fabricated test chip for a 6T-2MTJ nonvolatile TCAM. (a) Die photo. (b) Measured waveforms for the equality-search operation with power gating. (Reproduced from Ref. 29 with permission from IEEE.)

because one or more V_{CELL} values for the mismatched cells become low. Since the keeper clamps the match-line voltage to a certain level in accordance with the matched result, the result is easily detected by sensing the two kinds of voltage levels.

To verify the 6T-2MTJ nonvolatile TCAM, a test chip was fabricated using 90 nm CMOS and MTJ process technologies, as shown in Fig. 7.29a. Figure 7.29b shows measured waveforms that demonstrate the instant-on/off behavior of the fabricated chip. In these measurements, a periodic pulse was applied to V_{DD} to set up the power-on and power-off states. As shown by the waveforms, the same outputs were obtained before power-off and after power-on, which indicates the nonvolatility of the fabricated TCAM chip. Such instant-on/off operation in a nonvolatile TCAM is an important feature for reducing standby power dissipation.

To leverage the benefits of the nonvolatile TCAM, it is also important to achieve high-speed accessibility. For a high-speed-access word circuit, two types of cell circuits have been proposed: 9T-2MTJ (30,31) and 7T-2MTJ (32) (33,34). Figure 7.30 shows a design example of a high-speed TCAM word circuit using 7T-2MTJ cell circuits. A single-ended sense amplifier is embedded into the cell circuit to read out a matched result from the comparison logic circuit and to conduct the full swing output voltage to a single-pass transistor connected between ML and ground. Since the critical path for switching in a TCAM cell circuit is only a single MOS transistor, the switching delay of the TCAM word circuit is minimized.

Figure 7.31 summarizes the comparison results for 144-bit word circuits among the 6T-2MTJ, 7T-2MTJ, 9T-2MTJ, and CMOS types. The attractive feature of all the hybrid types is nonvolatility, which can eliminate the need for standby power. Since the 7T-2MTJ and 9T-2MTJ cells have not only a high driving capability via the full swing output voltage but also a shorter critical path between ML and ground, the search delay of the word circuits based on these types is

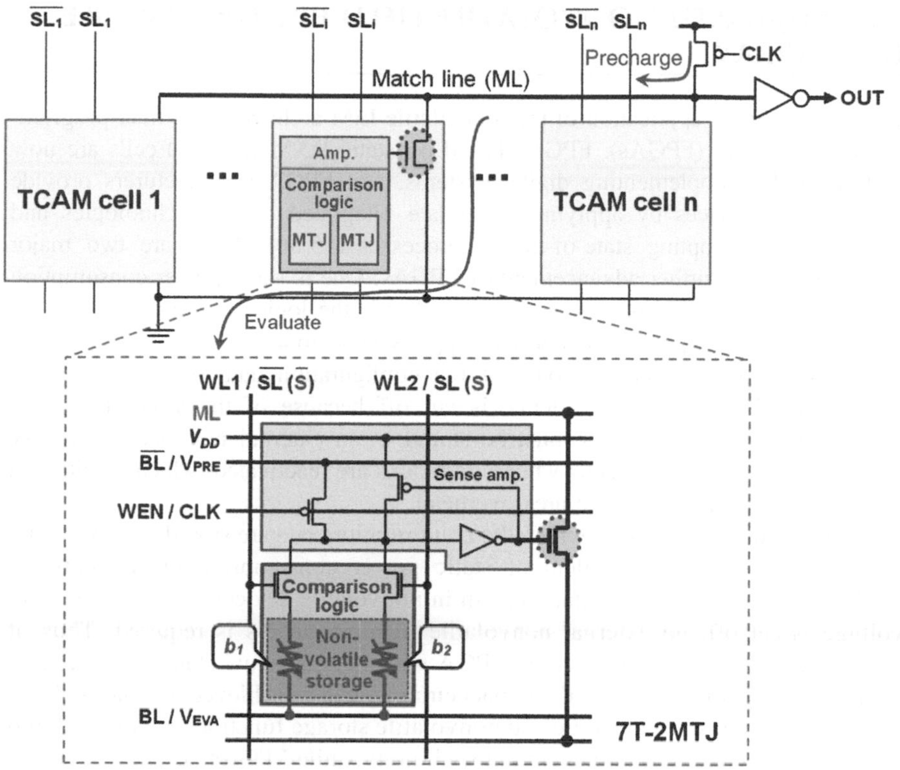

Figure 7.30 Nonvolatile TCAM word circuit with high-speed accessibility using a 7T-2MTJ cell.

reduced to less than 50% compared with the other types. In addition, the area cost of the proposed cells is less than that of conventional CMOS-based cells. As a result, the search speeds of the 7T-2MTJ-based and 9T-2MTJ-based TCAMs are enhanced to more than 200% while maintaining compactness.

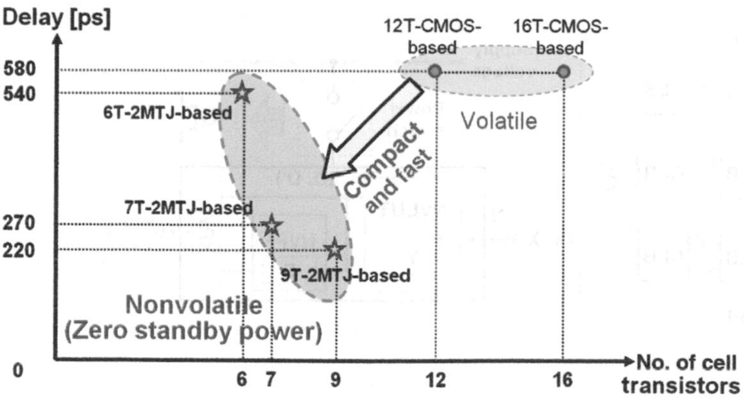

Figure 7.31 Comparison of results for nonvolatile and CMOS-based TCAMs.

7.6 MTJ-BASED NONVOLATILE FIELD-PROGRAMMABLE GATE ARRAY

The last important application of the nonvolatile LiM architecture is field-programmable gate arrays (FPGAs). FPGAs based on static RAM (SRAM) cells are now widely used in implementing digital systems, and FPGA manufacturers provide larger, faster devices by applying up-to-date integrated circuit technologies and sometimes by adopting state-of-the-art processes (35,36). There are two major problems for the further advancement of FPGAs: One is static power consumption due to nanometer-scale processes and the other is the logic capacity limitation. A power-gating technique that cuts off the supply voltage to idle circuit blocks (37) can reduce static power consumption, but the configuration data of the FPGA are eliminated when the supply voltage is cut off because of the volatility of the SRAM. Therefore, these data must be stored in an external nonvolatile memory and recalled to the SRAM cells when idle blocks are reactivated, which results in a long start-up latency and large area overhead.

Nonvolatile FPGAs, in which the configuration data are stored in nonvolatile devices, can solve the problem of static power consumption in conventional SRAM-based devices. Since data remain in nonvolatile devices even if the supply voltage is cut off, no external nonvolatile memory access is required. Thus, it should be possible to implement an FPGA with both ultralow standby power and instant-on capability. Moreover, compact circuitry can be achieved with nonvolatile LiM architecture, since the logic and nonvolatile storage functions are merged into nonvolatile devices. For implementing such nonvolatile FPGAs, an MTJ device is the most suitable candidate for nonvolatile memory because of its virtually unlimited endurance, high-speed read/write operations, write-energy scalability, and three-dimensional stacking capability (38,39).

Figure 7.32a shows the architecture of an MTJ-based nonvolatile FPGA. It consists of an array of configurable logic blocks (CLBs) composed of a cluster of nonvolatile logic elements (NVLEs) and connected through a network of switch matrices. As shown in Fig. 7.32b, an NVLE is composed of a nonvolatile flip-flop

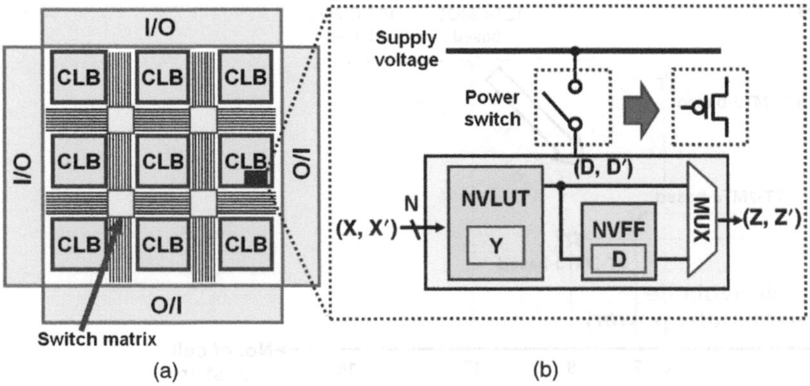

Figure 7.32 Nonvolatile FPGA. (a) Overall structure. (b) Nonvolatile LE.

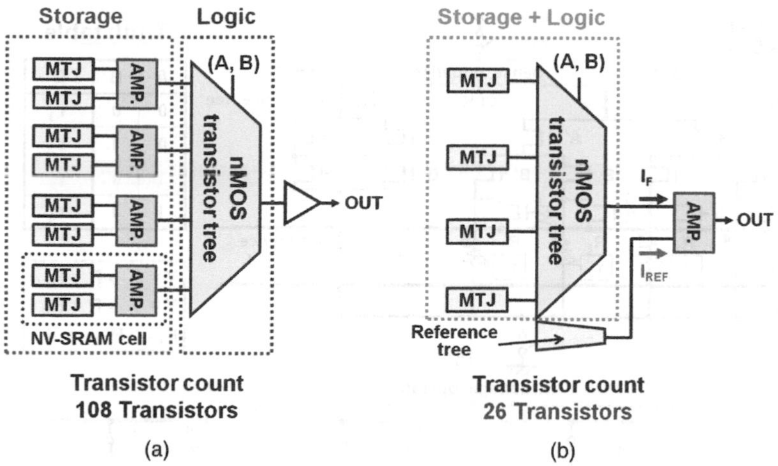

Figure 7.33 Block diagram of 2-input NVLUT circuits. (a) CMOS-based. (b) Proposed.

(NVFF) and an N-input nonvolatile lookup table (NVLUT) circuit. Any N-input combinational logic functions can be implemented by the NVLUT circuit, and sequential functions also can be implemented through combination of the NVLUT circuit and the NVFF. Since all data are stored in MTJ devices, fine-grained power gating can be achieved. Both the NVLUT circuit (39,40) and the NVFF (41) have already been implemented. To confirm the basic design concept of the proposed circuitry, we discuss a 2-input LUT circuit below.

Figure 7.33a shows a simplified block diagram for a conventional nonvolatile LUT circuit in which a CMOS-based volatile SRAM is replaced by an MTJ-based nonvolatile SRAM (41,42). A nonvolatile SRAM (NV-SRAM) cell is composed of one sense amplifier (SA) and two MTJ devices. The binary information stored in the MTJ devices is read by the SA. In this way, an NV-SRAM cell requires one SA per bit. Additionally, a bidirectional current control circuit to configure the MTJ devices is also required for each cell. The overhead for these devices increases with the number of inputs, since 2^N NV-SRAM cells are required for an N-input LUT circuit. Therefore, it is very important to consider how to maximally use the capabilities of the MTJ devices. Figure 7.33b shows the proposed LUT circuit. Since an MTJ device can be regarded as a variable resistor whose value is programmed to a low-resistance R_P or high-resistance R_{AP}, a 2-input LUT function can be achieved through the small difference in current level between an MTJ/MOS network and a reference cell with no signal amplification. As a result, the LUT circuit is simplified and the device count is greatly reduced.

Figure 7.34a shows a circuit diagram for the proposed 2-input NV-LUT circuit, which is composed of a sense amplifier, a LUT tree, and a reference tree, and two switches denoted as SW0 and SW1. The truth table for a 2-input logic function is stored in four MTJ devices (Y_0, Y_1, Y_2, Y_3). An MTJ device stores binary data Y as a resistance value: $Y=1$ corresponds to a low resistance R_P, and $Y=0$, to a high resistance R_{AP}. Figure 7.34b shows the basic behavior of the 2-input LUT circuit. In

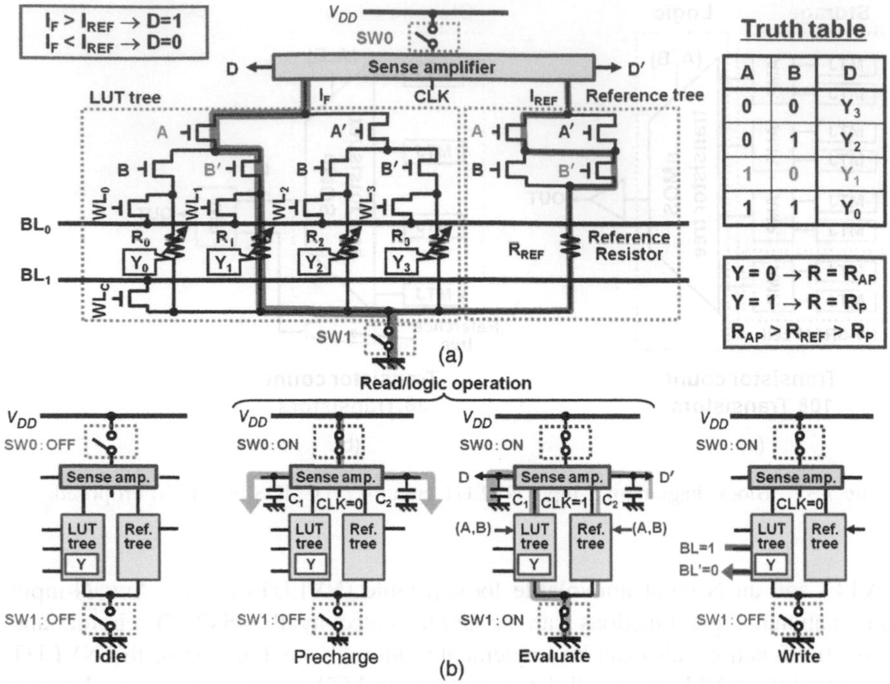

Figure 7.34 Circuit diagram of the proposed 2-input NVLUT circuit.

the idle mode, both SW0 and SW1 are turned off and the leakage current is completely cut off. The logic function is performed as follows: During a pre-charge phase ($CLK = 0$), two output node capacitances are charged to V_{DD}, while SW1 is turned off. During an evaluate phase ($CLK = 1$), SW1 turns on and the NMOS selector tree and NMOS reference tree are activated by complementary logic inputs (A, B), after which the currents I_F and I_{REF} pass through the corresponding MTJ device and the reference resistor, respectively. The difference between I_F and I_{REF} is sensed, and complementary full-swing outputs (D, D′) are generated by the sense amplifier. In the write operation, the MTJ devices in the NV-LUT circuit are configured by using word lines and bit lines, such as WL_0, WL_2, WL_2, WL_3, BL_0, and BL_1. The proposed LUT circuit can be extended to one with a higher number of inputs by using some "redundant" MTJ devices for variation compensation, while maintaining compact circuitry (note that the MTJ devices are stacked over the CMOS plane and do not affect the effective chip area) (40,42–44).

Figure 7.35a shows a die photo (chip size: 17.5 μm × 16.4 μm) of a fabricated test chip. The CMOS and metal layers were fabricated by a 0.14 μm CMOS process, and the MTJ devices were stacked over the third-metal layer via a back-end process (39). Figure 7.35b shows measured waveforms for the test chip. In this measurement, periodic voltage signals were applied to the inputs, while the power supply to the chip was cut off for a certain period of time. The results confirm that the output right after power-on was the same as it was just before power-off. This means that

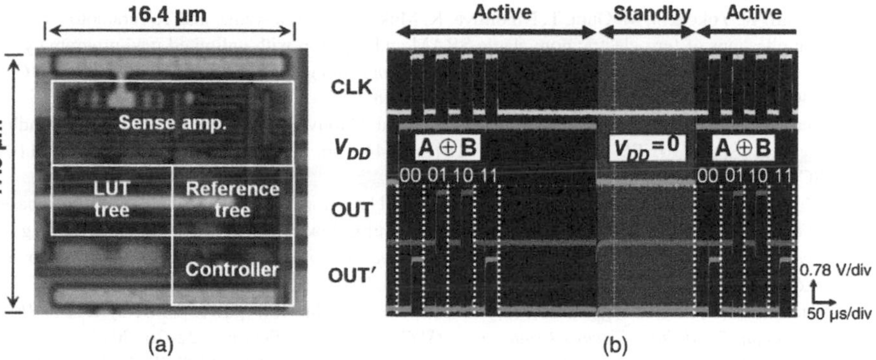

Figure 7.35 Fabricated test chip. (a) Die photo, (b) Measured waveforms. (Reproduced from Ref. 39 with permission from IEEE.)

the configuration information stored in the chip remains even if the power supply is cut off temporarily and then resumed. In this way, instant-on behavior was confirmed (44).

REFERENCES

1. S. Ikeda, J. Hayakawa, Y. M. Lee, F. Matsukura, Y. Ohno, T. Hanyu, and H. Ohno, "Magnetic tunnel junctions for spintronic memories and beyond," *IEEE Trans. Elect. Dev.* **54**, pp. 991–1002 (2007); doi: 10.1109/TED.2007.894617.
2. H. Honjo, S. Fukami, T. Suzuki, R. Nebashi, N. Ishiwata, S. Miura, N. Sakimura, T. Sugibayashi, N. Kasai, and H. Ohno, "Domain-wall-motion cell with perpendicular anisotropy wire and in-plane magnetic tunneling junctions," *J. Appl. Phys.* **111**, 07C903 (2012); doi: 10.1063/1.3671437.
3. H. Ohno, T. Endoh, T. Hanyu, N. Kasai, and S. Ikeda, "Magnetic tunnel junction for nonvolatile CMOS logic," *2010 IEEE International Electron Devices Meeting (IEDM), Technical Digest*, pp. 9.4.1–9.4.4 (2010); doi: 10.1109/IEDM.2010.5703329.
4. S. Ikeda, H. Sato, H. Honjo, E. C. I. Enobio, S. Ishikawa, M. Yamanouchi, S. Fukami, S. Kanai, F. Matsukura, T. Endoh, and H. Ohno, "Perpendicular-anisotropy CoFeB-MgO based magnetic tunnel junctions scaling down to 1X nm," *2014 IEEE International Electron Devices Meeting (IEDM)*, pp. 33.2.1–33.2.4 (2014); doi: 10.1109/IEDM.2014.7047160.
5. H. Ohno and S. Fukami, "Three-terminal spintronics memory devices with perpendicular anisotropy," *IEEE Intermag Conference*, Beijing, China, May 11–15, 2015, GA-02; doi: 10.1109/INTMAG.2015.7157427.
6. ITRS, "Process Integration, Devices & Structures," *International Technology Roadmap for Semiconductors* (2009).
7. ITRS, "2010 Future Memory Devices Workshop Summary," *International Technology Roadmap for Semiconductors*, July (2010).
8. T. Endoh, "Nonvolatile logic and memory devices based on spintronics," *IEEE International Symposium on Circuits and Systems* (2015).
9. ITRS, "Interconnect," *International Technology Roadmap for Semiconductors* (2009).
10. E. Pop, "Energy dissipation and transport in nanoscale devices," *Nano Res.* **3**, pp. 147–169 (2010); doi: 10.1007/s12274-010-1019-z.
11. W. H. Kautz, "Cellular logic-in-memory arrays," *IEEE Trans. Comput.* **C-18**, pp. 719–727 (1969); doi: 10.1109/T-C.1969.222754.
12. L. T. Clark, F. Ricci, and M. Biyani, "Low standby power state storage for sub-130-nm technologies," *IEEE J. Solid-State Circuits* **40**, pp. 498–506 (2005); doi: 10.1109/JSSC.2004.840987.

13. S. Masui, W. Yokozeki, M. Oura, T. Ninomiya, K. Mukaida, Y. Takayama, and T. Teramoto, "Design and applications of ferroelectric nonvolatile SRAM and flip-flop with unlimited read/program cycles and stable recall," *Proceedings of the IEEE Custom Integrated Circuits Conference (CICC)*, pp. 403–406 (2003); doi: 10.1109/CICC.2003.1249428.
14. N. Sakimura, T. Sugibayashi, R. Nebashi, and N. Kasai, "Nonvolatile magnetic flip-flop for standby-power-free SoCs," *IEEE J. Solid-State Circuits* **44**, pp. 2244–2250 (2009); doi: 10.1109/JSSC.2009.2023192.
15. T. Hanyu, D. Suzuki, N. Onizawa, S. Matsunaga, M. Natsui, and A. Mochizuki, "Spintronics-based nonvolatile logic-in-memory architecture towards an ultra-low-power VLSI computing paradigm," *Proceedings of the Design, Automation & Test in Europe Conference* (2015), pp. 1006–1011; http://dl.acm.org/citation.cfm?id=2757048.
16. T. Hanyu, "Nonvolatile logic-in-memory architecture for ultra-low power VLSI systems," *IEEE International Solid-State Circuits Conference (ISSCC)*, Forum 4, February 22–26, 2015.
17. T. Hanyu, D. Suzuki, A. Mochizuki, M. Natsui, N. Onizawa, T. Sugibayashi, S. Ikeda, T. Endoh, and H. Ohno, "Challenge of MOS/MTJ-hybrid nonvolatile logic-in-memory architecture in dark-silicon era," *2014 IEEE International Electron Devices Meeting (IEDM)*, December 15–17, 2014, pp. 28.2.1–28.2.3; doi: 10.1109/IEDM.2014.7047124.
18. N. Sakimura, T. Sugibayashi, R. Nebashi, and N. Kasai, "Nonvolatile magnetic flip-flop for standby-power-free SoCs," *Proceedings of the IEEE Custom Integrated Circuits Conference (CICC)*, pp. 355–358 (2008); doi: 10.1109/CICC.2008.4672095.
19. H. Honjo, R. Nebashi, T. Suzuki, S. Fukami, N. Ishiwata, T. Sugibayashi, and N. Kasai, "Performance of write-line inserted MTJ for low-write-current magnetic random access memory cell," *J. Appl. Phys.* **103**, p. 07A711 (2008); doi: 10.1063/1.2839288.
20. T. Endoh, S. Togashi, F. Iga, Y. Yoshida, T. Ohsawa, H. Koike, S. Fukami, S. Ikeda, N. Kasai, N. Sakimura, T. Hanyu, and H. Ohno, "A 600 MHz MTJ-based nonvolatile latch making use of incubation time in MTJ switching," *IEEE International Electron Devices Meeting (IEDM), Technical Digest*, pp. 4.3.1–4.3.4 (2011); doi: 10.1109/IEDM.2011.6131487.
21. H. Kimura, M. Ibuki, and T. Hanyu, "TMR-based logic-in-memory circuit for low-power VLSI," *International Technical Conference on Circuits/Systems, Computers and Communications (ITC-CSCC)*, pp. 8C3L-3-1–8C3L-3-4 (2004).
22. A. Mochizuki, H. Kimura, M. Ibuki, and T. Hanyu, "TMR-based logic-in-memory circuit for low-power VLSI," *IEICE Transactions on Fundamentals of Electronics, Communications and Computer Sciences*, E88-A, pp. 1408–1415 (2005); doi: 10.1093/ietfec/e88-a.6.1408.
23. S. Matsunaga, J. Hayakawa, S. Ikeda, K. Miura, H. Hasegawa, T. Endoh, H. Ohno, and T. Hanyu, "Fabrication of a nonvolatile full adder based on logic-in-memory architecture using magnetic tunnel junctions," *Appl. Phys. Express* **1**, 091301 (2008); doi: 10.1143/APEX.1.091301.
24. H. Ohno, T. Endoh, T. Hanyu, N. Kasai, and S. Ikeda, "Magnetic tunnel junction for nonvolatile CMOS logic," *IEEE International Electron Devices Meeting (IEDM), Technical Digest*, pp. 9.4.1–9.4.4 (2010); doi: 10.1109/IEDM.2010.5703329.
25. M. Natsui, D. Suzuki, N. Sakimura, R. Nebashi, Y. Tsuji, A. Morioka, T. Sugibayashi, S. Miura, H. Honjo, K. Kinoshita, S. Ikeda, T. Endoh, H. Ohno, and T. Hanyu, "Nonvolatile logic-in-memory LSI using cycle-based power gating and its application to motion-vector prediction," *IEEE J. Solid-State Circuits* **50**, pp. 476–489 (2015); doi: 10.1109/JSSC.2014.2362853.
26. R. Nebashi, N. Sakimura, Y. Tsuji, S. Fukami, H. Honjo, S. Saito, S. Miura, N. Ishiwata, K. Kinoshita, T. Hanyu, T. Endoh, N. Kasai, H. Ohno, and T. Sugibayashi, "A content addressable memory using magnetic domain wall motion cells," *Symposium on VLSI Circuits, Digest of Technical Papers*, pp. 300–301 (2011); http://ieeexplore.ieee.org/xpls/abs_all.jsp?arnumber=5986430
27. S. Matsunaga, K. Hiyama, A. Matsumoto, S. Ikeda, H. Hasegawa, K. Miura, J. Hayakawa, T. Endoh, H. Ohno, and T. Hanyu, "Standby-power-free compact ternary content-addressable memory cell chip using magnetic tunnel junction devices," *Appl. Phys. Express*, **2**, 023004 (2009); doi: 10.1143/APEX.2.023004.
28. S. Matsunaga, M. Natsui, K. Hiyama, T. Endoh, H. Ohno, and T. Hanyu, "Fine-grained power-gating scheme of a metal–oxide–semiconductor and magnetic-tunnel-junction-hybrid bit-serial ternary content-addressable memory," *Japan. J. Appl. Phys.* **49**, 04DM05 (2010); doi: 10.1143/JJAP.49.04DM05.

29. S. Matsunaga, A. Katsumata, M. Natsui, S. Fukami, T. Endoh, H. Ohno, and T. Hanyu, "Fully parallel 6T-2MTJ nonvolatile TCAM with single-transistor-based self match-line discharge control," *IEEE Symposium on VLSI Circuits*, 28-2, pp. 298–299 (2011); http://ieeexplore.ieee.org/xpls/abs_all.jsp?arnumber=5986429.
30. S. Matsunaga, A. Katsumata, M. Natsui, T. Endoh, H. Ohno, and T. Hanyu, "High-speed-search nonvolatile TCAM using MTJ devices," *International Conference on Solid State Devices and Materials (SSDM)*, pp. 454–455 (2011).
31. S. Matsunaga, A. Katsumata, M. Natsui, T. Endoh, H. Ohno, and T. Hanyu, "Design of a nine-transistor/two-magnetic-tunnel-junction-cell-based low-energy nonvolatile ternary content-addressable memory," *Jpn. J. Appl. Phys.* **51**, 02BM06 (2012); doi: 10.1143/JJAP.51.02BM06.
32. S. Matsunaga, A. Katsumata, M. Natsui, T. Endoh, H. Ohno, and T. Hanyu, "Design of a 270 ps-access 7-transistor/2-magnetic-tunnel-junction cell circuit for a high-speed-search nonvolatile ternary content-addressable memory," *J. Appl. Phys.* **111**, p. 07E336 (2012); doi: 10.1063/1.3677875.
33. S. Matsunaga, S. Miura, H. Honjou, K. Kinoshita, S. Ikeda, T. Endoh, H. Ohno, and T. Hanyu, "A 3.14 μm^2 4T-2MTJ-cell fully parallel TCAM based on nonvolatile logic-in-memory architecture," *Symposium VLSI Circuits, Digest of Technical Papers*, pp. 44–45 (2012); doi: 10.1109/VLSIC.2012.6243781.
34. S. Matsunaga, N. Sakimura, R. Nebashi, Y. Tsuji, A. Morioka, T. Sugibayashi, S. Miura, H. Honjo, K. Kinoshita, H. Sato, S. Fukami, M. Natsui, A. Mochizuki, S. Ikeda, T. Endoh, H. Ohno, and T. Hanyu, "Fabrication of a 99%-energy-less nonvolatile multi-functional CAM chip using hierarchical power gating for a massively-parallel full-text-search engine," *Symposium on VLSI Technology, Digest of Technical Papers*, pp. 106–107 (2013); http://ieeexplore.ieee.org/xpls/abs_all.jsp?arnumber=6576611.
35. S. Brown and J. Rose, "FPGA and CPLD architectures: a tutorial," *IEEE Des. Test Comput.* **12**, pp. 42–57 (1996); doi: 10.1109/54.500200.
36. A. Amara, F. Amiel, and T. Ea, "FPGA vs. ASIC for low power applications," *Microelectron. J.* **37**, pp. 669–677 (2006); doi: 10.1016/j.mejo.2005.11.003.
37. T. Tuan, A. Rahman, S. Das, S. Trimberger, and S. Kao, "A 90-nm low-power FPGA for battery-powered applications," *IEEE Trans. Comput. Aided Des. Int. Circuits Syst.* **26**, pp. 296–300 (2007); doi: 10.1109/TCAD.2006.885731.
38. W. Zhao, E. Belhaire, C. Chappert, and P. Mazoyer, "Spin transfer torque (STT)-MRAM-based runtime reconfiguration FPGA circuit," *ACM Trans. Embedded Comput. Syst.* **9**, pp. 1–14 (2009); doi: 10.1145/1596543.1596548.
39. D. Suzuki, M. Natsui, S. Ikeda, H. Hasegawa, K. Miura, J. Hayakawa, T. Endoh, H. Ohno, and T. Hanyu, "Fabrication of a nonvolatile lookup-table circuit chip using magneto/semiconductor-hybrid structure for an immediate-power-up field programmable gate array," *Symposium on VLSI Circuits, Digest of Technical Papers*, pp. 80–81 (2009); http://ieeexplore.ieee.org/xpl/articleDetails.jsp?arnumber=5205282
40. D. Suzuki, M. Natsui, H. Ohno, and T. Hanyu, "Design of a process-variation-aware nonvolatile MTJ-based lookup-table circuit," *International Conference on Solid State Devices and Materials (SSDM)*, K-9-4, pp. 1146–1147 (2010).
41. D. Suzuki, M. Natsui, T. Endoh, H. Ohno, and T. Hanyu, "A compact nonvolatile logic element using an MTJ/MOS-hybrid structure," *International Conference on Solid State Devices and Materials (SSDM)*, N-8-2, pp. 1464–1465 (2011).
42. D. Suzuki, M. Natsui, T. Endoh, H. Ohno, and T. Hanyu, "Six-input lookup table circuit with 62% fewer transistors using nonvolatile logic-in-memory architecture with series/parallel-connected magnetic tunnel junctions," *J. Appl. Phys.* **111**, p. 07E318 (2012); doi: 10.1063/1.3672411.
43. D. Suzuki, M. Natsui, T. Endoh, H. Ohno, and T. Hanyu, "50%-transistor-less standby-power-free 6-input LUT circuit using redundant MTJ-based nonvolatile logic-in-memory architecture," *Conference on Magnetism and Magnetic Materials*, GD-07, p. 480 (2011).
44. D. Suzuki, M. Natsui, A. Mochizuki, S. Miura, H. Honjo, K. Kinoshita, H. Sato, S. Ikeda, T. Endoh, H. Ohno, and T. Hanyu, "Fabrication of a magnetic tunnel junction-based 240-tile nonvolatile field-programmable gate array chip skipping wasted write operations for greedy power-reduced logic applications," *IEICE Electron. Express*, **10**, 20130772 (2013); doi: 10.1587/elex.10.20130772.

APPENDIX

UNITS FOR MAGNETIC PROPERTIES

Symbol	Quantity	Conversion from Gaussian and cgs emu to SI
Φ	Magnetic flux	$1\ \text{Mx} \rightarrow 10^{-8}\ \text{Wb} = 10^{-8}\ \text{V} \cdot \text{s}$
B	Magnetic flux density, magnetic induction	$1\ \text{G} \rightarrow 10^{-4}\ \text{T} = 10^{-4}\ \text{Wb/m}^2$
H	Magnetic field strength	$1\ \text{Oe} \rightarrow 10^3/(4\pi)\ \text{A/m}$
m	Magnetic moment	$1\ \text{erg/G} = 1\ \text{emu} \rightarrow 10^{-3}\ \text{A} \cdot \text{m}^2 = 10^{-3}\ \text{J/T}$
M	Magnetization	$1\ \text{erg/(G} \cdot \text{cm}^3) = 1\ \text{emu/cm}^3 \rightarrow 10^3\ \text{A/m}$
$4\pi M$	Magnetization	$1\ \text{G} \rightarrow 10^3/(4\pi)\ \text{A/m}$
σ	Mass magnetization, specific magnetization	$1\ \text{erg/(G} \cdot \text{g}) = 1\ \text{emu/g} \rightarrow 1(\text{A} \cdot \text{m}^2)/\text{kg}$
j	Magnetic dipole moment	$1\ \text{erg/G} = 1\ \text{emu} \rightarrow 4\pi \times 10^{-10}\ \text{Wb} \cdot \text{m}$
J	Magnetic polarization	$1\ \text{erg/(G} \cdot \text{cm}^3) = 1\ \text{emu/cm}^3 \rightarrow 4\pi \times 10^{-4}\ \text{T}$
χ, κ	Susceptibility	$1 \rightarrow 4\pi$
χ_ρ	Mass susceptibility	$1\ \text{cm}^3/\text{g} \rightarrow 4\pi \times 10^{-3}\ \text{m}^3/\text{kg}$
μ	Permeability	$1 \rightarrow 4\pi \times 10^{-7}\ \text{H/m} = 4\pi \times 10^{-7}\ \text{Wb/(A} \cdot \text{m)}$
μ_r	Relative permeability	$\mu \rightarrow \mu_\text{r}$
w, W	Energy density	$1\ \text{erg/cm}^3 \rightarrow 10^{-1}\ \text{J/m}^3$
N, D	Demagnetizing factor	$1 \rightarrow 1/(4\pi)$

Gaussian units are the same as cgs emu for magnetostatics; Mx = maxwell, G = gauss, Oe = oersted; Wb = weber, V = volt, s = second, T = tesla, m = meter, A = ampere, J = joule, kg = kilogram, H = henry.

Introduction to Magnetic Random-Access Memory, First Edition. Edited by Bernard Dieny, Ronald B. Goldfarb, and Kyung-Jin Lee.

INDEX

Introduction to Magnetic Random-Access Memory, First Edition. Edited by Bernard Dieny, Ronald B. Goldfarb, and Kyung-Jin Lee.

Printed and bound by CPI Group (UK) Ltd, Croydon, CR0 4YY
27/10/2024
14580278-0001